Studies in Organic Chemistry 5

COMPREHENSIVE CARBANION CHEMISTRY

PART C

Studies in Organic Chemistry

Other titles in this series:

1 **Complex Hydrides** by A. Hajós
2 **Proteoglycans — Biological and Chemical Aspects in Human Life** by J.F. Kennedy
3 **New Trends in Heterocyclic Chemistry** edited by R.B. Mitra, N.R. Ayyangar, V.N. Gogte, R.M. Acheson and N. Cromwell
4 **Inositol Phosphates: Their Chemistry, Biochemistry and Physiology** by D.J. Cosgrove
5 **Comprehensive Carbanion Chemistry. Part A. Structure and Reactivity** edited by E. Buncel and T. Durst
Comprehensive Carbanion Chemistry. Part B. Selectivity in Carbon-Carbon Bond Forming Reactions edited by E. Buncel and T. Durst
6 **New Synthetic Methodology and Biologically Active Substances** edited by Z.-I. Yoshida
7 **Quinonediazides** by V.V. Ershov, G.A. Nikiforov and C.R.H.I. de Jonge
8 **Synthesis of Acetylenes, Allenes and Cumulenes: A Laboratory Manual** by L. Brandsma and H.D. Verkruijsse
9 **Electrophilic Additions to Unsaturated Systems** by P.B.D. de la Mare and R. Bolton
10 **Chemical Approaches to Understanding Enzyme Catalysis: Biomimetic Chemistry and Transition-State Analogs** edited by B.S. Green, Y. Ashani and D. Chipman
11 **Flavonoids and Bioflavonoids 1981** edited by L. Farkas, M. Gábor, F. Kállay and H. Wagner
12 **Crown Compounds: Their Characteristics and Applications** by M. Hiraoka
13 **Biomimetic Chemistry** edited by Z.-I. Yoshida and N. Ise
14 **Electron Deficient Aromatic- and Heteroaromatic-Base Interactions. The Chemistry of Anionic Sigma Complexes** by E. Buncel, M.R. Crampton, M.J. Strauss and F. Terrier
15 **Ozone and its Reactions with Organic Compounds** by S.D. Razumovskii and G.E. Zaikov
16 **Non-benzenoid Conjugated Carbocyclic Compounds** by D. Lloyd
17 **Chemistry and Biotechnology of Biologically Active Natural Products** edited by Cs. Szántay, A. Gottsegen and G. Kovács
18 **Bio-Organic Heterocycles: Synthetic, Physical Organic and Pharmacological Aspects** edited by H.C. van der Plas, L. Ötvös and M. Simonyi
19 **Organic Sulfur Chemistry: Theoretical and Experimental Advances** edited by F. Bernardi, I.G. Czismadia and A. Mangini
20 **Natural Products Chemistry 1984** edited by R.I. Zalewski and J.J. Skolik
21 **Carbocation Chemistry** by P. Vogel
22 **Biocatalysts in Organic Syntheses** edited by J. Tramper, H.C. van der Plas and P. Linko
23 **Flavonoids and Bioflavonoids 1985** edited by L. Farkas, M. Gábor and F. Kállay
24 **The Organic Chemistry of Nucleic Acids** by Y. Mizuno
25 **New Synthetic Methodology and Functionally Interesting Compounds** edited by Z.-I. Yoshida
26 **New Trends in Natural Products Chemistry 1986** edited by A.-ur-Rahman and P.W. Le Quesne
27 **Bio-Organic Heterocycles 1986. Synthesis, Mechanisms and Bioactivity** edited by H.C. van der Plas, M. Simonyi, F.C. Alderweireldt and J.A. Lepoivre
28 **Perspectives in the Organic Chemistry of Sulfur** edited by B. Zwanenburg and A.J.H. Klunder
29 **Biocatalysis in Organic Media** edited by C. Laane, J. Tramper and M.D. Lilly
30 **Recent Advances in Electroorganic Synthesis** edited by S. Torii
31 **Physical Organic Chemistry 1986** edited by M. Kobayashi
32 **Organic Solid State Chemistry** edited by G.R. Desiraju

Studies in Organic Chemistry 5

COMPREHENSIVE CARBANION CHEMISTRY

Edited by

E. Buncel
Department of Chemistry, Queen's University, Kingston, Ontario, Canada K7L 3N6

T. Durst
Department of Chemistry, University of Ottawa, Ottawa, Ontario, Canada K1N 6N5

PART C

GROUND AND EXCITED STATE REACTIVITY

ELSEVIER
Amsterdam — Oxford — New York — Tokyo 1987

ELSEVIER SCIENCE PUBLISHERS B.V.
Sara Burgerhartstraat 25
P.O. Box 211, 1000 AE Amsterdam, The Netherlands

Distributors for the United States and Canada:

ELSEVIER SCIENCE PUBLISHING COMPANY INC.
52, Vanderbilt Avenue
New York, NY 10017, U.S.A.

ISBN 0-444-42869-0 (Vol. 5)
ISBN 0-444-41737-0 (Series)

Printed in The Netherlands

Contributors to Part C

M. R. Bryce — Department of Chemistry, University of Durham, Durham, DH1 3LE, U.K.

R. D. Chambers — Department of Chemistry, University of Durham, Durham, DH1 3LE, U.K.

Marye Anne Fox — Department of Chemistry, University of Texas, Austin, Texas 78712, U.S.A.

Erling Grovenstein, Jr. — School of Chemistry, Georgia Institute of Technology, Atlanta, Georgia 30332, U.S.A.

Heinz F. Koch — Department of Chemistry, Ithaca College, Ithaca, New York 14850, U.S.A.

Wai-Kee Li — Department of Chemistry, The Chinese University of Hong Kong, Shatin, N.T., Hong Kong.

Ross H. Nobes — Research School of Chemistry, Australian National University, Canberra, A.C.T. 2601, Australia.

Dieter Poppinger — CIBA-GEIGY Limited, CH 4002 Basle, Switzerland

Leo Radom — Research School of Chemistry, Australian National University, Canberra, A.C.T. 2601, Australia.

Laren M. Tolbert — School of Chemistry, Georgia Institute of Technology, Atlanta, Georgia 30332, U.S.A.

COMPREHENSIVE CARBANION CHEMISTRY

Part A. Structure and Reactivity

R. B. Bates University of Arizona	Dianions and Polyanions
J. I. Brauman and M. J. Pellerite Stanford University	Gas-Phase Acidities of Carbon Acids
E. Buncel and B. Menon Queen's University	Spectrophotometric Investigations of Arylmethyl Carbanions
J. Corset CNRS, Thiais	Vibrational Infrared and Raman Spectroscopy of Carbanions
R. D. Guthrie University of Kentucky	Electron-Transfer Reactions of Carbanions and Related Species
D. H. O'Brien Texas A&M University	The Nuclear Magnetic Resonance of Carbanions
A. Streitwieser, Jr., E. Juaristi and L. L. Nebenzahl University of California, Berkeley	Equilibrium Carbon Acidities in Solution

Part B. Selectivity in Carbon-Carbon Bond Forming Reactions

T. Durst University of Ottawa	Techniques in Carbanion Chemistry
R. Epsztein Universite Paris-Sud	The Formation and Transformation of Allenic-α-acetylene Carbanions
R. R. Fraser University of Ottawa	The Syn Effect and the Use of Enolate Equivalents in Synthesis
C. H. Heathcock University of California	Stereoselective Aldol Condensations
L. S. Hegedus Colorado State University	Formation of Carbon-Carbon bonds via π-allyl Complexes of Transition Metals

FOREWORD

The impetus to the study of the structure, stability and reactivity of carbanions was provided to a large extent by Professor Donald J. Cram, through his own work, and in the publication in 1965 of the classic monograph "Fundamentals of Carbanion Chemistry". Since that time, the study of carbanion chemistry has intensified, in part as new techniques such as ^{13}C and hetero-nuclear magnetic resonance spectroscopy have become available, and in part as new theoretical concepts have come to light. A particularly interesting development is that of ion cyclotron resonance and related techniques for the gas phase investigation of carbanion reactivity. Such studies have increased our awareness of medium effects and have caused a re-evaluation of the theoretical basis of structural and electronic effects.

Carbanion chemistry plays a central role in modern synthetic organic chemistry. This relationship has developed at an extremely rapid rate since the 1960's due to the commercial availability of a variety of alkyllithiums, and the development of a number of potent non-nucleophilic bases such as lithium diisopropylamide, lithium 2,2,6,6-tetramethylpiperidide and sodium hexamethyldisilazide. The availability of these reagents has made accessible many structural types of carbanions which are extremely valuable as synthetic units since they generally undergo efficient carbon-carbon bond formation with typical electrophiles.

In this series, which has been entitled "Comprehensive Carbanion Chemistry", it is hoped to discuss in detail key aspects of carbanion chemistry, via contributed chapters written by active investigators in the field. Several volumes are in progress, each devoted to related aspects of carbanion chemistry. Thus, Part A, already published, is concerned with the physical aspects of carbanions, namely the acidity of carbon acids both in solution and in the gas phase, the structure of mono- and dianions as determined by proton and carbon nuclear magnetic resonance, ultraviolet, infrared and Raman spectroscopy, and the involvement of electron transfer processes in carbanion reactions. Part B, also published, emphasizes regio- and stereoselective processes, for example in aldol condensations, the use of enolate equivalents in synthesis, and some reactions of allenic-alpha-acetylenic carbanions; other topics discussed include the formation of carbon-carbon bonds via π-complexes of transition metals, as well as the techniques used in preparative carbanion chemistry. Part C, the current volume, continues on different aspects of ground state chemistry of carbanions but considers, as well, their excited state reactivity. Thus the topics treated include the molecular orbital theory of carbanions, their electrochemistry, the structures of dianions, the chemistry of fluoro-carbanions, the role of hydrogen-bonded carbanion intermediates and the photochemistry of resonance-stabilized carbanions.

We would like to thank the authors of the individual chapters in the present volume for their excellent contributions and the extra effort which they undertook in preparing camera-ready copy.

June, 1987

E.B.
T.D.

CONTENTS

Foreword vii

Chapter 1. Molecular orbital theory of carbanions, by Wai-Kee Li, R. H. Nobes, D. Poppinger and L. Radom 1

Chapter 2. Electrochemistry of carbanions, by Marye Anne Fox 93

Chapter 3. Structures of organodialkali metal compounds ("Dianions"), by E. Grovenstein, Jr. 175

Chapter 4. The photochemistry of resonance-stabilized anions, by L. M. Tolbert 223

Chapter 5. Fluoro-carbanions, by R.D. Chambers and M.R. Bryce 271

Chapter 6. Reactions of hydrogen-bonded carbanion intermediates, by Heinz F. Koch 321

Index 361

Chapter 1

MOLECULAR ORBITAL THEORY OF CARBANIONS

ROSS H. NOBES, DIETER POPPINGER,[1] WAI-KEE LI[2] AND LEO RADOM
Research School of Chemistry, Australian National University, Canberra, A.C.T. 2601, Australia

CONTENTS

I. INTRODUCTION 3
II. SURVEY OF THEORETICAL METHODS 4
A. POTENTIAL ENERGY SURFACES 4
B. QUANTUM CHEMICAL METHODS 5
1. AB INITIO METHODS 6
a. BASIS SETS 6
b. THE HARTREE-FOCK METHOD 8
c. CORRELATED WAVEFUNCTIONS 8
2. SEMIEMPIRICAL METHODS 9
III. PREDICTIVE CAPABILITIES OF QUANTUM CHEMICAL METHODS 10
A. GEOMETRIES 10
B. PROTON AFFINITIES 14
C. INVERSION BARRIERS 16
D. ELECTRON AFFINITIES 20
E. VIBRATIONAL FREQUENCIES 23
IV. APPLICATIONS TO SPECIFIC CARBANIONS 25
A. THE METHYL ANION AND ITS DERIVATIVES 25
1. THE PARENT SYSTEM CH_3^- 25
2. SUBSTITUTED METHYL ANIONS XCH_2^- 25
a. GENERAL CONSIDERATIONS 25
b. NITROMETHYL ANION 29
c. CYANOMETHYL AND SUBSTITUTED CYANOMETHYL ANIONS 30
d. ISOCYANOMETHYL AND SUBSTITUTED ISOCYANOMETHYL ANIONS 31
e. CYCLOPROPYLMETHYL ANION 31
f. FLUORO-SUBSTITUTED METHYL ANIONS 32
g. MERCAPTOMETHYL ANION 33
h. SILYLATED METHYL ANIONS 34
i. DIPHENYLMETHYL ANION 35
B. THE ETHYL ANION AND ITS DERIVATIVES 35
1. THE PARENT SYSTEM $CH_3CH_2^-$ 35

2. β-SUBSTITUTED ETHYL ANIONS. HYPERCONJUGATION 39
3. α-SUBSTITUTED ETHYL ANIONS 45
C. HIGHER ALKYL ANIONS 47
D. CYCLOALKYL ANIONS 47
1. CYCLOPROPYL ANION 48
2. CYCLOBUTYL ANION 49
3. OTHER CYCLOALKYL ANIONS 49
E. THE VINYL ANION AND ITS DERIVATIVES 49
1. THE PARENT SYSTEM $CH_2{=}CH^-$ 50
2. SUBSTITUTED VINYL ANIONS 52
3. ANALOGUES OF VINYL ANION 53
F. CARBONYL ANIONS 53
G. ALKYNYL ANIONS 54
1. THE PARENT SYSTEM $HC{\equiv}C^-$ 54
2. SUBSTITUTED ETHYNYL ANIONS 55
3. ANALOGUES OF THE ETHYNYL ANION 55
a. CYANIDE ANION 55
b. SILA-ANALOGUES OF HCC^- 56
4. PROPYNYL AND LARGER ALKYNYL ANIONS 56
H. π-DELOCALIZED CARBANIONS 57
1. THE ALLYL ANION $CH_2CHCH_2^-$ 57
2. ANALOGUES OF THE ALLYL ANION 58
a. ENOLATE ANIONS 58
b. METHANIMIDAMIDE ANION 59
3. SUBSTITUTED ALLYL AND POLYENYL ANIONS AND THEIR ANALOGUES 59
4. CYCLOALKENYL ANIONS 61
5. CYCLOHEXADIENYL AND SUBSTITUTED CYCLOHEXADIENYL ANIONS 62
6. BENZYL ANIONS 64
7. THE PROPARGYL ANION AND SUBSTITUTED PROPARGYL ANIONS 64
8. ANALOGUES OF PROPARGYL ANION 65
I. AROMATIC, ANTIAROMATIC AND HOMOAROMATIC ANIONS 66
1. CYCLOPROPENYL ANION 66
2. 6π- AND 8π-ELECTRON CARBANIONS 67
3. HOMOAROMATIC ANIONS 68
J. MULTIPLY-CHARGED CARBANIONS 71
1. GENERAL CONSIDERATIONS 71
2. METHYLENE DIANION CH_2^{2-} 72
3. ETHYLENE DIANION $CH_2CH_2^{2-}$ 72
4. ACETYLENE DIANION $HCCH^{2-}$ 72
5. ACETYLIDE DIANION CC^{2-} 72

6. DIDEPROTONATED VINYLACETYLENE 73
7. Y DELOCALIZATION 73
8. AROMATIC MULTIPLY-CHARGED ANIONS 73
V. CONCLUDING REMARKS 74
ACKNOWLEDGEMENTS 75
NOTE ADDED IN PROOF 75
REFERENCES 76

I. INTRODUCTION

Carbanions are among the most important synthetic organic intermediates, largely because of the efficient manner in which they can undergo carbon-carbon bond formation through reaction with appropriate electrophiles.[3-6] Since Cram's classic book[3] appeared in 1965, there has been a vigorous growth in the carbanion literature, mainly associated with synthetic applications. However, despite recent advances (see, for example, refs. 5 and 7), detailed physical characterization of free carbanions remains a difficult experimental task.

Under such circumstances, theory has the opportunity to play an important role. Quantum chemical calculations can be used *in principle* to study reactive species such as ions with no more difficulty than to study stable, neutral molecules (see, for example, ref. 8). They can, for example, be used to determine both structures and thermodynamic and kinetic stabilities. Such calculations may help in the interpretation of experimental results or may serve in a predictive capacity, thus providing a valuable impetus for experimental research.

Although these applications of quantum chemical methods are now well developed for small neutral molecules and for cations, anion calculations, which had long been regarded as difficult, have until recently received relatively little attention. The topic of anion calculations was last reviewed in 1977.[9] Because of considerable recent progress, a new assessment would seem appropriate. In this article, we survey the current status of quantum chemical calculations on carbanions.

We adopt the definition[3] of a carbanion as the conjugate base of a carbon acid, i.e. as a species formally derived from an organic molecule by heterolytic fission of a carbon-hydrogen bond, and restrict our coverage to anions encompassed by this definition. This is in general equivalent to a restriction to polyatomic organic anions in which the negative charge is localized on a carbon atom in at least one of the normal resonance structures. Radical anions and systems in excited states are excluded from our treatment, as is the very important topic of organometallic analogues of carbanions such as appropriate organolithium compounds (see, for example, refs. 10 and 11). Another conscious omission is treatment of the S_N2 reaction; the pentavalent intermediates in such reactions do not fall within the definition of carbanions presented above. One final restriction concerns the theoretical methods covered. Only those methods which can in principle handle molecules of any shape and composition and which can be expected to predict both molecular geometries and energies are considered. Methods such as Hückel π-electron theory are not included.

We begin in Sect. II with a brief description of the theoretical methods. An important aspect of this article is the evaluation in Sect. III of the performance of these procedures with respect to anion calculations. Section IV discusses the application of theory to specific carbanions. Finally, the present status of theoretical work on carbanions is summarized in Sect. V.

II. SURVEY OF THEORETICAL METHODS

(A) Potential energy surfaces

Important goals in theoretical chemistry include the prediction of structures of molecules and mechanisms of chemical reactions. Knowledge of the potential energy surface, the potential energy of the molecule or assembly of molecules as a function of geometry, represents a vital step in achieving these goals. For most purposes, it is usually sufficient to consider only two types of points on the surface: minima, which correspond to stable structures, and first-order saddle points, which according to absolute rate theory correspond to transition structures through which reactions proceed.

Given a theoretical method to generate the potential energy surface, locating minima is a straightforward application of mathematical optimization techniques (see, for example, refs. 12-16). A reasonable guess at the molecular geometry, obtained for example from tabulations of average bond lengths and angles, is fed into an energy-minimizing algorithm, which generally returns the exact location of the minimum closest to the starting point. The coordinates of this minimum represent the theoretical geometry of the molecule, as defined by the method used to generate the surface. Similar, but less straightforward, methods exist for locating saddle points, i.e. for calculating geometries of transition structures.[14-21]

The relative energies ΔE_0 at these stationary points refer to molecular systems at 0 K in a hypothetical vibrationless state. From the curvature of the surface in the vicinity of minima and saddle points, entropies as well as thermodynamic corrections to ΔE_0, and hence equilibrium constants and reaction rates, can be calculated.

A potential energy surface can of course be constructed not only for an isolated molecule, but equally well (although at much greater computational expense) for an assembly of molecules, such as a solvated species. This would be the appropriate approach for treating carbanions, which are usually associated with counterions and solvent molecules. The counterion problem has been examined to some extent (see, for example, refs. 11 and 22) but, with a few exceptions, theoretical work on carbanions themselves has so far only been carried out for isolated, unsolvated molecules. Therefore, the theoretical results should strictly only be compared with gas-phase experimental data. Direct comparison with results from solution-phase experiments may be dangerous.

(B) Quantum chemical methods

Quantum chemistry provides a number of methods of varying accuracy to calculate the energy of a molecule as a function of its geometry. Most quantum chemical methods used for the calculation of energies construct the wavefunction Ψ of the molecule from a set of simple atom-centred basis functions ϕ_μ. Basis functions are usually either of the Slater type (STFs, radial dependence $e^{-\zeta r}$) or, more commonly, of the Gaussian type (GTFs, radial dependence $e^{-\zeta r^2}$). The simpler sets of basis functions bear some relation to the atomic orbitals (1s, 2s, 2p, 3d, etc.) which play such a central role in qualitative electronic structure theory, but this relation becomes obscured for larger sets.

From the basis functions, one constructs the so-called molecular orbitals (MOs) ψ_i:

$$\psi_i = \sum_\mu c_{\mu i} \phi_\mu . \tag{1}$$

The square of ψ_i describes the probability distribution of an electron occupying this particular molecular orbital in the field of the nuclei and all other electrons.

The molecular wavefunction Ψ can be written as a linear combination of configurations Ψ_k:

$$\Psi = \sum_k a_k \Psi_k. \quad (2)$$

The configurations Ψ_k are themselves constructed as antisymmetrized products of occupied molecular orbitals. Equation 2 forms the basis of the configuration interaction approach. Usually, the configuration Ψ_0 in which the molecular orbitals of lowest one-electron energy are occupied will dominate the wavefunction, and other configurations (which differ from Ψ_0 by single, double, triple, ... electronic excitations) represent small correction terms.

The total electronic energy associated with the wavefunction Ψ is then a function of many adjustable parameters (ζ, $c_{\mu i}$, a_k), which are determined by using the variational method, i.e. by minimizing the electronic energy with respect to these parameters. In the *ab initio* approach, all mathematical expressions arising in this task are fully evaluated.

In principle, one can approximate the molecular wavefunction to any desired accuracy by employing ever larger basis sets (eq. 1) and longer configuration expansions (eq. 2). There are practical limits, however, even with the present generation of very fast computers, and approximations are necessary to reduce the amount of numerical work. Such approximations can be made within the *ab initio* framework: they consist of selecting a particular (usually small) set of basis functions with fixed exponents ζ and a particular (usually severely truncated) type of configuration expansion. Alternatively, some or most of the cumbersome mathematical expressions may be simply neglected, replaced by simpler functional forms, or estimated from experimental data; this leads to the various *semiempirical* methods.

(1) *Ab initio methods*

(a) *Basis sets*. The first step in an *ab initio* calculation is the selection of a particular set of basis functions. A large number of different basis sets have been developed over the past twenty five years. For the purpose of this review, they can be classified broadly as minimal, extended, polarization and diffuse-augmented.

A *minimal* basis set includes one function (STF, GTF or fixed linear combination of GTFs) for each orbital which is occupied in the electronic ground state of the atom while maintaining spherical symmetry. For example, for carbon, 1s, 2s, $2p_x$, $2p_y$ and $2p_z$ functions are used, while for hydrogen a single 1s function is employed. A well-tested and popular basis set of this type is STO-3G.[23-27]

Extended basis sets employ more than one basis function per atomic orbital. A common practice is to have two functions per atomic orbital, leading to a so-called double-zeta basis set. An alternative is to employ two or more functions for the valence atomic orbitals (2s, $2p_x$, $2p_y$, $2p_z$ for C, 1s for H) only, while the inner-shell atomic orbitals are still represented by a single function each. Such a basis set is termed split-valence. Examples include the 3-21G[28,29], 4-31G[30-32], 6-31G[33-36] and 6-311G[37] sets.

Polarization basis sets are usually derived from extended sets by adding basis functions with higher orbital angular momentum quantum numbers (for example, d, f, ... on C, p, d, ... on H). Simple basis sets of this kind include 6-31G(d)[35,36,38] (sometimes written 6-31G*), formed from the 6-31G representation by adding a single set of d functions to non-hydrogen atoms, 6-31G(d,p) (sometimes written 6-31G**) which has, in addition, a single set of p functions on hydrogen atoms, and 6-311G(d,p)[37] (sometimes written 6-311G**).

Diffuse-augmented basis sets are usually derived from extended or polarization sets by adding one or several diffuse functions (with small exponents ζ). As we shall see, these extra functions are especially important for the accurate description of negatively charged systems (for an early recognition of this requirement, see ref. 39) and, in particular, as demonstrated by Schleyer and co-workers, for energy comparisons (see, for example, ref. 40 and references therein). Examples include 3-21+G[40], derived from the 3-21G representation by adding a set of diffuse s and p functions to non-hydrogen atoms, and 6-31+G(d)[41], derived from 6-31G(d) in the same manner.

Basis sets are usually developed for free atoms or small representative molecules and then used without further modification for larger systems.

As a general rule, estimates of the energy and geometry of a system can be improved by either optimizing a basis set for the problem at hand, or by simply using a larger basis set. It is generally more profitable to take the latter approach and to employ larger and more flexible sets. It is important to note that different basis sets of the same type (e.g. two different split-valence basis sets) usually lead to similar structures and relative energies. This is especially true for large (extended and larger) sets. However, predictions made with basis sets of different types are often quite different.

The computational effort involved in *ab initio* calculations increases rapidly with the number of basis functions used to expand the wavefunction. Calculations with around one hundred functions are now considered routine. This allows reasonably accurate calculations for organic molecules with three or four first- or second-row atoms. For large molecules (e.g. benzene and larger) one is, however, still forced to use rather small basis sets, and results obtained may be of limited reliability.

(b) *The Hartree-Fock method*. The second problem one has to face is the selection of configurations in eq. 2. In a first approximation, one simply neglects all but the dominant configuration Ψ_0, which, in the closed-shell case, corresponds to the n molecular orbitals of lowest energy being pairwise occupied by the 2n electrons. This strategy leads to a relatively straightforward iterative method for optimizing wavefunctions, the so-called *Hartree-Fock* (HF) method. It is well established that the HF method is generally quite successful (at least if adequate basis sets are used) in predicting both the shape of potential energy surfaces in the vicinity of minima and the relative energies of isomers. However, this method is less reliable (and does in fact sometimes fail quite spectacularly) for energy comparisons between species with different bonding characteristics, such as molecules in different electronic states, dissociation processes, diradical systems, and so on. By virtue of its basic simplicity and modest computational requirements, the HF method is nevertheless widely used, and most calculations discussed in Sect. IV are of this type.

(c) *Correlated wavefunctions*. That part of the total electronic energy which is neglected by truncating eq. 2 after the first term, i.e. by using the HF method, is called the *correlation energy*. It can be evaluated approximately in a variety of ways, all of which add significantly to the computational effort involved in *ab initio* calculations (for a recent review, see ref. 42).

A configuration expansion which is sometimes used includes Ψ_0 and all doubly excited configurations; this is termed configuration interaction with doubles (CID). More commonly, singly excited configurations are also admitted into the wavefunction, leading to the CISD method (for a discussion of the methods of configuration interaction, see ref. 43). An alternative and, in some respects, more convenient method of accounting for the correlation energy is via Møller-Plesset perturbation theory.[44] Møller-Plesset methods are denoted by the order at which the perturbation expansion is truncated, the most common being second[45] (MP2), third[45] (MP3) and fourth[46,47] (MP4) orders. Other schemes for accounting for electron correlation are also available, the pair-correlation methods IEPA and CEPA being examples.[48]

(2) Semiempirical methods. There are basically two different classes of semiempirical methods, both of which seek to reduce the computational expense of *ab initio* methods by neglecting or approximating certain difficult mathematical expressions. Empirical parameters are often introduced to compensate the resulting errors. Although some semiempirical methods do include correlation effects, most are of the single-configuration HF type.

In the CNDO/2 and related INDO methods, the parameters are chosen so as to reproduce the results of *ab initio* HF calculations with minimal basis sets.[49]. These semiempirical schemes are known to account reasonably well for the geometries of most molecules, but they are expected and indeed known to be less successful for energy comparisons. With the recent advances in computer technology, the CNDO-type methods have largely been superseded by *ab initio* methods and are now rarely used except for very large systems, such as molecules of biological interest.

In the MINDO, MNDO and, most recently, AM1 methods, the aim is to compensate not only for the mathematical approximations, but also for the use of small basis sets and the neglect of electron correlation. Such methods are therefore parametrized to reproduce experimental structural data and heats of formation.[50,51] Lack of appropriate experimental data often hampers the extension of these methods to new areas, and the optimization of parameters is a difficult task. However, if used with discretion, methods of this type can be very efficient and satisfactory tools for the exploration of potential energy surfaces.

The extended Hückel method represents the simplest kind of semiempirical theory which can be applied to molecules of any type and composition.[52] The assumptions inherent in this method are, however, so severe that it is nowadays best viewed as a tool to support qualitative arguments rather than as a method to generate potential energy surfaces. In particular, it is well established that the extended Hückel method fails for polar or charged systems and cannot reliably predict molecular geometries.

III. PREDICTIVE CAPABILITIES OF QUANTUM CHEMICAL METHODS

All the quantum chemical methods, both *ab initio* and semiempirical, that we have mentioned above are *approximate*, the former because of the limited basis sets and configuration expansions used, and the latter, in addition, because of their simplifying assumptions and empirical parametrizations. They may or may not give results in good agreement with experiment, and must therefore be evaluated thoroughly before they can be used as predictive tools. Experience shows that methods which perform well for certain properties and certain classes of molecules quite often fail when applied to other properties or different molecules. For example, a method which yields satisfactory geometries and relative energies of neutral hydrocarbons may well produce quite poor hydrocarbon spectra or quite poor geometries for ionic systems. It is therefore important to test any theoretical method in the intended area of application.

Semiempirical methods must obviously be tested against empirical data such as structural information, heats of reaction, excitation energies, activation parameters, and so forth. *Ab initio* methods are usually tested in the same way. However, in the absence of experimental data, they can also be tested purely theoretically, by examining the convergence of calculated properties with respect to increasing sophistication (size of basis set, configuration expansion) of the method used. This aspect represents a major advantage of *ab initio* over semiempirical methods in areas where few reliable experimental data are available for comparison.

(A) Geometries

Relative energies of molecules can be calculated at experimental, assumed (model), or theoretically determined geometries. Use of experimental geometries appears to be a reasonable prescription for reducing the computational effort involved in geometry optimizations, especially if used in conjunction with theoretical methods which would have predicted geometries close to the experimental structures. However, this would not allow application of theory to molecules for which experimental structures are not available (such as transient species) or not obtainable as a matter of principle (such as transition structures). On the other hand, to use model structures is quite suspect if the aim of the calculations is to predict relative energies, since these are clearly dependent on whatever geometrical assumptions are made. As a rule, relative energies are best calculated at the theoretically determined equilibrium geometries. The ability to reproduce known experimental geometries is therefore one of the more important criteria of a good theoretical model, even though the geometries themselves may not be the main point of interest in many investigations.

Both semiempirical and *ab initio* methods have been tested quite thoroughly for small neutral molecules for which structural data are available. We only summarize the main results here and refer the reader to the extensive literature[8,49,51,53-55] for further details.

Extended Hückel theory is generally incapable of predicting reasonable geometries, even for simple hydrocarbon molecules for which almost every other method gives reasonable results. For example, extended Hückel theory yields CC bond lengths of 1.92 and 0.85 Å for ethane and acetylene, respectively,[52] whereas the experimental values are 1.54 and 1.21 Å.

CNDO/2 performs reasonably well for hydrocarbons; compared with experimental results, bond lengths are usually reproduced to within 0.03 Å and bond angles to within 3°. Errors are often unacceptably large in molecules containing heteroatoms. Bonds between electronegative atoms in particular are not handled satisfactorily; for example, the NN bond length in hydrazine is underestimated by 0.12 Å.[49] MINDO/3 and MNDO are generally more reliable than CNDO/2. Bond lengths in hydrocarbons are usually accurate to 0.01 Å; no marked improvement over CNDO/2 is apparent, however, for bond angles, which are often overestimated by about 5°. Bonds between electronegative atoms are handled somewhat better than in CNDO/2, but problems still remain.[53] The more recent AM1 method[51] appears to be superior to its predecessors MINDO and MNDO.

Ab initio calculations with minimal basis sets give results of comparable quality to MINDO/3; bond angles are often too small by several degrees. Again, bonds between electronegative elements are not handled satisfactorily.[8] Some improvement in bond lengths is obtained at the extended basis set level. However, bonds between electronegative atoms remain troublesome, and bond angles are usually overestimated by about 5°, in particular at atoms having lone pairs. Polarization functions are important for accurate (within 2°) predictions of bond angles. They also improve bond length predictions for the "difficult" bonds. With very few exceptions, bond lengths are accurate to 0.04 Å (usually too short).[54] Inclusion of electron correlation at the MP2 level usually has the effect of increasing bond lengths so that many bonds are now longer than experiment by up to 0.03 Å.[54] Excellent agreement with experiment is obtained with polarization basis sets at the MP3 level, typical deviations being <0.01 Å.[8,55]

From a practical point of view, both MNDO/AM1 and ab initio methods with extended basis sets can be used to predict adequate geometries for most neutral organic molecules. However, this does not necessarily imply that good predictions can also be made for anions. In Tables 1 and 2 we compare calculated geometries for a small set of anions with experimentally determined structures.[56-62] It should be kept in mind when comparing experimental with theoretical structures (which refer to isolated anions in the gas phase) that some of the experimental data come from X-ray diffraction measurements in the solid state.

Shown in Table 1 are theoretical geometries determined at the semiempirical MNDO and AM1 and ab initio Hartree-Fock levels. In general, the MNDO and AM1 geometries are in reasonable agreement with experiment, an exception being the bond length in BH_4^- (too short by ≈0.07 Å). The ab initio STO-3G bond lengths in NH_2^-, BH_4^- and NO_2^- are poorly described, while the HNH angle in NH_2^- is underestimated by ≈9°. In most cases, improved geometries are obtained with the 3-21G basis set. Diffuse functions seem to be necessary to describe well the essentially localized anions OH^- and NH_2^-, although the importance of adding these additional diffuse functions diminishes as the size of the underlying sp basis increases. For example, the bond length in OH^- and the HNH angle in NH_2^- are changed substantially on adding diffuse functions to the 3-21G basis set (3-21G → 3-21+G), but are hardly altered on adding diffuse functions to the large triple-zeta set of Lee and Schaefer.[66] Of greater importance in some cases is the addition of polarization functions to the basis set. For example, the NO distance in NO_2^- is reduced by 0.06 Å on going from 3-21G to 6-31G(d).

The effect of inclusion of electron correlation on anion geometries is demonstrated by the results in Table 2. The trends

TABLE 1

Semiempirical and *ab initio* Hartree-Fock geometries of anions[a]

Level[b]	OH^-	BH_4^-	FHF^-	CN^-	N_3^-	NH_2^-		NO_2^-	
	r_{OH}	r_{BH}	r_{FH}	r_{CN}	r_{NN}	r_{NH}	<HNH	r_{NO}	<ONO
MNDO[c]	0.939	1.184[d]	1.144[d]	1.178	1.168	1.013	100.9	1.215	116.5
AM1[d]	0.949	1.186	1.084	1.179	1.171	1.006	102.6	1.212	121.3
STO-3G[e]	1.068	1.176	1.111	1.162	1.202	1.080	95.2	1.294	114.3
3-21G[f]	1.029	1.241	1.145[g]	1.166	1.176[g]	1.066	98.0	1.286	116.4
3-21+G[g]	0.978	1.244	1.152	1.171	1.177	1.029	105.9	1.285	116.5
6-31G(d)[f]	0.962	1.243	1.127[g]	1.161	1.156[g]	1.030	99.3	1.229	116.7
6-31+G(d)[f]	0.953	1.245[d]	1.134[d]	1.162[d]	1.156[d]	1.018	103.5	1.225[d]	117.3[d]
TZ[h]	0.968			1.166		1.028	105.2		
TZ+diff[i]	0.967			1.167		1.025	106.1		
TZ+pol[j]	0.948			1.154		1.107	102.2		
NHFL[k]	0.944			1.152		1.012	103.1		
Expt[l]	0.97	1.25	1.14	1.18	1.18	1.03	104	1.24	115

[a] Bond lengths in ångströms, bond angles in degrees.
[b] See Sect. IIB1a for a description of the basis sets.
[c] Ref. 63 unless otherwise stated.
[d] Present work.
[e] Ref. 64.
[f] Ref. 65 unless otherwise stated.
[g] Ref. 40.
[h] Triple-zeta basis set. Ref. 66.
[i] Triple-zeta basis plus diffuse functions. Ref. 66.
[j] Triple-zeta basis plus polarization functions. Ref. 66.
[k] Near-Hartree-Fock-limit results. Ref. 66.
[l] Refs. 56-62, respectively.

here seem to parallel those for neutral systems. In particular, Hartree-Fock bond lengths with the polarization basis set used here are, in general, too short relative to the experimental values. Calculations at the MP2 level lead in many cases to bond lengths which overestimate the experimental values, while at the MP3 level good agreement with experiment is obtained (at least within the limited reliability of the comparisons presented here).

TABLE 2

Effect of electron correlation on calculated geometries of anions[a]

Level[b]	OH^-	BH_4^-	FHF^-	CN^-	N_3^-	NH_2^-		NO_2^-	
	r_{OH}	r_{BH}	r_{FH}	r_{CN}	r_{NN}	r_{NH}	<HNH	r_{NO}	<ONO
HF[c]	0.948	1.245	1.125	1.162	1.156	1.016	103.4	1.225	117.3
MP2	0.970	1.232	1.149	1.201	1.220	1.029	103.0	1.278	116.0
MP3	0.964	1.234	1.139	1.182	1.184	1.027	102.5	1.256	116.3
Expt[d]	0.97	1.25	1.14	1.18	1.18	1.03	104	1.24	115

[a] Bond lengths in ångströms, bond angles in degrees. All results have been obtained in the present study with the 6-31++G(d,p) basis set.

[b] Correlation introduced via Møller-Plesset perturbation theory terminated at second (MP2) or third (MP3) order.

[c] Hartree-Fock results.

[d] Refs. 56-62, respectively.

In summary, it would appear that, in general, theory is equally as successful in predicting the geometries of anions as it is for neutral systems. Localized anions require special care and for such systems the use of a large basis set or the inclusion of diffuse functions is recommended.

(B) Proton affinities

Experimental gas-phase proton affinities are now available for many anions and can be used to assess the accuracy of theoretical methods. The experimental proton affinity (PA) generally refers to $-\Delta H^{\circ}_{298}$ for the reaction

$$A^- + H^+ \rightarrow AH, \quad (3)$$

and therefore the experimental values should be corrected for zero-point-vibrational and thermal energies before being compared with calculated values which refer to motionless molecules at 0 K. Because of the gain in vibrational degrees of freedom upon protonation, these corrected PA values are usually higher than the directly measured experimental values.

Proton affinities for a number of small anions, calculated at the MNDO/AM1 and *ab initio* Hartree-Fock levels, are compared with experimental values in Table 3. In general, the MNDO and AM1

TABLE 3

Semiempirical and *ab initio* Hartree-Fock proton affinities of anions[a]

Level[b]	CH_3^-	NH_2^-	OH^-	F^-	CN^-
MNDO[c]	1823	1760	1766	1721	1619
AM1[d]	1594	1566	1505	1641	1370
STO-3G[e]	2342	2289	2365	2519	1936
3-21G[c]	1936	1933	1883	1807	1586
3-21+G[c]	1806	1748	1637	1503	1470
6-31G(d)[e]	1912	1860	1830	1713	1548
6-31+G(d)[d]	1816	1763	1684	1566	1480
6-311++G(d,p)[f]	1814	1767	1699	1592	1486
6-311++G(3df,3pd)[f]	1814	1766	1705	1604	1489
Expt[g]	1782±4	1729±9	1663±2	1574±3	1486±8

[a] In kJ mol^{-1}.

[b] See Sect. IIB1a for a description of the basis sets.

[c] Ref. 67.

[d] Data from ref. 65 and present work.

[e] Data from ref. 65.

[f] Ref. 68. Geometries optimized at the CID/6-31G(d) or CID/6-31+G(d) levels.

[g] Experimental proton affinities at 298 K corrected for zero-point vibrational energy and to 0 K using data from ref. 68.

results are poor, with errors of up to 200 kJ mol^{-1}. *Ab initio* calculations with minimal, extended and polarization basis sets fare even worse, the largest errors being 945 (STO-3G), 233 (3-21G) and 167 (6-31G(d)) kJ mol^{-1}. Addition of diffuse functions to the basis set is essential in order to obtain reasonable proton affinities for anions. With the diffuse-function-augmented 3-21+G and 6-31+G(d) sets, the largest errors drop to just 71 and 34 kJ mol^{-1}, respectively. With the exception of CN^-, the two largest basis sets yield proton affinities which are uniformly too large by 30-40 kJ mol^{-1}.

The effect on proton affinities of inclusion of electron correlation is demonstrated by the results in Table 4. At the MP4 level, excellent agreement is found between theory and experiment. Interestingly, it has been noted[68] that the even orders of perturbation theory (MP2 and MP4) are in general superior to the odd orders (HF and MP3) for calculating proton affinities.

TABLE 4

Effect of electron correlation on calculated proton affinities of anions[a]

Level[b]	CH_3^-	NH_2^-	OH^-	F^-	CN^-
HF[c]	1814	1766	1705	1604	1489
MP2	1776	1719	1659	1577	1490
MP3	1790	1744	1690	1600	1496
MP4	1781	1725	1664	1580	1491
Expt[d]	1782±4	1729±9	1663±2	1574±3	1486±8

[a] Proton affinities in kJ mol^{-1}. Based on calculations using the 6-311++G(3df,3dp) basis set and CID/6-31G(d) or CID/6-31+G(d) geometries. Taken from ref. 68.

[b] Correlation introduced via Møller-Plesset perturbation theory terminated at second (MP2), third (MP3) or fourth (MP4) order.

[c] Hartree-Fock results.

[d] See footnote g to Table 3.

In summary, diffuse-function-augmented basis sets are essential for reliable *ab initio* calculation of the proton affinities of anions. Provided that a large enough basis set is used and that electron correlation effects are taken into account, results of comparable accuracy to those for neutrals and in excellent agreement with experiment can be obtained.[68] The MP2/6-311++G(d,p) level has been suggested[68] as a useful level at which to carry out anion proton affinity calculations.

(C) *Inversion barriers*

Reliable experimental values for inversion barriers in carbanions are not available. As mentioned above, the performance of *ab initio* methods in this area can nevertheless be tested by examining the convergence of calculated barriers with respect to increasing sophistication of the method used. Tables 5 and 6 illustrate this procedure for pyramidal inversion in the methyl anion, CH_3^-, and in the isoelectronic ammonia, NH_3, where an experimental value is available. The semiempirical methods MINDO/3, MNDO and AM1 perform rather poorly for CH_3^-; compared with the most accurate *ab initio* values, MNDO and AM1 underestimate the inversion barrier (actually predicting a planar methyl anion) while MINDO/3 overestimates it. *Ab initio* calculations with small basis sets (STO-3G, 3-21G) also perform poorly. The diffuse-augmented 3-21+G basis leads to an

TABLE 5

Calculated semiempirical and *ab initio* Hartree-Fock inversion barriers for methyl anion and ammonia[a]

Level[b]	CH_3^-	NH_3
MINDO/3[c]	74	25
MNDO	0[d]	48[e]
AM1[e]	0	18
STO-3G[f]	100	47
3-21G[g]	38	7
3-21+G[g]	4	2
6-31G(d)[g]	54	27
6-31+G(d)[g]	13	23
6-31++G(d,p)[e]	11	18
6-311++G(d,p)[e]	13	19
6-311++G(df,p)[e]	12	17
Best[h]	7-9[i]	22[j]
Expt		22[k]

[a] In kJ mol^{-1}.
[b] See Sect. IIB1a for a description of the basis sets.
[c] Ref. 69.
[d] Ref. 63.
[e] Present work.
[f] Ref. 65.
[g] Ref. 40.
[h] Extended basis sets with two sets of diffuse functions and either one or two sets of d polarization functions.
[i] Refs. 70, 71 and present work.
[j] Near-Hartree-Fock-limit value. Ref. 72.
[k] Ref. 73.

underestimation of the barrier in ammonia, but gives a barrier in CH_3^- which is (possibly fortuitously) close to the most reliable values. The polarization 6-31G(d) basis set performs quite well for NH_3, but grossly overestimates the barrier in CH_3^-. Clearly, diffuse functions are necessary to describe well the barrier in the anion: the diffuse-augmented 6-31+G(d) basis gives a result far superior to that obtained with 6-31G(d). The best calculations shown in Table 5, which include in particular a second set of diffuse basis functions, result in a further small

TABLE 6

Effect of electron correlation on calculated inversion barriers for methyl anion and ammonia[a]

Level[b]	CH_3^-	NH_3
HF[c]	9	19
MP2	9	21
MP3	8	22
MP4	8	23
Best	0-8[d]	23[e]
Expt		22[f]

[a] Inversion barriers in kJ mol^{-1}. Unless otherwise stated, all results were obtained in the present work using the 6-311++G(d,p) basis set for NH_3 and the 6-311++G(2d,p) basis set with an extra diffuse sp shell on carbon for CH_3^-.

[b] Electron correlation introduced via Møller-Plesset perturbation theory terminated at second (MP2), third (MP3) or fourth (MP4) order.

[c] Hartree-Fock results.

[d] Correlated calculations with large basis sets. Refs. 71, 74-76 and present work.

[e] MP3 calculation with large basis set. Ref. 72.

[f] Ref. 73.

decrease in the computed barrier for CH_3^-. The effect of electron correlation on the pyramidal inversion barriers in CH_3^- and NH_3, as demonstrated by the results in Table 6, is small.

A very recent paper[76] which deals with inversion in CH_3^- exposes an important general problem for anion calculations. It happens that the experimental electron affinity for CH_3^- is very small and positive but that, at the highest theoretical levels employed to date, the calculated electron affinity is negative (see Sect. IIID). Under these circumstances, the best theoretical description of CH_3^- corresponds to that of the methyl radical with an additional electron located in a very diffuse orbital. The consequence of relevance to this section is that with a sufficiently flexible, high-level theoretical procedure, the methyl anion emerges with a planar geometry (i.e. that of the methyl radical).[76] It would seem (paradoxically) that this is a less reliable result than those obtained at slightly lower

TABLE 7

Calculated barriers to linear inversion in the vinyl anion and in formaldimine[a]

Level[b]	CH_2CH^-	CH_2NH
MNDO	8	112
AM1	28	91
STO-3G	211	172[c]
3-21G	152	109[c]
3-21+G	127[d]	101
6-31G(d)	168	136[c]
6-31+G(d)	143[d]	130
6-31++G(d,p)	141	128
MP2/6-31+G(d)[e]	126	127
MP3/6-31+G(d)[e]	133	130
MP4/6-31+G(d)[e]	132	130

[a] In kJ mol^{-1}. All results from present work unless otherwise specified.
[b] See Sect. IIB1a for a description of the basis sets. MP2, MP3 and MP4 refer to Møller-Plesset calculations terminated at second, third and fourth orders, respectively.
[c] Ref. 65.
[d] Ref. 77.
[e] Using HF/6-31+G(d) geometries.

theoretical levels.

Calculated barriers to linear inversion in the vinyl anion, CH_2CH^-, and in the isoelectronic formaldimine molecule, CH_2NH, are shown in Table 7. The trends here follow closely those for pyramidal inversion in CH_3^- and NH_3. MNDO/AM1 and *ab initio* calculations with small basis sets fare poorly: addition of polarization functions and, in the case of vinyl anion, diffuse functions to the basis set has a large effect. In the case of formaldimine, inclusion of electron correlation changes the barrier by an insignificant amount; for the vinyl anion, inclusion of correlation at the MP4 level lowers the barrier by 11 kJ mol^{-1}.

In summary, accurate *ab initio* calculation of anion inversion barriers requires the use of diffuse-function-augmented polarization basis sets.

(D) Electron affinities

Gas-phase anions are often only weakly bound with respect to loss of an electron to give the corresponding neutral radical:

$$A^- \rightarrow A^\bullet + e^-. \quad (4)$$

The ability to calculate accurately the enthalpy change of reaction 4 (i.e. the electron affinity of the radical $A^\bullet$) is therefore of great importance in determining whether a particular anion might be an observable species in the gas phase.

Unfortunately, the calculation of accurate electron affinities has proved to be a very difficult problem for theoretical chemistry. Most *ab initio* studies have focussed on the simplest of systems, the first-row atoms. Even with extremely large basis sets and long configuration expansions, errors of up to 30 kJ mol^{-1} are common (see, for example, ref. 78 and references therein). Possibly the best calculations are those of Raghavachari,[79] who performed fourth-order Møller-Plesset calculations with a large sp basis augmented with two sets of diffuse functions and four d-type and two f-type polarization functions. Even here, errors of ±10 kJ mol^{-1} were found for the electron affinities of the boron, carbon, oxygen and fluorine atoms. Similar results have been found[80] for the hydrides of the first-row atoms.

The difficulty for theory arises from the fact that the number of electron pairs is not conserved in reaction 4. The anion A^- has one more pair than the radical $A^\bullet$, and hence has a greater correlation energy. This difference in correlation energies makes it very difficult to treat the anion and radical on an equal footing.

Given the difficulties encountered in treating atoms, it is not surprising that there have been very few systematic studies of the electron affinities of molecules. Shown in Table 8 are some selected results for a small number of systems in which the anion has a closed-shell electron configuration. MNDO and AM1 in general fare poorly, particularly in cases where the negative charge is essentially localized on a single atom.[63] *Ab initio* Hartree-Fock calculations, even with large basis sets, yield very poor results, with errors of up to 200 kJ mol^{-1}. Introduction of electron correlation via Møller-Plesset perturbation theory to fourth order improves the situation considerably; however, large errors (up to 60 kJ mol^{-1}) still remain. It is particularly disturbing that the MP4 calculations *overestimate* the electron

TABLE 8

Calculated adiabatic electron affinities of doublet radicals[a]

Level[b]	$F^{\cdot}$	$OH^{\cdot}$	$NH_2^{\cdot}$	$CH_3^{\cdot}$	$CH_3O^{\cdot}$	$C_2H^{\cdot}$	$CN^{\cdot}$
MNDO[c]	151	26	-42	-129	166	243	310
AM1[d]	65	63	-59	-111	175	249	294
HF	119	32[e]	-108	-151	-36[f]	149	283
MP2	329	169[e]	64	-18	133[f]	331	447
MP3	276	113[e]	28	-29	90[f]	309	424
MP4	311	158	52	-17		320	427
CISD(Q)[g]	287	132	32	-29		261	344
EOM[h]	343	177	48	-1		357	431
Expt[i]	328	177	73	8	153	284	369

[a] In kJ mol^{-1}. Unless otherwise stated, all calculations performed with 6-311++G(2d,2p) basis and taken from ref. 81.

[b] HF refers to a restricted Hartree-Fock calculation for the anion and an unrestricted Hartree-Fock calculation for the radical. Electron correlation introduced via Møller-Plesset perturbation theory terminated at second (MP2), third (MP3) or fourth (MP4) order.

[c] Ref. 63.

[d] Present work.

[e] Triple-zeta basis plus two diffuse p shells and two sets of d-type polarization functions. Ref. 82.

[f] 6-31+G(d) basis set. Ref. 82.

[g] Spin-restricted singles-and-doubles configuration interaction calculation with correction for quadruple excitations.

[h] Equations-of-motion calculation with 6-311++G(d,p) basis set. Ref. 81.

[i] All experimental values from ref. 83.

affinities of $C_2H^{\cdot}$ and $CN^{\cdot}$; this has been ascribed[81] to the presence of spin contamination in the unrestricted Hartree-Fock reference configuration for the radical and the slow convergence of the Møller-Plesset perturbation series under such circumstances. This is supported by the spin-restricted configuration interaction results shown in Table 8, these results uniformly underestimating the experimental electron affinities. Other approaches, such as the Green's function[84] and equations-of-motion[85] methods, may lead to better results for electron affinities; a thorough assessment of these methods is clearly desirable.

The use of thermodynamic cycles[86] and isogyric reactions[87] (in which the number of unpaired spins is conserved during the reaction) has been proposed in an attempt to avoid the difficult problem of calculating the difference in correlation energy of the anion and radical. For example, the electron affinity of the ethyl radical might be estimated from the calculated enthalpy change of the isogyric electron-transfer reaction

$$CH_3^- + C_2H_5^\bullet \rightarrow CH_3^\bullet + C_2H_5^- \quad (5)$$

together with the known experimental electron affinity of the methyl radical. Such an approach, which relies on cancellation of errors, may prove useful when the two radicals have similar electronic structures. For example, it appears to hold reasonably well for the set of radicals $F^\bullet$, $OH^\bullet$, $NH_2^\bullet$ and $CH_3^\bullet$, the electron affinities at the MP4 level being consistently ≃20 kJ mol^{-1} less than the experimental values. However, an examination of Table 8 indicates that it may break down when systems with markedly different electronic characteristics are compared. For example, at the MP4 level, ΔE for the reaction

$$OH^\bullet + CN^- \rightarrow CN^\bullet + OH^- \quad (6)$$

is calculated to be 269 kJ mol^{-1}, whereas the experimental value is 192 kJ mol^{-1}. Again, this may be traced to the presence of spin contamination in the unrestricted Hartree-Fock treatment of the cyano radical; use of the CISD(Q) results in Table 8 leads to a more reasonable ΔE of 212 kJ mol^{-1}. In summary, care must be exercised in the use of isogyric reactions to estimate electron affinities, especially when comparing species of greatly differing electronic structure. If such an approach is adopted, however, the CISD(Q) level of theory, for which there is the greatest constancy of errors in the calculated electron affinities, would appear to be the method of choice. UMP calculation would only be satisfactory for cases in which spin contamination is unimportant.[81]

It should be emphasized that a calculated <u>negative</u> electron affinity implies that the anion is unstable towards spontaneous ejection of an electron to yield the corresponding radical; under such circumstances, it is only some inadequacy in the theoretical treatment (such as the imposition of a closed-shell electron configuration) which prevents it from doing so. This is exemplified by a recent study[76] of the methyl anion: as noted in Sect. IIIC, when a sufficiently flexible basis set was used in conjunction with a two-configuration SCF procedure, an equilibrium geometry identical to that of the methyl radical was obtained. In other words, an electronic structure corresponding to an almost unperturbed methyl radical with an additional electron in a very diffuse molecular orbital was obtained. However, as pointed out in ref. 76, if the anion is in reality stable, then use of such a two-configuration reference wavefunction does not necessarily provide a better zero-order approximation to the exact solution than does a single closed-shell determinant.

(E) Vibrational frequencies

Over the past five years, methods have been developed to measure the high-resolution infrared spectra of gas-phase cations (see, for example, refs 88 and 89). Guided by high-level *ab initio* calculations,[90] Rosenbaum *et al.*[91] very recently applied such a method to measure the first direct infrared absorption spectrum of a gas-phase negative ion, the hydroxide anion. Accurate theoretical predictions of vibrational frequencies are likely to prove to be of great assistance to spectroscopists in finding vibrational bands of other gas-phase anions.

Harmonic vibrational frequencies for a number of small anions are shown in Table 9. The trends here closely parallel those for anion geometries noted earlier. In particular:

(i) Addition of diffuse functions to the basis set becomes less important as the size of the underlying sp basis increases.

(ii) Diffuse functions are more important for the localized anions (OH^-, NH_2^- and CH_3^-) than for the delocalized systems (CN^- and C_2H^-).

(iii) Addition of polarization functions to the triple-zeta basis of Lee and Schaefer[66] is generally of greater importance than addition of diffuse functions.

(iv) For neutral systems, *ab initio* calculations generally lead to vibrational frequencies which overestimate the experimental values by ≈10%.[93] This parallels the result that, at the Hartree-Fock level, bond lengths are generally predicted to be too short. Similarly, for the anions in Table 9, the near-Hartree-Fock-limit frequencies

TABLE 9

Ab initio harmonic vibrational frequencies for anions[a]

Species	Mode	6-31G(d)[b]	6-31+G(d)[c]	TZ[d]	TZ+diff[e]	TZ+pol[f]	NHFL[g]	Best
OH^-	ν_1	3739	3969	3694	3736	3998	4057	3731[h,i]
CN^-	ν_1	2360	2330	2223	2215	2323	2314	2052[j]
C_2H^-	ν_1			3573	3566	3553	3546	
	ν_2			706	699	667	620	
	ν_3			1988	1980	2033	2029	
NH_2^-	ν_1	3299	3481	3288	3334	3455	3509	
	ν_2	1747	1663	1612	1584	1609	1612	
	ν_3	3328	3555	3359	3427	3513	3574	
CH_3^-	ν_1	2852	3045	2979	3066	2959	3019	
	ν_2	1230	880	856	599	963	800	
	ν_3	2849	3109	3048	3179	3008	3096	
	ν_4	1630	1577	1585	1560	1544	1532	

[a] In cm^{-1}.
[b] Extended basis plus polarization functions. Ref. 68.
[c] Extended basis plus polarization and diffuse functions. Ref. 68.
[d] Triple-zeta basis set. Ref. 66.
[e] Triple-zeta basis plus diffuse functions. Ref. 66.
[f] Triple-zeta basis plus polarization functions. Ref. 66.
[g] Near-Hartree-Fock-limit results. Ref. 66.
[h] MCSCF SCEP calculation. Ref. 90.
[i] Experimental value 3738 cm^{-1}. Ref. 91.
[j] CEPA calculation. Includes anharmonic effects. Ref. 92.

might be expected to be too large. This is indeed the case for the hydroxide ion, the experimental[91] harmonic frequency being ≈8% less than the near-Hartree-Fock-limit value of Lee and Schaefer[66] and very close to the best correlated *ab initio* calculation.[90] Similarly, for CN^-, the best *ab initio* calculation[92] predicts a frequency ≈11% lower than the near-Hartree-Fock-limit value.

IV. APPLICATIONS TO SPECIFIC CARBANIONS

(A) The methyl anion and its derivatives

(1) The parent system CH_3^-. Simple hybridization theory leads one to expect a pyramidal (C_{3v}) rather than a planar (D_{3h}) equilibrium geometry for the methyl anion. Theoretical work[40,41,63,66,68-71,74-76,94-123] generally confirms this qualitative argument. However, as mentioned in Sect. IIIC, the predicted height of the inversion barrier is very sensitive to details of the method used, and values ranging from zero to greater than 100 kJ mol^{-1} have been obtained. Semiempirical methods perform rather badly: either a planar geometry is predicted (CNDO,[98] MNDO[63], AM1[123]) or the barrier height is overestimated by 40 to 80 kJ mol^{-1} (CNDO/2,[98,101,110] INDO,[95,110] MINDO[97]). Ab initio calculations are equally unsatisfactory unless diffuse and polarization functions are included in the basis set (cf. Sect. IIIC). The best current theoretical values[71,74-76,123] for the inversion barrier in the methyl anion are in the range 0-8 kJ mol^{-1}. An early indirect experimental determination of the inversion barrier[124] led to a value greater than 25 kJ mol^{-1} which now appears too high. More recenty, the vibronic structure of the photoelectron spectrum of CH_3^- has been found to be consistent with a much smaller barrier (<9 kJ mol^{-1}).[125]

A high-level calculation[68] of the proton affinity of the methyl anion yielded a value of 1747 kJ mol^{-1}, in excellent agreement with the experimental value of 1744±4 kJ mol^{-1}.

The methyl radical is known experimentally[125] to have a small positive electron affinity (8±3 kJ mol^{-1}). A configuration interaction calculation with a large basis set[71] gave a value of -11 kJ mol^{-1}. Other recent ab initio estimates are -13 kJ mol^{-1} (via Møller-Plesset calculations),[81] -1 kJ mol^{-1} (via an equations-of-motion approach)[81] and -4 kJ mol^{-1} (via a multiconfigurational-SCF configuration-interaction procedure).[76]

The vibrational frequencies of the methyl anion have been calculated using large basis sets at the Hartree-Fock level.[66]

(2) Substituted methyl anions XCH_2^-

(a) General considerations. The influence of substituents on the stability, geometry and inversion barrier of the methyl anion

TABLE 10.

Influence of substituents X on stabilization energies, pyramidalization angles and inversion barriers in XCH_2^-

X	Stabilization energy[a] (kJ mol^{-1})		Angle (°)[b]		Inversion barrier (kJ mol^{-1})	
	4-31G[c]	"Best"[d]	4-31G[c]	4-31+G[e]	4-31G[f]	"Best"[g]
H	0	65	38	34	4-8[h]	
Li	73	34	0	19		13
BeH	169	138	0	0		0
BH_2	283	240	0	0		0
CH_3	9	-13	60	49	34[i]	10[i,j]
NH_2	22	-3	62	52	42[i]	
OH	66	23	71	66	85[i]	
F	103	38	76	58	99	58
SiH_3		106		11[k]		
PH_2		93		32[k]		
SH		61		73[k]		
Cl		68		88[k]		
CN	256		0	2[l]	0	0[l]
NC	156[m]		55[m]		12[m]	
NO_2	410		0		0	
CF_3	238		31			
CH_2CH_3	23		59			
$CHCH_2$	157		0		0	
CCH	185		1		0	
CHO	299		0		0	
$COCH_3$	284[f]		0[f]		0	
C_6H_5	186[n]		0[n]			

[a] ΔE for the reaction $XCH_2^- + CH_4 \rightarrow XCH_3 + CH_3^-$.

[b] Angle between the C-X bond and the HCH plane. For planar XCH_2^- this angle is 0°; for "sp^3" XCH_2^- it is 54.8°.

[c] Ref. 118 unless otherwise indicated.

[d] MP2/6-31+G(d) for first-row substituents and MP2/6-21+G(d) for second-row substituents using geometries optimized with either the 3-21+G or 4-31+G basis set. From refs. 11 and 41.

[e] From ref. 41 unless otherwise indicated.

[f] From ref. 116 unless otherwise indicated.

[g] MP2/6-31+G(d)//4-31+G unless otherwise indicated. From ref. 41.

[h] See Sect. IVA1.

[i] Using model geometry for transition structure.

[j] MP2/4-31+G//4-31+G. From ref. 41.

[k] 3-21+G values. From ref. 11.

[l] HF/6-31+G(d) calculation. Present work.

[m] Double-zeta basis set. From ref. 129.

[n] Partially optimized STO-3G geometry.

has been the subject of numerous theoretical studies, often with a view towards disentangling the various effects which have traditionally been held responsible for the conformational stability or instability of carbanions in solution. We first discuss the effects of substituents in general terms and then turn to some systems which have been studied in greater detail.

Systematic _ab initio_ studies of substituted methyl anions XCH_2^- have been carried out by several groups.[11,41,116,118,122,126-128] Some of this work has been performed using basis sets which, as we have seen in Sect. III, perform poorly in the calculation of certain properties of anionic systems. The ability of such basis sets to describe satisfactorily the effect of substituents, i.e. relative rather than absolute quantities, may be assessed by comparisons with results from better calculations (using diffuse-function-augmented basis sets) as shown in Table 10. In general, the simpler calculations are useful in a qualitative sense but there are exceptional cases, as noted below.

The results in Table 10 generally confirm expectations based on qualitative molecular orbital arguments. When the substituent X is a π acceptor (X=Li, BeH, BH_2, CN, NO_2, $CHCH_2$, CCH, CHO, $COCH_3$ and C_6H_5), the anion is strongly stabilized. In such cases, stabilization of XCH_2^- can take place through interaction of the CH_2^- lone-pair-type orbital with an empty π-type orbital on X. This stabilizing two-electron interaction (Fig. 1) is accompanied by a reduction in the C-X bond length, which facilitates orbital overlap. Interaction of the high-lying orbital of the CH_2^- fragment with the vacant orbital at X is also facilitated by flattening of the CH_2^- pyramid. Such a flattening increases the p character of the lone-pair-type orbital making it a better donor and also allows improved overlap with the orbital at X. Indeed, for the π-electron-accepting substituents X, the XCH_2^- anions are found in nearly all cases to be planar or almost planar.

Despite the electropositive nature of Li, Be and B, stabilizing effects are found for the Li, BeH and BH_2

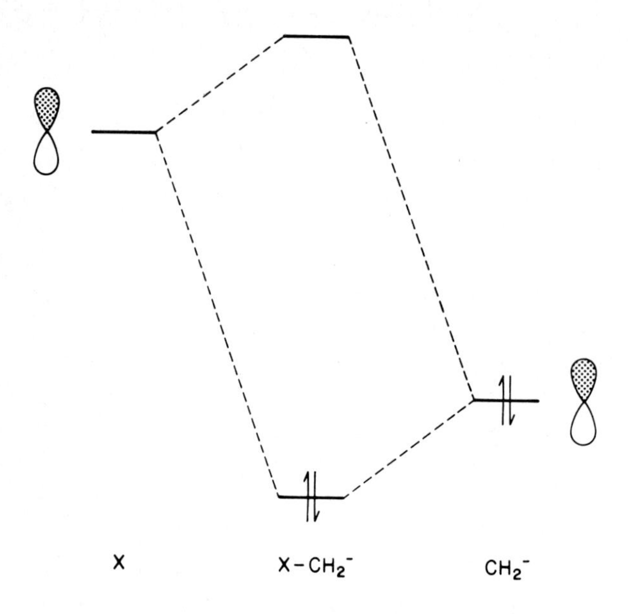

Fig. 1. Two-electron stabilizing interaction between the lone-pair orbital on CH_2^- and a vacant p orbital on X in a planar XCH_2^- anion.

substituents as a result of their low-lying empty orbitals. The best calculations,[41] however, indicate a pyramidal structure for (singlet) $LiCH_2^-$ with an inversion barrier of 13 kJ mol^{-1}. It has been suggested[41] that the electronic ground state of $LiCH_2^-$ may be a triplet, as one would expect for a loose complex of methylene and Li^-. $NaCH_2^-$ is found[126] to be pyramidal, at least at the minimal-basis level.

The trifluoromethyl substituent also leads to strong stabilization and a decreased pyramidalization angle, although again the carbanion centre retains its nonplanar conformation. That CF_3 acts as a π-electron acceptor can be thought of as a manifestation of (negative) hyperconjugation. This topic is dealt with in more detail in Sect. IVB2.

Calculations with small basis sets have suggested that the methyl and ethyl substituents are capable of favourable hyperconjugative interaction with an adjacent lone pair: at the 4-31G level, for example, the carbanionic centre is flattened by a few degrees and the inversion barrier is reduced marginally.[116] In addition, the C-H bond anti to the lone pair is weakened, as one would expect if an electron pair is partially delocalized into an antibonding σ^*_{C-H} acceptor orbital. However, improved calculations which include diffuse basis functions[41,130] find that the methyl substituent *destabilizes* the methyl anion and leads to an increase in the pyramidalization angle; the weakening of the *anti* C-H bond is, however, still observed.

Substituents with lone pairs, such as NH_2, OH and F, stabilize the pyramidal relative to the planar conformation of XCH_2^- and thus increase the pyramidalization. The OH and F substituents also lead to a stabilization of the methyl anion, this time through σ-electron withdrawal. $NH_2CH_2^-$ has a *gauche* conformation similar to that of the isoelectronic hydrazine molecule, and in $HOCH_2^-$ (isoelectronic with hydroxylamine) the sp^2 lone pair on oxygen and the carbanionic lone pair are *anti*. These effects are readily rationalized as the result of a four-electron destabilizing interaction of the two lone pairs involved.[118] An alternative interpretation has also been proposed.[131]

The second-row substituents (SiH_3, PH_2, SH and Cl) stabilize the methyl anion much more effectively than their first-row counterparts. This has been ascribed to the enhanced polarizability of the second-row atoms and/or the availability of low-lying σ^* orbitals available for negative hyperconjugation (see Sects. IVA2f and IVA2g).

(b) *Nitromethyl anion*. The nitromethyl anion has been studied at various semiempirical[67,132,133] and *ab initio*[40,67,116,118,122,133-136] levels. It is uniformly predicted to adopt a planar C_{2v} conformation (1). Rotation about the formal C-N single bond does not proceed via the perpendicular structure (2), but rather via (3), which has a pyramidal configuration at carbon.

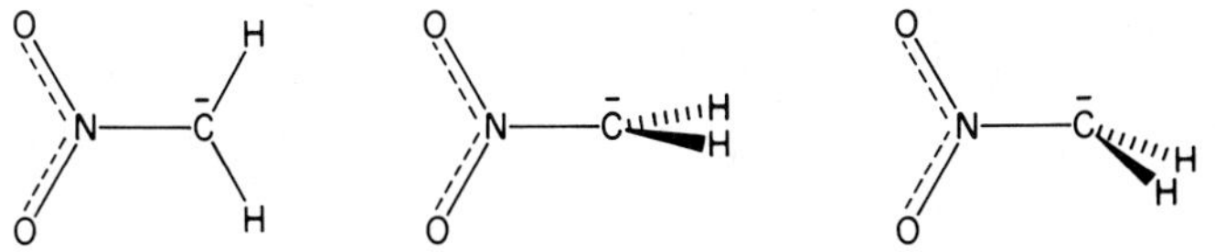

(1) (2) (3)

A barrier of 67 kJ mol^{-1} has been calculated[134] for the rotation-inversion process. On the other hand, the di- and tri-nitromethyl anions are predicted to be nonplanar, the latter species having a "propeller-like" C_3 geometry.[122]

Since the nitromethyl radical $O_2NCH_2^{\bullet}$ has a relatively large dipole moment (calculated value[136] >4 debyes), the nitromethyl anion is expected to be stable towards electron loss.[137,138] However, the calculated[136] electron affinity of the radical, 238 kJ mol^{-1}, is in poor agreement with an experimental estimate[139] of ≈50 kJ mol^{-1}.

Protonation of the nitromethyl anion can occur at C or O, leading to nitromethane or its aci tautomer. The experimental gas-phase proton affinity[140] is PA(C) = 1500±10 kJ mol^{-1}. Ab initio PA(C) values are dependent on the particular method used (minimal basis: 1989,[141] double-zeta: 1485,[116] split-valence: 1510-1530,[116,118,134,135] diffuse-augmented split-valence: 1432-1466,[40,67,141] polarized: 1596[118]). Theory clearly has not yet converged here; calculations with diffuse-augmented polarization basis sets and incorporating electron correlation would probably be needed to obtain quantitative agreement with experiment (see Sect. IIIB).

(c) Cyanomethyl and substituted cyanomethyl anions. In the cyanomethyl anion, there seems to be an almost perfect balance between the σ- and π-accepting effects of the cyano substituent. The potential energy surface is very shallow along the pyramidalization coordinate. Semiempirical calculations predict a planar (C_{2v}) species.[63,142,143] The results of ab initio calculations show a marked qualitative dependence on the type of basis set used: $NCCH_2^-$ is found to be pyramidal at the minimal-basis-set level,[120,122,144,145] planar if split-valence or double-zeta basis sets are used[116,118,130,144,146] and pyramidal again if polarization functions[144] or polarization and diffuse functions[123] are added. The role of electron correlation has not yet been established. It appears likely, however, that the isolated (gas-phase) cyanomethyl anion is quasi-planar and that the inversion barrier is at most of the order of a few kilojoules per mole. On the other hand, the di- and tri-cyanomethyl anions are planar even at the minimal-basis-set ab initio level.[122] Stereochemical studies of cyano-substituted carbanions in solution indicate that these species are intrinsically planar.[3]

Both σ induction and π resonance stabilize the negative charge in $NCCH_2^-$. Acetonitrile is a much stronger acid than CH_4, both in solution (pK_a = 31.3)[147] and in the gas phase (PA = 1557 or 1563 kJ mol^{-1}).[140,148] While the theoretical proton affinities reproduce this trend qualitatively,[40,67,116,118,120,129,141] the calculated values show the expected (cf. Sect. IIIB) strong basis-set dependency. Calculations at a level sufficient to reproduce the proton affinity quantitatively have not yet been reported.

The relationship between π delocalization and stability in the cyanomethyl anion has been studied using a valence-bond analysis and perturbation molecular orbital arguments.[149]

Methyl substituents have been predicted[129] to destabilize the cyanomethyl anion: the proton affinities calculated with a diffuse-augmented basis set for $NCCHCH_3^-$ and $NCC(CH_3)_2^-$ are 24 and 38 kJ mol^{-1}, respectively, higher than that for $NCCH_2^-$.

As in other carbanion systems, α chlorine substitution is expected to increase the inversion barrier. However, a prediction[145] that $NCCHCl^-$ is pyramidal should be viewed with caution because of the small basis set used. $NCCH(C_6H_5)^-$ has been calculated[145] to be planar, in apparent agreement with infrared and proton NMR data.[150]

Crystal structural data are available for the tricyanomethide anion $C(CN)_3^-$. Small-basis-set *ab initio* calculations are quite successful in reproducing the experimental bond lengths for this ion.[64]

(d) Isocyanomethyl and substituted isocyanomethyl anions. According to theory,[120,129] the isocyano substituent stabilizes an adjacent carbanion centre less than CN but more than F (Table 10). For example, the calculated exothermicities (120-160 kJ mol^{-1}) for the isomerization reactions

$$R_1R_2\bar{C}NC \rightarrow R_1R_2\bar{C}CN \qquad R_1,\ R_2 = H,\ CH_3, \tag{7}$$

are considerably greater than those (70-85 kJ mol^{-1}) for the corresponding neutral isomerization

$$R_1R_2HCNC \rightarrow R_1R_2HCCN. \tag{8}$$

In addition, the gas-phase acidities of alkyl isocyanides are computed to be somewhat lower than those of alkyl cyanides.[120] At the levels of theory employed to date, the isocyanomethyl anion has a small but distinct (12-23 kJ mol^{-1}) inversion barrier.

(e) Cyclopropylmethyl anion. As part of a study[151] of conformational preferences in substituted cyclopropanes, the cyclopropylmethyl anion has been investigated in two conformations, bisected (4) and perpendicular (5).

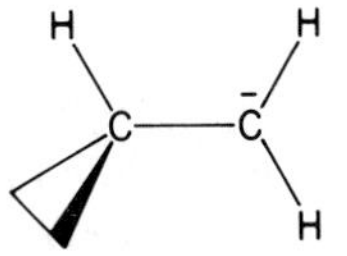

(4) 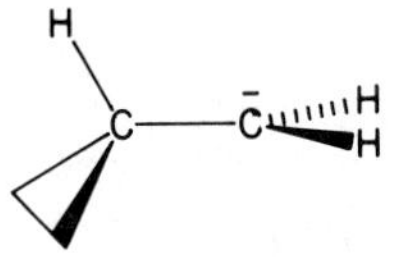 (5)

Using a diffuse-augmented double-zeta basis set, and after partial geometry optimization, the bisected conformation (4) is favoured by 5 kJ mol^{-1}. A substantially greater preference (130 kJ mol^{-1}) for the bisected conformation has been calculated[152] for the corresponding cation, the much-studied cyclopropylcarbinyl cation. The calculated C-C bond lengths around the ring in (4) are slightly longer (by 0.01-0.02 Å) than those in cyclopropane, and this has been taken as evidence for the delocalization of charge from the anionic lone pair into the lowest unoccupied (a_2') Walsh orbital[153] of the cyclopropyl fragment (6).

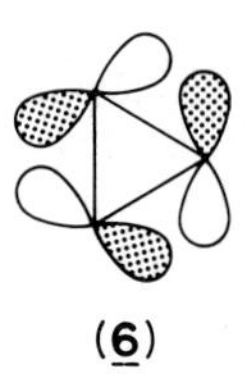

(6)

Judging by the computed proton affinities[151] of the cyclopropylmethyl and ethyl anions (1791 and 1833 kJ mol^{-1}, respectively), it appears that the cyclopropyl substituent stabilizes an adjacent methyl anion by ≈40 kJ mol^{-1} (relative to a methyl substituent). For comparison, the stabilization by a vinyl group (calculated with the same basis set)[151] amounts to 190 kJ mol^{-1}.

An isomer of the cyclopropylmethyl anion, the trimethylenemethyl anion (7), has a 4σ-electron Möbius orbital topology[154] and could therefore be expected to enjoy aromatic stabilization.[155] However, (7) is calculated to be about 200 kJ mol^{-1} less stable than the classical isomer (4).[151]

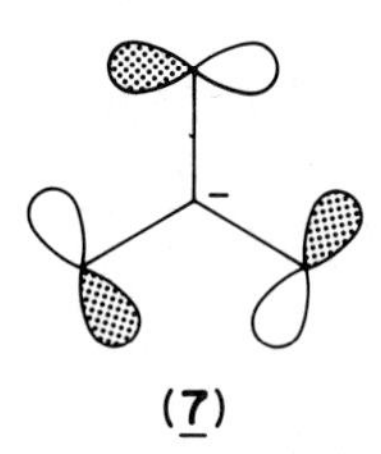

(7)

(f) Fluoro-substituted methyl anions. Because of the σ-inductive and π-donating effects of the fluorine substituent, the fluoromethyl anion is expected to have a higher inversion barrier and to be more pyramidal than CH_3^-. This is confirmed by *ab* *initio* calculations with various basis sets.[41,115,116,118,123,126,128,134,156] The most reliable value of the inversion barrier to date,[41] obtained at the MP2 level with a polarized and diffuse-augmented basis set, is 58 kJ mol^{-1}. The C-F bond lengthens markedly (by about 0.1 Å) upon pyramidalization. This has been interpreted[41] in terms of delocalization of the anion lone pair into an antibonding σ^*_{C-F} orbital, which is only possible in the pyramidal conformation.

While the σ and π effects work in unison to increase the inversion barrier in CH_2F^-, they show a competing influence on the anion stability. The net result is a (calculated) stabilization of 38 kJ mol^{-1}, as measured[41] by the isodesmic proton-transfer reaction

$$CH_2F^- + CH_4 \rightarrow CH_3F + CH_3^-. \qquad (9)$$

In the gas phase, CH_3F is therefore expected to be a stronger acid than CH_4, but a weaker acid than CH_3-CH=CH_2. The best current estimate[41] of the proton affinity of CH_2F^- is 1747 kJ mol^{-1}; a corresponding calculation for CH_3^- reproduces the experimental proton affinity (corrected for zero-point vibrational effects) to within 5 kJ mol^{-1}. An experimental estimate[157] of the gas-phase proton affinity of CH_2F^- is >1690 kJ mol^{-1}.

Semiempirical[63] and minimal-basis *ab initio*[122] calculations predict a pyramidal C_s structure for the difluoromethyl anion. The trifluoromethyl anion, CF_3^-, is predicted to adopt a pyramidal C_{3v} structure by *ab initio* calculations with a variety of basis sets.[122,123,158,159] The most reliable calculations[123,159] lead to an estimate for the inversion barrier in CF_3^- of 500 kJ mol^{-1}, more than eight times greater than the barrier in CH_2F^-. Since this is only slightly lower than the average C-F bond dissociation energy in CF_4 (506-531 kJ mol^{-1}), it has been argued[159] that the planar trifluoromethyl anion may not be bound.

Within the series CH_3^-, CH_2F^-, CHF_2^- and CF_3^-, each successive fluoro substituent is found to lower the proton affinity *more* than the previous one.[141] This is in contrast to the analogous nitro- and cyano-substituted anions, where the opposite trend is observed.[141]

(g) *Mercaptomethyl anion.* The SH substituent strongly stabilizes an adjacent carbanion centre, significantly more so than does the OH substituent (see Table 10). Traditionally, this stabilization was thought to arise from conjugation via $(p \rightarrow d)_\pi$ bonding. However, some ten years ago, it was proposed[160,161] that it is the high polarizability of the sulphur atom which is responsible for this stabilization. Another model, involving hyperconjugative delocalization of the carbon lone pair into a low-lying S–H antibonding orbital, was also proposed.[162] Although the latter model accounts well for the observed stereochemistry of carbanion formation adjacent to sulphur,[163] it has been pointed out[164] that the high polarizability of sulphur must be invoked to account for the large stabilization of the mercaptomethyl anion.

In all of these early studies, the unimportance of $(p \rightarrow d)_\pi$ bonding between carbon and sulphur was stressed. In a more recent study,[165] however, it has been shown that $(p \rightarrow d)_\pi$ bonding is important, and that earlier studies failed to detect this because of the use of only partially optimized geometries or geometries optimized without d functions in the basis set. Other work has confirmed that d functions are important in determining the geometry and inversion barrier of the mercaptomethyl anion,[128] but that the presence of d orbitals has a negligible effect on the stabilization energy.[11,128] Recent calculations[166] have also demonstrated the importance of d orbitals in determining the geometries of α-sulphinyl and α-sulphonyl carbanions.

The mercaptomethyl anion may exist in two possible conformations, the W form (8) and the Y form (9).

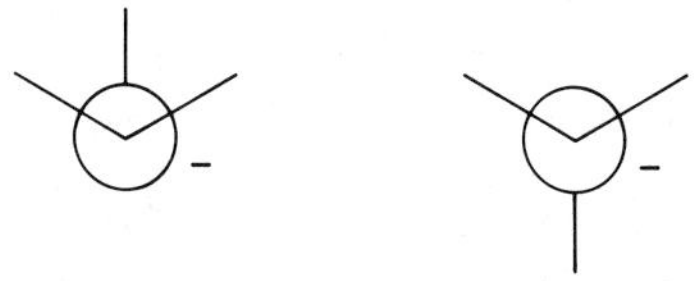

(8) (9)

In contrast to the hydroxymethyl anion, which prefers the Y conformation,[128] the W conformation is favoured for the mercaptomethyl anion,[11,128,160,165] although the energy difference between (8) and (9) is quite sensitive to the theoretical method employed.

(h) Silylated methyl anions. Trialkylsilyl groups are also known to stabilize adjacent carbanion centres (see, for example, ref. 167). As with the mercaptomethyl anion (see previous section), this effect was originally explained in terms of a (p →

d)$_\pi$ conjugative interaction:

$$R_3SiCH_2^- \longleftrightarrow R_3\bar{Si}=CH_2. \quad (10)$$

Calculations[168] with d orbitals on carbon and silicon atoms yield a stabilization energy for the SiH_3 substituent of 123 kJ mol^{-1} via the isodesmic reaction

$$SiH_3CH_2^- + CH_4 \rightarrow SiH_3CH_3 + CH_3^-. \quad (11)$$

However, virtually the same result (a stabilization energy of 126 kJ mol^{-1}) is obtained if d orbitals on Si are not included in the basis set. It has been concluded, therefore, that the stabilization of carbanions by α-silyl groups cannot be due to d-orbital participation; as with the mercaptomethyl anion, negative hyperconjugation and the high polarizability of the second-row atom have been invoked.[11,168] The best value for the stabilization energy of the silylmethyl anion is 106 kJ mol^{-1} (Table 10).

$H_3SiCH_2^-$ is calculated to be less stable than its tautomer $CH_3SiH_2^-$ by about 66 kJ mol^{-1}. The barrier to interconversion is large (272 kJ mol^{-1}), as one would expect for anionic 1,2-hydrogen shifts.[168]

Di- and tri-silylated methyl anions, $(SiH_3)_2CH^-$ and $(SiH_3)_3C^-$, have been examined[169] for comparison with the isoelectronic amines, $(SiH_3)_2NH$ and $(SiH_3)_3N$. As with the amines, both these carbanions are found to have a planar arrangement of bonds at the central atom.

(i) Diphenylmethyl anion. Semiempirical and minimal-basis-set *ab initio* calculations[170] have been used to study the diphenylmethyl anion. Although the prevailing view has been that this ion is substantially twisted from planarity, the MINDO/3, MNDO and STO-3G calculations point to a less twisted structure with a widened central bond angle to alleviate steric interactions.

(B) The ethyl anion and its derivatives

(1) The parent system $CH_3CH_2^-$. The ethyl anion has not yet been observed as an isolated species. A most important question regarding C_2H_5 is then whether it can be expected to exist at all. There are two obvious ways in which $C_2H_5^-$ can fragment. The

first is by hydride loss:

$$CH_3\text{-}CH_2^- \rightarrow CH_2\text{=}CH_2 + H^- \quad (12)$$

while the second is by electron loss:

$$CH_3\text{-}CH_2^- \rightarrow CH_3\text{-}CH_2^{\cdot} + e^-. \quad (13)$$

The energy of fragmentation into ethylene and a hydride ion (eq. 12) has been calculated with medium-sized and large basis sets.[22,130,171] The most reliable value[22] is $\Delta E = 61$ kJ mol^{-1} (or $\Delta H = 31$ kJ mol^{-1} after correction for zero-point energies). The barrier to hydride loss has been estimated[171] to be about 80 kJ mol^{-1}. The ethyl anion is therefore expected to be stable, both kinetically and thermodynamically, towards hydride loss.

From the discussion in Sect. IIID, reliable direct calculation of the electron affinity (EA) of the ethyl radical (ΔH for reaction 13) might be expected to be difficult because the number of electron pairs is not conserved in this process and a large change in correlation energy is expected. The best directly calculated _ab initio_ value reported to date[77] is $EA(C_2H_5^{\cdot}) = -83$ kJ mol^{-1}. The semiempirical MNDO method[63] yields a value of -61 kJ mol^{-1}.

Better estimates of the electron affinity of $C_2H_5^{\cdot}$ might be expected if use is made of the isogyric electron-transfer reaction

$$C_2H_5^- + CH_3^{\cdot} \rightarrow C_2H_5^{\cdot} + CH_3^- \quad (14)$$

(see Sect. IIID). This approach leads to electron affinities for the ethyl radical of -16[22] and -19[77] kJ mol^{-1} if an experimental value[125] of 8 kJ mol^{-1} for the electron affinity of the methyl radical is used.

A related approach makes use of the thermochemical cycle

$$EA(C_2H_5^{\cdot}) = PA(CH_3^-) - PA(C_2H_5^-) + DE(C_2H_6) - DE(CH_4) + EA(CH_3^{\cdot}). \quad (15)$$

Using calculated proton affinities[41] for $C_2H_5^-$ and CH_3^- and experimental values for the C-H bond dissociation energy DE[172] and for $EA(CH_3^{\cdot})$[125] leads to an electron affinity for the ethyl radical of -30 kJ mol^{-1}.

The ethyl anion is thus predicted to be unstable (by 16-30 kJ mol^{-1}) with respect to electron loss. It is predicted to be less

TABLE 11

Methyl stabilization energies in $CH_3CH_2^-$ calculated using different basis sets

Basis set[a]	Stabilization energy[b]	
	No correlation[c]	With correlation[d]
STO-3G (minimal)	38	
4-31G (extended)	9	
6-31G(d) (polarized)	6	14
4-31+G (ext.+diffuse)	-24	-20
6-31+G(d) (ext.+pol.+diff.)	-17	-13

[a] For a description of the basis sets, see Sect. IIB1a.

[b] Energy change (in kJ mol^{-1}) for the isodesmic proton-transfer reaction

$$CH_3CH_2^- + CH_4 \rightarrow CH_3CH_3 + CH_3^-.$$

[c] Hartree-Fock calculations.

[d] Correlation introduced via Møller-Plesset perturbation theory to second order.

stable than CH_3^- (which itself is only marginally bound) because the additional methyl substituent *destabilizes* the carbanion ($PA(C_2H_5^-) > PA(CH_3^-)$) but *stabilizes* the radical ($DE(C_2H_6) < DE(CH_4)$). These conclusions rest critically on the theoretical prediction that the methyl group in $CH_3CH_2^-$ is destabilizing, a prediction quite sensitive to the method used (cf. Table 11). It is noteworthy, however, that those *ab initio* methods which account well for the absolute value of $PA(CH_3^-)$, namely those employing diffuse-augmented basis sets, all support the view that the methyl substituent in $CH_3CH_2^-$ is *destabilizing*.

Properties of $C_2H_5^-$ other than its stability have been investigated by a number of authors.[77,100,106,111,173-175] Because it is likely that $C_2H_5^-$ cannot exist as an isolated species, the practical significance of these results appears somewhat dubious and they will be mentioned only briefly here.

The ethyl anion is calculated to be pyramidal at the carbanion centre and to prefer a conformation in which the lone pair and a β C-H bond are *antiperiplanar* (10).

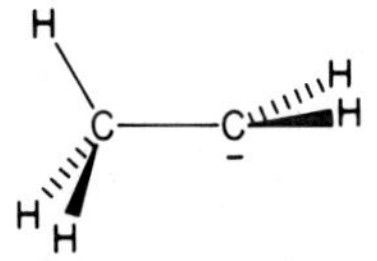

(10)

This conformational preference, as well as the lengthening of the *antiperiplanar* C-H bond, are generally ascribed to negative hyperconjugation.[41,118,175] The conformational dependence of secondary β-deuterium isotope effects in anions has been interpreted in a similar manner.[174] The inversion-rotation potential energy surface has been discussed in detail;[106] the inversion barrier in $C_2H_5^-$ was originally found to be less than that in CH_3^-. Other calculations,[41,176] however, disagree. The most reliable estimate[77] of the inversion-rotation barrier in the ethyl anion is 15 kJ mol^{-1}, only slightly greater than that calculated for CH_3^- at the same level of theory.

The proton affinity of $C_2H_5^-$ has been calculated as a function of the HCH angle of the methyl group. Decreasing this angle from its normal approximately tetrahedral value induces greater s character in the lone pair and decreases the calculated proton affinity. These results[173] provide a possible rationalization for the experimental observation (solution data) that a hydrogen adjacent to strained systems has enhanced acidity.

The addition of a hydride ion to ethylene has been investigated as a model for nucleophilic addition to alkenes.[171,177] The calculated[171] transition structure (11) has a *trans*-bent ethylene fragment; this has been rationalized by using simple orbital interaction arguments similar to those noted in Sect. IVE1 for the approach of hydride to acetylene. A value of 69 kJ mol^{-1} has been obtained[171] for the activation energy at a fairly low level of theory.

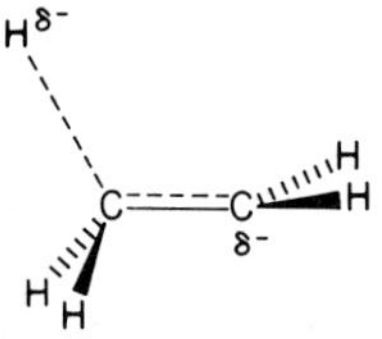

(11)

The degenerate 1,2-hydrogen shift in the ethyl anion has recently been studied.[22,77] The transition structure (12) is found to prefer C_2 symmetry, and the barrier to rearrangement is calculated to be 202 kJ mol^{-1}. This large barrier is consistent with the absence of known examples of 1,2-hydrogen migration along a carbon chain[178] and contrasts with the ease of such rearrangements in carbocations.

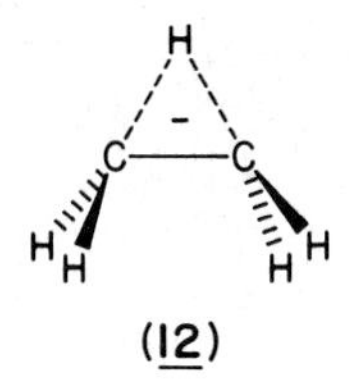

(12)

(2) <u>β-Substituted ethyl anions. Hyperconjugation</u>. The ability of groups such as CH_3 to interact with double and triple bonds was first recognized about fifty years ago. This phenomenon was termed <u>hyperconjugation</u> by Mulliken,[179,180] and has been the subject of vigorous debate and continuing controversy (for reviews, see refs. 181-185).

The particular concept of <u>negative</u> (or anionic) <u>hyperconjugation</u>, as exemplified by the β-fluoroethyl anion,

$$\mathrm{F{-}CH_2{-}\bar{C}H_2} \longleftrightarrow \mathrm{F^-}\quad \mathrm{CH_2{=}CH_2}$$

was introduced by Roberts[186] to explain reactivity patterns in fluoro-substituted systems. The importance of negative hyperconjugation has been questioned,[183,184,187,188] but the weight of recent theoretical evidence is very strongly in favour of negative hyperconjugation as a phenomenon with significant energetic and structural consequences.[185,189-198]

Substituted ethyl anions have provided the prototype systems for the theoretical investigation of negative hyperconjugation. Particular advantage has been taken of examining systems in artificially constrained geometries so as to distinguish, for example, <u>conformationally dependent</u> hyperconjugative interactions from <u>conformationally independent</u> inductive interactions.

An important early paper in this area was that of Hoffmann <u>et al.</u>,[189] which pointed out that for substituted ethyl anions $XCH_2CH_2^-$ (with idealized geometries, locally tetrahedral at XCH_2 and trigonal planar at CH_2^-), the eclipsed conformation (<u>13</u>) is preferred if X is less electronegative than H while the perpendicular conformation (<u>14</u>) is preferred if X is more electronegative than H. These conformational preferences are readily rationalized through a consideration of the principal orbital interactions (Fig. 2). For example, if X is electronegative, the π_x* orbital is both lowered in energy and concentrated on C, leading to a decreased energy gap and

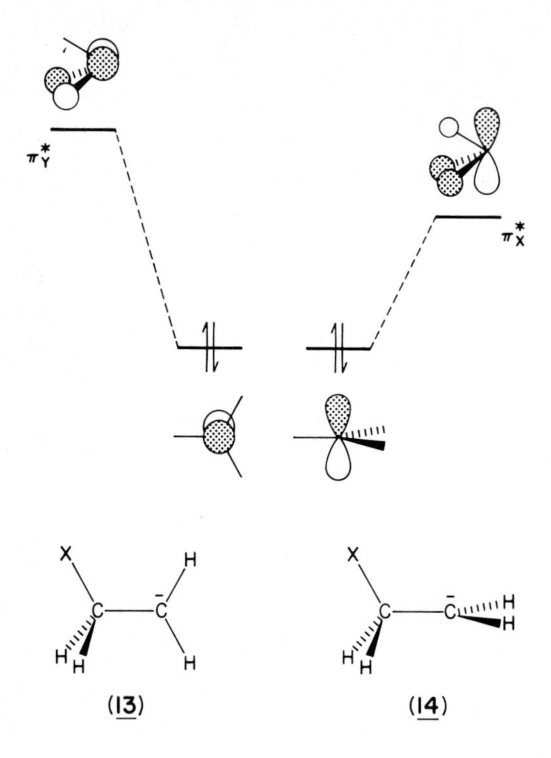

Fig. 2. Comparison of hyperconjugative interactions in eclipsed (13) and perpendicular (14) conformations of substituted ethyl anions $XCH_2CH_2^-$ (X electronegative).

increased overlap. Both of these effects lead to improved hyperconjugative interaction in the perpendicular conformation compared with the eclipsed conformation in which the interaction (with π_y*) is relatively unaffected by the substituent. For an electropositive substituent X, the reverse considerations apply. Interaction in the perpendicular conformation is made worse through an increased energy gap and poorer overlap, leading to the expectation, in the absence of other interactions, of a preference for the eclipsed conformation.

These predictions have been tested by calculations on a series of model $XCH_2CH_2^-$ anions with a variety of electropositive and electronegative substituents X.[22,188,189,191-196] Calculated stabilization energies, i.e. energies for the reaction

$$XCH_2CH_2^- + CH_3CH_3 \rightarrow XCH_2CH_3 + CH_3CH_2^- \qquad (16)$$

for the eclipsed conformation (13) provide a rough measure of the inductive effect, whereas the changes in going to the perpendicular conformation (14) provide a measure of hyperconjugative and other conformationally dependent interactions. It is found[193] that (i) the model ethyl anions $XCH_2CH_2^-$ are generally stabilized by both electropositive and electronegative substituents; (ii) electronegative substituents are most effective in stabilizing $XCH_2CH_2^-$ through hyperconjugative electron acceptance in the perpendicular conformation; and (iii) for electropositive substituents containing an appropriately oriented vacant orbital, the perpendicular conformation is again preferred but the dominant interaction involves direct 1,3-transfer rather than hyperconjugation (15).

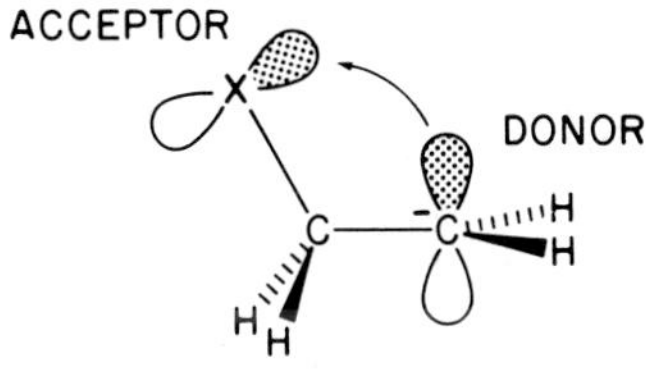

ACCEPTOR

DONOR

(15)

Calculations on the $XCH_2CH_2^-$ anions using standard geometrical parameters but assuming tetrahedral (rather than trigonal) angles at the carbanionic centre[22] yield similar qualitative conclusions. These calculations are able to distinguish, in addition, between *anti* (16) and *syn* (17) variants of the perpendicular conformation (14).

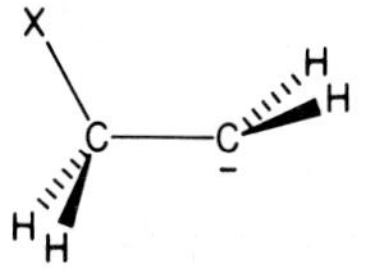

(16) 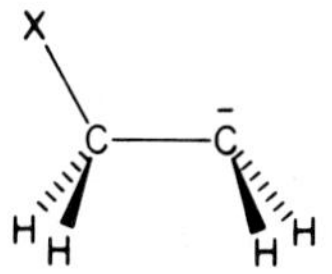 (17)

For all but one of the substituents examined, the stability ordering is *anti* > *syn* > *gauche*.

An extensive set of calculations has been carried out for β-substituted ethyl anions with full geometry optimization.[22] Such calculations are clearly necessary for the investigation of structural changes, particularly if the main point of interest is the ethyl anions *per se* (as opposed to an interest in larger molecules which might be *modelled* by the ethyl anions, or an interest in the testing of concepts such as the conformational dependence of hyperconjugative interactions).

The optimization studies[22] show that, in most cases, there are major distortions from classical ethyl anion structures. Thus, anions with electropositive substituents (X=Li, BeH, BH_2, MgH, AlH_2) prefer three-membered cyclic structures, consistent with (15) above. The classical structure of $SiH_3CH_2CH_2^-$ is a local minimum but the cyclic structure, a transition structure for 1,2-SiH_3 migration, lies only 37 kJ mol^{-1} higher in energy. Ethyl anions with electronegative β substituents are generally unstable with respect to elimination. Thus, for X=F, PH_2, SH and Cl, there is essentially complete transfer of negative charge to X and the global minimum on the $C_2H_4X^-$ surface resembles a hydrogen-bonded complex of ethylene with X^-. For example, the classical β-fluoroethyl anion is unstable with respect to dissociation to F^- + C_2H_4, whereas the adduct (18) is a stable species.[22,199]

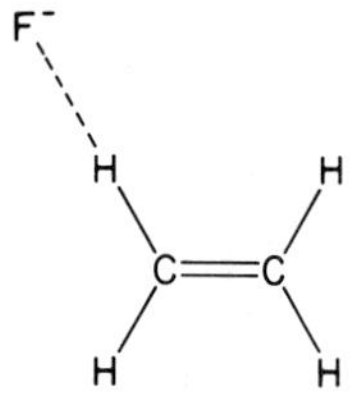

F^- ··· H–C(H)=C(H)–H

(18)

Both $HOCH_2CH_2^-$ and $NH_2CH_2CH_2^-$ are local minima, but favour conformations which involve interaction between a hydrogen of the substituent and the carbanionic lone pair.

The calculations[22] suggest that the $SiH_3CH_2CH_2^{\bullet}$ radical should have only a marginally positive electron affinity. Much larger electron affinities are predicted for the strongly stabilized anions with X=BeH, BH_2, MgH, AlH_2 and OH. As discussed in Sect. IVC, the n-propyl anion is predicted to be unstable with respect to electron loss.

One of the central arguments of the opponents of the concept of negative hyperconjugation has been based on the similarity of the observed effects of β-F and β-CF_3 substituents[183,184] compared with the expectation that the hyperconjugative effects of these groups should differ. However, calculations[194,195] show that β-F and β-CF_3 groups stabilize an adjacent carbanionic centre to similar extents and in both cases there is a substantial hyperconjugative component to the stabilization. The slightly smaller hyperconjugative effect of the CF_3 substituent is fully compensated by a greater inductive effect.

More recently, Streitwieser et al.[188] have found from electron density calculations that integrated spatial electron populations at fluorine in the eclipsed (19) and perpendicular (20) conformations of the β-fluoroethyl anion show only small differences. They conclude that polarization of electrons rather than charge transfer is important and that the traditional fluorine anionic hyperconjugation mechanism is essentially insignificant.

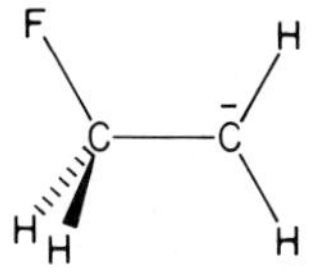

(19) 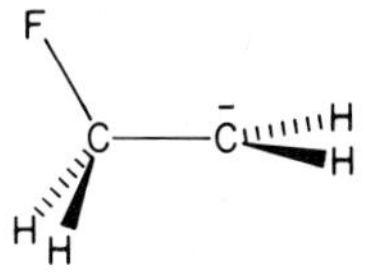 (20)

This view has been (strongly) criticized by Schleyer et al.,[22,197] who point out that the definition of hyperconjugation involves orbital interactions, not charge transfer. Both charge transfer as well as polarization are possible consequences of hyperconjugation and the dominance of one or the other varies with different molecular systems.[200] In addition, as noted above, full geometry optimization of the β-fluoroethyl anion leads to steady elongation of the C-F bond without a barrier.[192,197,199] In the extreme, this corresponds to complete dissociation to ethylene + F^-, i.e. to complete charge transfer to fluorine.

The 2,2,2-trifluoroethyl anion does not split off F^- on geometry optimization and therefore allows a more meaningful examination of hyperconjugation. It is found[118,197] that the C-F bond in (21), which is ideally oriented for hyperconjugative interaction with the lone pair, is much longer than the other two C-F bonds. The C-C bond is shortened considerably, reflecting substantial double-bond character. Schleyer and Kos[197] conclude that "negative hyperconjugation is not controversial; it is firmly established."

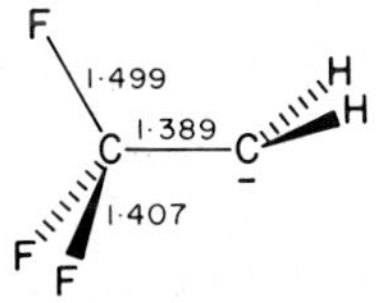

1·499 1·389 1·407

(21)

β-Substituted carbanions have been invoked in the mechanisms of a variety of organic reactions.[22,195,196] These reactions include, in the first place, Elcb elimination[201] (eq. 17+20), β-proton exchange[140] (eq. 17+18) and nucleophilic addition to alkenes[202] (eq. 19+18).

The rate and stereochemistry of these reactions is influenced by the effect of the substituent X on the stability and conformation of the anion and of the appropriate transition structures.

The stereochemical preference anti > syn > gauche in elimination reactions is often attributed to the desirability of maximizing the overlap of the developing p orbitals on adjacent carbon atoms (see, for example, ref. 192). The ordering is, however, the same as the stability ordering for different conformations (defined by the relative orientation of the C-X bond and the carbanion lone pair) of the $XCH_2CH_2^-$ anions.[22] It also coincides with the ordering of the lengths (and presumably strengths) of the C-X bond in $XCH_2CH_2^-$ when calculated as a function of the orientation of the C-X bond with respect to the lone-pair orbital. These results suggest that the preferred stereochemistry of Elcb eliminations, proceeding via carbanion intermediates, should be the same as that found for E2 eliminations.[22]

Likewise, the preferred anti stereochemistry for the addition of an electronegative nucleophile to an alkene[202] is consistent with the preferred anti conformation of the $XCH_2CH_2^-$ anion.[22]

The conformational properties of β-substituted ethyl anions have also been used to rationalize[195,196] the stereochemistry of a second set of related reactions, nucleophilic vinylic substitution[203,204] (eq. 21), epoxidation (eq. 22) and cyclopropanation[196] (eq. 23).

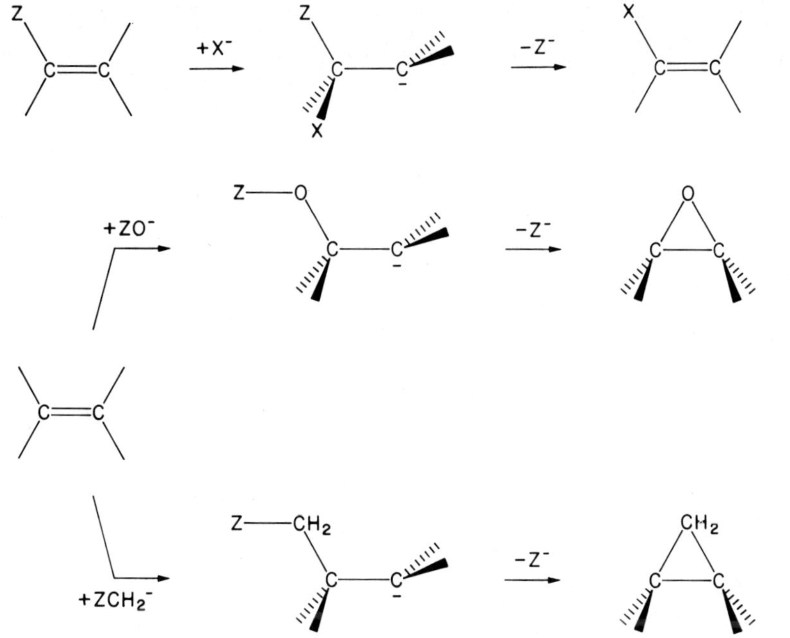

(21)

(22)

(23)

The preference for retention of configuration in nucleophilic vinylic substitution[203,204] has been rationalized[191] in terms of a preference for internal rotation to occur in the ethyl anion intermediate in a direction which maximizes hyperconjugative interaction with the β substituents. It is concluded in addition[196] that the greater the hyperconjugative interaction of the bond at the β carbon, the greater should be the stereospecificity of reactions 21-23. As a consequence, cyclopropanation should be less stereospecific than epoxidation.

(3) <u>α-Substituted ethyl anions</u>. The effect of substituents in α-substituted ethyl anions, as given by the energy change in the reaction

$$CH_3CHX^- + CH_4 \rightarrow CH_3^- + CH_3CH_2X, \qquad (24)$$

is compared with the effect in substituted methyl anions (XCH_2^-), calculated at the same level of theory,[22] in Table 12. The differences are quite small and, in fact, correspond to the effect of a methyl substituent on the stability of XCH_2^- anions, i.e. the energy change for

$$CH_3CHX^- + CH_3X \rightarrow CH_2X^- + CH_3CH_2X. \qquad (25)$$

TABLE 12

Calculated stabilization and isomerization energies for substituted ethyl anions[a]

X	$SE(CH_3CHX^-)$[b]	$SE(CH_2X^-)$[c]	ΔSE[d]	ΔE[e]
H	-23	0	-23	0
Li	25	10	15	80
BeH	123	133	-10	36
BH_2	230	229	1	-24
CH_3	-25	-23	-1	13
NH_2	-7	-8	1	33
OH	23	32	-10	19
F	55	65	-10	
Na	44	24[f]	20[f]	
SiH_3	77	95[f]	18	60
Cl	138	147[f]	8	

[a] In kJ mol^{-1}. 4-31+G//4-31+G calculations unless otherwise noted. From ref. 22.
[b] Energy change for reaction (24).
[c] Energy change for the reaction
$$XCH_2^- + CH_4 \rightarrow XCH_3 + CH_3^-.$$
[d] Energy change for reaction (25).
[e] E(α-substituted)-E(β-substituted).
[f] 3-21+G//3-21+G.

The effect of the methyl substituent is variable, ranging from a 23 kJ mol^{-1} destabilization when X=H (cf. 13 kJ mol^{-1} destabilization for this reaction at the highest level of theory of Table 10) to a 20 kJ mol^{-1} stabilization when X=Na.

Also shown in Table 12 are the calculated relative energies of α- and β-substituted ethyl anions.[22] The β-substituted anions are favoured in all cases except $X=BH_2$ (for which the α-substituted anion is isoelectronic with propene and benefits from extensive charge delocalization). For electronegative substituents, the inductive effect should favour α substitution over β. The observed preference for β substitution suggests that the σ-acceptance is less important than the favourable hyperconjugative interaction in the β form together with π-electron destabilization in the α form.

(C) Higher alkyl anions

Calculations on the *n*-propyl anion[130] yield a proton affinity 13 kJ mol^{-1} lower than that of the ethyl anion. This contrasts with the result, noted in Sect. IVB1, that the ethyl anion has a proton affinity greater than that of methyl anion, and suggests that hyperconjugation of the lone pair with a CC bond is more effective than with a CH bond. This suggestion is supported by the preferred conformation of *n*-propyl anion, which has the lone pair antiperiplanar to the CC bond (22).

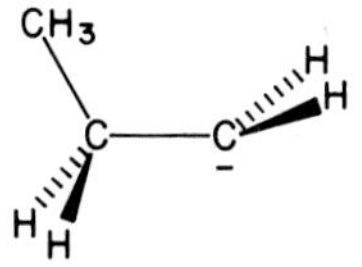

CH_3 H H C C H H

(22)

Nevertheless, the *n*-propyl anion is unlikely to be stable with respect to electron loss.[22,130]

The isopropyl anion lies about 13 kJ mol^{-1} higher in energy than the *n*-propyl anion[22,130] and should also be unstable with respect to electron loss. Its inversion barrier has been calculated[130] to be about 17 kJ mol^{-1} greater than that of the methyl anion.

(D) Cycloalkyl anions

Angular constraints in three- and four-membered rings are expected to destabilize the planar relative to the pyramidal geometry of AX_3 molecules and thereby increase the inversion barrier.[205] This effect is well known in the amine series; the nitrogen inversion barriers in aziridines, for example, are much higher than in corresponding acyclic amines. A similar effect is anticipated in carbanions:[206] cyclobutyl anions should be configurationally more stable than secondary acyclic carbanions, and cyclopropyl anions even more so. Indeed, the cyclopropyl anion appears experimentally to be configurationally stable.[207-209]

The effect of angular constraints on the inversion barrier in carbanions has been probed by calculating the barrier in CH_3^- as a function of the HCH angle. As this angle is gradually decreased to 90° (the value appropriate for cyclobutyl anion), the calculated inversion barrier increases from ≈35 to 85 kJ mol^{-1} (split-valence basis),[113] in accordance with the simple hybridization angle picture. This energy change can be traced back to the energetic behaviour of the highest occupied molecular orbital in CH_3^-.[113]

(1) Cyclopropyl anion. The cyclopropyl anion has been studied at various levels of theory, both semiempirical and *ab initio*, and is usually found to be nonplanar around the anionic carbon (23).

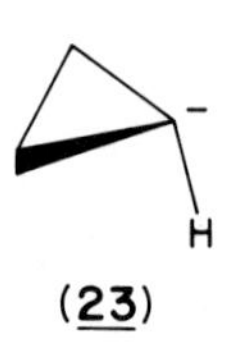

(23)

Early semiempirical[97] and *ab initio*[210,211] calculations involved quite arbitrary assumptions about the molecular geometry and are of doubtful significance. More recent *ab initio* calculations have yielded values for the inversion barrier in the cyclopropyl anion of 73,[212] 63[67] and 81[206] kJ mol^{-1}. These calculations are in qualitative accord with the experimentally observed configurational stability of cyclopropyl anions.[207-209] MNDO, which wrongly predicts the methyl anion to be planar, also produces a planar carbanionic centre in the cyclopropyl anion.[213]

Substitution in the α position by fluorine has the expected effect: the anionic centre becomes more pyramidal and the inversion barrier increases to 175[208] or 220[206] kJ mol^{-1}. The isocyano substituent increases the barrier slightly (from 81 kJ mol^{-1} in cyclopropyl anion to 90 kJ mol^{-1} in the α-isocyanocyclopropyl anion[206]). The lack of configurational stability of the cyanocyclopropyl anion[214] is reflected in the low calculated barrier to inversion of 24 kJ mol^{-1} for this species.[206]

In the case of cyclopropane and fluoro- and isocyanocyclopropane, anion formation is predicted to lead to a considerable relief in ring strain.[206] Consequently, these molecules are more acidic than their substituted propane analogues. Cyanocyclopropyl anion and cyanocyclopropane have similar strain energies; cyanocyclopropane and 2-cyanopropane are of almost equal acidity.[206]

Minimal-basis-set *ab initio* calculations predict that the nitrocyclopropyl anion has a triplet ground electronic state, in contrast to the formyl-, carboxyl- and cyano-cyclopropyl anions, all of which prefer a singlet ground state.[215] Higher-level calculations would be required to test the validity of this prediction.

The electrocyclic transformation of the cyclopropyl to the allyl anion is uniformly predicted, as expected, to follow a conrotatory course.[210,213,216-221] . MNDO[213] predicts an activation barrier of 78 kJ mol^{-1}. Earlier <u>ab initio</u> values are higher, but are of doubtful reliability in view of the use of assumed geometries,[210] small basis sets[220,221] and non-optimum molecular orbitals[219] in these studies.

(2) <u>Cyclobutyl anion</u>. The cyclobutyl anion is calculated, like the neutral parent molecule, to have a puckered structure, with the lone pair preferring the axial (<u>24</u>) rather than the equatorial (<u>25</u>) position.[222] The barrier between these two conformations has not been determined.

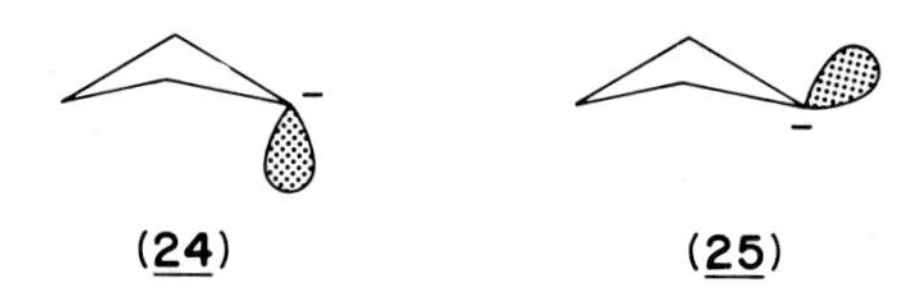

(<u>24</u>) (<u>25</u>)

(3) <u>Other cycloalkyl anions</u>. The kinetic acidities of cycloalkanes decrease in the order cyclopropane > cyclobutane > cyclopentane > cyclohexane, as one would expect from hybridization arguments. This trend is not reproduced by CNDO/2, INDO and NDDO calculations, casting some doubt on the applicability of CNDO-type methods to anions in general.[223]

The calculated (MINDO/2) structures of anions (<u>26</u>)-(<u>28</u>) in the 7-norbornyl series are slightly nonplanar at C-7[218] (H being 3 to 4° out of plane in these structures).

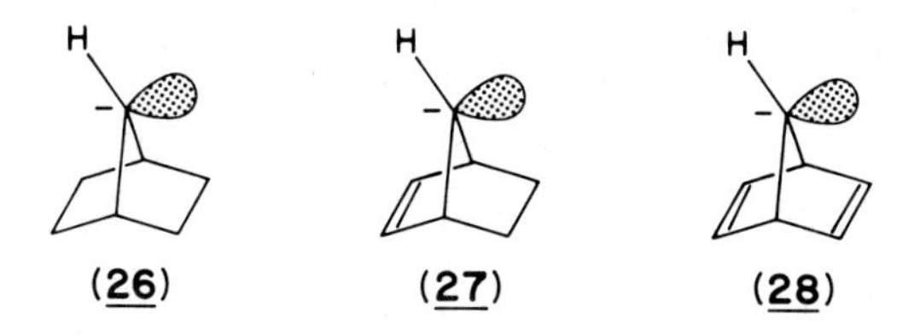

(<u>26</u>) (<u>27</u>) (<u>28</u>)

(E) <u>The vinyl anion and its derivatives</u>

Vinyl anions, in which the lone pair is localized in an sp^2 hybrid orbital, are expected to be configurationally more stable than sp^3 hybridized carbanions. Vinyl anions are believed to be involved in nucleophilic additions to triple bonds, and their configurational preference and stability are thought to be responsible for the E stereoselectivity often observed in these reactions.[224,225]

(1) The parent system CH_2=CH^-. The parent vinyl anion (29) and its barrier to inversion through a linear transition structure (30) have been investigated theoretically by several groups.[77,97,111,123,171,226-228] The most reliable estimate[123] of the inversion barrier is 132 kJ mol^{-1} (Sect. IIIC). An experimental estimate[229] based on solution-phase kinetic data is 125±20 kJ mol^{-1}.

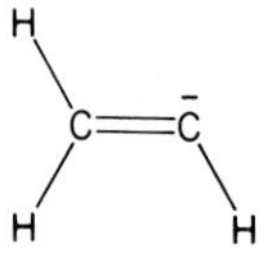

(29) 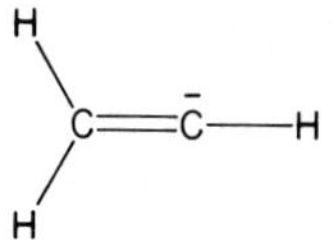 (30)

A recent calculation[228] of the proton affinity of the vinyl anion yielded a value of 1738 kJ mol^{-1}, consistent with the experimental value[230] of >1690 kJ mol^{-1}.

Calculated values of the electron affinity of the vinyl radical have been obtained[77] in the range 38-102 kJ mol^{-1}, suggesting that the vinyl anion is a stable species and should be observable experimentally. An indirect experimental estimate of 125 kJ mol^{-1} has been reported;[231] other experiments[232,233] have yielded lower and upper limits of 40 and 125 kJ mol^{-1}, respectively.

The addition of H^- to acetylene is the simplest example of a nucleophilic addition to the CC triple bond and has been investigated in detail.[171,227,234] The calculated transition structure (31) for this process is "early" in the sense that the developing C-H bond is relatively long (≃2.1 Å) and the CC bond length is halfway between triple- and double-bond values.

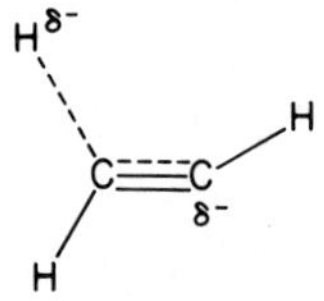

(31)

Calculated values for the heat of reaction and activation energy are −109 and 67 kJ mol^{-1}, respectively.[227] It is noteworthy that the incoming nucleophile induces a trans bending of the acetylene skeleton. This can be rationalized by noting that a stabilizing interaction of the lowest unoccupied molecular orbital $\pi^*(C\equiv C)$ (32) with the highest occupied $\sigma(C-H)$ orbital (33) is, for symmetry reasons, possible upon *trans* but not *cis* bending.[171] The trajectory of the approaching hydride makes an angle of ≈125° with the CC bond.[171,227,234] This does not agree, however, with Baldwin's suggestion[235] that nucleophiles Nu attack acetylenes RCCR with an RCNu angle of 120°, leading to a rather narrow CCNu angle of 60°.

(32) (33)

Conjugating groups adjacent to the lone pair of the vinyl anion should lower the inversion barrier by an amount related to their ability to increase the gas-phase acidity of methane, i.e. CHO > CO_2CH_3 > CN.[140] Calculations[236] show that the inversion barrier in $CH_2{=}CX^-$ does in fact decrease in the expected order X=H > CN > CO_2CH_3 > CHO. The 1-formylvinyl anion (34) is predicted to have a linear carbon skeleton.

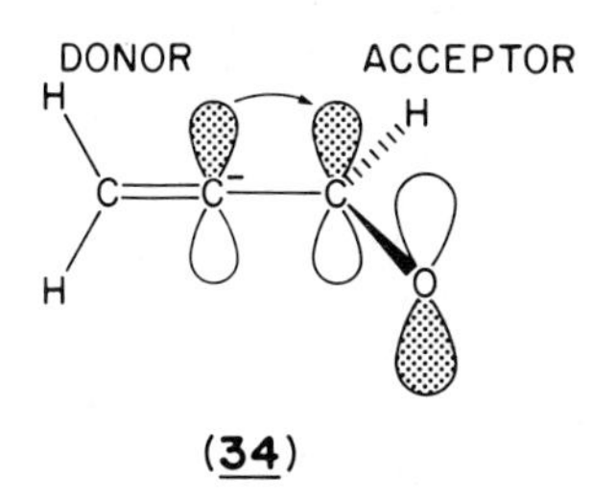

(34)

While the numerical values of inversion barriers calculated in this study appear suspect because rather small basis sets were used, the qualitative trend is probably correct and accounts for the low or variable stereochemistry of nucleophilic additions to activated acetylenes.[225]

The degenerate rearrangement of CH_2CH^- via a 1,2-hydrogen shift has also been studied.[77,227] The most recent study[77] reports transition structures of C_2 and C_s symmetries, corresponding respectively to migration of the hydrogens *syn* and *anti* with respect to the α hydrogen in CH_2CH^-. The C_2 transition structure (35) is found to lie lower in energy, the calculated barrier via

(35) being 235 kJ mol^{-1}. This is in sharp contrast to the cationic species, where the barrier to degenerate rearrangement is close to zero.[237]

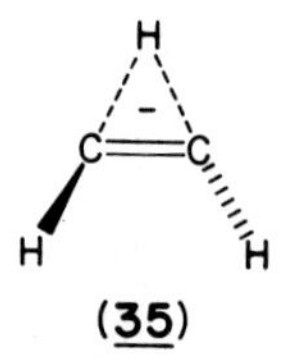

(**35**)

(2) Substituted vinyl anions. Reaction of n-BuLi with substituted alkoxyalkenes produces vinyl rather than allyl anions.[238] In qualitative terms (supported by *ab initio* calculations), this may be attributed to the fact that the carbanion-stabilizing effect of alkoxy is greater in (36) than in (37) because the charge is more localized in the vinylic anion.[239]

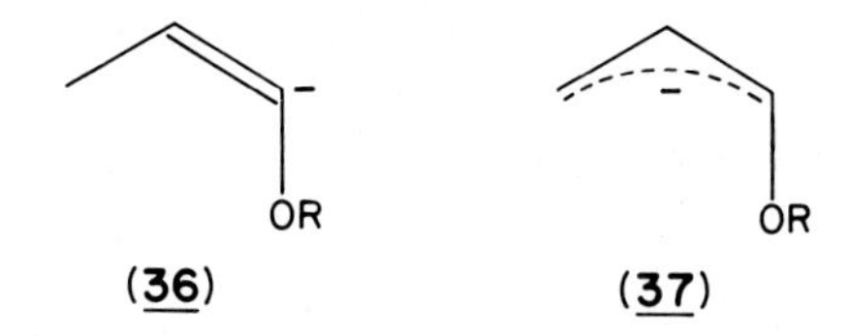

(**36**) (**37**)

However, since the absolute value of carbanion stabilization by an adjacent OH group is only ≈23 kJ mol^{-1} (Table 10), it seems surprising that this would suffice to shift the equilibrium between allyl anion and 1-propenyl anion in favour of the latter. Kinetic control may be important here (as in the metallation of certain propenes and alkyl benzenes[240,241]), and the role of the counterion is, as usual, unclear.

Ethynyl substituents stabilize adjacent lone pairs substantially; as expected, the 1-ethynylvinyl anion (38) is calculated[242] to be more stable than the 2-isomer (39).

$H_2C{=}\bar{C}{-}C{\equiv}CH$ (**38**)

$H\bar{C}{=}C(H){-}C{\equiv}CH$ (**39**)

Ab initio calculations[197,199] indicate that the most stable $C_2H_2F^-$ isomer is the hydrogen-bonded complex of acetylene and F^- (40). The classical α- and β-fluorovinyl anions lie somewhat higher in energy.

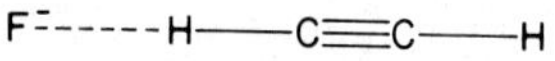

F^{-}-----H—C≡C—H

(40)

The computed greater stability of *cis* $CHF{=}CF^{-}$ compared with the *trans* isomer (by a few kilojoules per mole) has been rationalized in terms of interacting fragment orbitals.[243]

The rates of base cleavage of substituted 2-thienyltrimethylsilanes increase with the relative stability of the thienyl anions (41), and this correlation suggests that the thienyl anions are formed in the rate-determining step.[244]

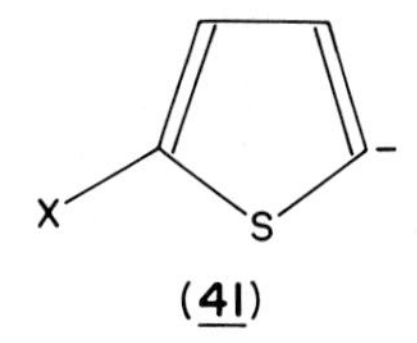

(41)

The proton affinities of a number of substituted vinyl anions have been determined in an attempt to explain the highly regioselective lithiation of β-functionally substituted acrylic acid derivatives.[245]

(3) Analogues of vinyl anion. Deprotonation at the nitrogen site of formaldimine, leading to (42), is found to be favoured by 50 kJ mol^{-1} over deprotonation at the carbon site.[228] In the analogous phosphorus-containing system, the difference in acidities of the hydrogens is even greater:[228] deprotonation to yield (43) is favoured by 113 kJ mol^{-1}.

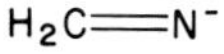

$H_2C{=}N^{-}$ (42) $H_2C{=}P^{-}$ (43)

(F) Carbonyl anions

Carbonyl anions (44) have been studied using MNDO and *ab initio* methods.[246]

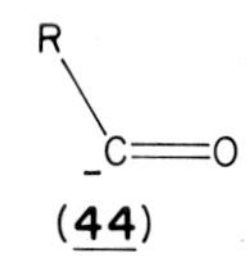

(44)

They are generally found to have elongated R-C bonds, with the exception of the vinyl- and phenyl-substituted species which have ketene-like structures, such as (45b).

$$H_2C{=}\overset{H}{C}{-}\overset{-}{C}{=}O \longleftrightarrow H_2\overset{-}{C}{-}\overset{H}{C}{=}C{=}O$$

(45a) (45b)

The formyl anion (HCO^-), which is marginally stable in the gas phase,[247] is bound with respect to H^- and CO by just 18 kJ mol^{-1}. Other carbonyl anions, especially $CH_3{-}\bar{C}{=}O$ and $H_2N{-}\bar{C}{=}O$, are much more stable with respect to decarbonylation. Formaldehyde is calculated to be a stronger acid than ethylene but weaker than acetylene; formyl fluoride is predicted to be of comparable acidity to mercaptans and phenols.[246]

(G) Alkynyl anions

(1) The parent system $HC{\equiv}C^-$. The ethynyl anion $HC{\equiv}C^-$ has been investigated at a variety of theoretical levels.[40,66,77,111,228,242,248-261] The predicted structure is linear.

A high-level theoretical calculation of the proton affinity yielded a result (1583 kJ mol^{-1} after inclusion of zero-point vibrational and thermal corrections)[255] which compares well with the experimental value (1570±10 kJ mol^{-1}).[140] However, the predicted equilibrium constant (log K_p value of 11.9 at 298 K) for the reaction

$$C_2H_2 + OH^- \rightleftharpoons C_2H^- + H_2O, \quad (26)$$

calculated at the same level of theory, is in poor agreement with experiment (log K_p = 5.4). This has been attributed to inadequate treatment of electron correlation effects and occurs despite the isogyric nature of reaction 26.[255]

The most recent experimental study of the electron affinity of the ethynyl radical[260] gave a value of 284±10 kJ mol^{-1}. Theoretically derived EA values vary widely.[77,81,111,256,259,260] An eighth-order perturbation treatment[260] gave a value of 307±25 kJ mol^{-1}, while in a more recent study[81] an estimate of 261 kJ mol^{-1} was derived by direct calculation of the energy difference between the radical and anion which, after incorporating a suggested empirical correction, yields an electron affinity of 293 kJ mol^{-1}. By using the isogyric reaction

$CH_3^- + C_2H_2 \rightarrow CH_4 + C_2H^-$ (27)

in conjunction with data from ref. 81, an estimate of 298 kJ mol^{-1} may be derived.

The barrier to the degenerate 1,2-hydrogen shift in the ethynyl anion has been calculated to be 73 kJ mol^{-1}, substantially less than the barriers for the analogous rearrangements in the vinyl and ethyl anions (235 and 202 kJ mol^{-1} respectively).[77] The transition structure has C_{2v} symmetry and may be formally considered as a complex between the acetylide dianion and a proton (46).

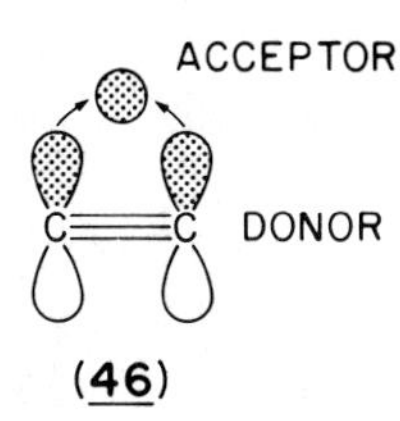

(**46**)

The binding of ethynyl anion to water and acetylene in the gas phase has been studied.[261] The strengths of the resulting hydrogen bonds are calculated to be 75 and 44 kJ mol^{-1}, respectively.

The bonding in HCC^- has been compared with that in HCCH and HCCLi, using density difference maps and localized orbitals.[253]

(2) Substituted ethynyl anions. In an *ab initio* study of twenty-one monosubstituted acetylenes, a good correlation was found between the calculated acidities and the rates of tritium isotope exchange in solution.[262]

(3) Analogues of the ethynyl anion

(a) Cyanide anion. The cyanide anion CN^- has been the subject of a large number of theoretical studies.[39,40,66-68,81,92,228,263-276] Only the more recent results are summarized here.

Large-basis-set Hartree-Fock calculations[66,271] yield an equilibrium bond length of 1.152-1.157 Å for the cyanide ion. More accurate estimates of the bond length, obtained in calculations incorporating electron correlation, are in the range 1.173-1.198 Å.[92,123,273,274] The most reliable value is likely to be that of Botschwina[92] (1.180 Å). This value coincides with the average of lengths derived from an X-ray crystallographic study of a set of cyanide salts.[59]

High-quality calculations[68] of the proton affinity of CN^- yielded a value of 1467 kJ mol^{-1}, in excellent agreement with the experimental value of 1462±8 kJ mol^{-1}.

Spectroscopic properties of CN^- have been studied in a number of papers.[92,272-274] The fundamental vibrational band has been predicted[92] to lie at 2052±6 cm^{-1}.

The electron affinity (EA) of the cyano radical has aroused considerable interest, possibly because the cyanide ion is exceptionally stable with respect to detachment of an electron. The experimental EA[277] is 369±2 kJ mol^{-1}. The EA at the Hartree-Fock limit has been estimated[271] to be 317±5 kJ mol^{-1}. Transition-operator[275] and equations-of-motion[269] methods yield values of 372 and 357 kJ mol^{-1}, respectively; however, the latter study has been criticized.[270] Configuration interaction calculations with moderately large basis sets[81,273,276] lead to values of 350-362 kJ mol^{-1}.

Other studies of the cyanide ion include its binding to BX_3 and AlX_3 species[278] and the formation of hydrogen-bonded complexes with hydrogen fluoride.[279]

(b) Sila-analogues of HCC^-. The hypothetical sila derivative $H\text{-}Si{\equiv}C^-$ of the ethynyl anion has been calculated to be less stable than its tautomer $H\text{-}C{\equiv}Si^-$ and to rearrange to $H\text{-}C{\equiv}Si^-$ without activation.[176]

(4) Propynyl and larger alkynyl anions. Propyne is known to be a weaker acid than acetylene, both in solution and in the gas phase.[280] This destabilizing effect of the methyl group, as measured by the energy change (ΔH_{298} = −18 kJ mol^{-1})[140] for the proton-transfer reaction

$$CH_3C{\equiv}C^- + HC{\equiv}CH \rightarrow CH_3C{\equiv}CH + HC{\equiv}C^-, \qquad (28)$$

is reproduced well by *ab initio* calculations.[118,249,250] On the other hand, the methyl group is stabilizing in the methoxide anion.[258,281-284] A rationalization of this dual behaviour in terms of orbital interactions has been presented.[258] Larger alkyl groups also destabilize the ethynyl anion, but less so than methyl. The experimental order of acidities of terminal alkynes RCCH, R=H > t-Bu > n-Pr > Me, is reproduced surprisingly well by minimal-basis-set *ab initio* calculations,[249,250] but not by the semiempirical CNDO/2 method.[280]

A vinyl substituent slightly stabilizes the ethynyl anion, as reflected in a decrease of 11 kJ mol^{-1} in the computed proton affinity[242] in going from $HC{\equiv}C^-$ to $H_2C{=}CH{-}C{\equiv}C^-$. The stabilities of the deprotonated vinyl acetylenes decrease in the order $H_2C{=}CH{-}C{\equiv}C^-$ > $H_2C{=}\bar{C}{-}C{\equiv}CH$ > $H\bar{C}{=}CH{-}C{\equiv}CH$.

(H) π-Delocalized carbanions

(1) The allyl anion $CH_2CHCH_2^-$. Theoretical interest in this species[63,67,99,116,118,123,142,208,213,216,219-221,239,285-300] has focussed mainly on the question of allylic resonance and its relation to nonplanar distortions, rotational barriers, and proton affinities. Another point of interest has been the electrocyclic interconversion of the allyl and cyclopropyl anions (cf. Sect. IVD1).

The allyl anion is one of the textbook examples of a strongly delocalized and therefore presumably planar (C_{2v}) π system. Early semiempirical calculations[99,286] somewhat disconcertingly predicted the allyl anion to be nonplanar, the calculated inversion barrier at C_1 being 65-70 kJ mol^{-1}. A more recent minimal-basis-set *ab initio* investigation with partial geometry optimization[291] was in qualitative agreement, but predicted a very small inversion barrier. Better *ab initio* calulations, however, with full geometry optimization, predict a perfectly planar C_{2v} equilibrium structure for the allyl anion,[123,293,298] as does the semiempirical MNDO method.[63]

Photoelectron spectroscopic measurements have been used to deduce a bond angle of 140±4° and CC bond lengths of 1.505 Å for the allyl anion.[301] These results have been convincingly challenged by Schleyer,[299] who suggests that the 6-31+G(d) results of 132.2° and 1.388 Å are unlikely to be in error by more than ±2° and ±0.02 Å, respectively. The widening of the CCC angle in the allyl anion (132.2°) compared with allyl radical (124.6°)[299] is consistent with 1,3-repulsion between the out-of-phase p orbitals in the highest occupied molecular orbital (47).

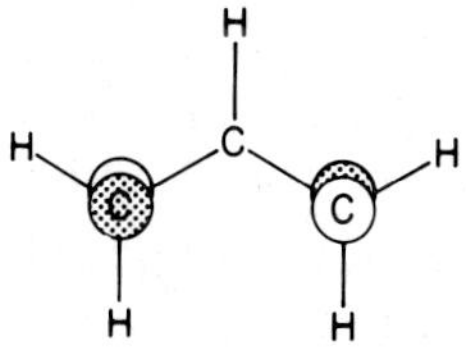

(**47**)

As expected from Hückel theory, the CC bond lengths are similar for the allyl cation (1.373 Å), radical (1.390 Å) and anion (1.388 Å).

The barrier to rotation about a CC bond is expected to reflect the allylic resonance stabilization of the planar equilibrium structure. Experimental rotational barriers for the free anion are not available, but barriers have been reported for the organometallic species CH_2CHCH_2M in tetrahydrofuran: 45, 70 and 75 kJ mol^{-1} for M=Li, K and Cs, respectively. The last value has been considered to be a lower bound for the barrier in the free anion.[302] Calculations[67,220,221,289,290,292,295,298,300] lead to values from 45 to 184 kJ mol^{-1} depending on the method used. The highest-level calculations give 93[67] and 81[300] kJ mol^{-1}.

The proton affinity calculated at the MP2/4-31+G level[67] coincides exactly with the experimental value[293] of 1636 kJ mol^{-1}. Experimental electron affinities of 53±5[142] and 35±2[301] kJ mol^{-1} have been reported for the allyl radical. A theoretical value obtained through the isogyric comparison with the methyl system[299] is 33 kJ mol^{-1}.

(2) Analogues of the Allyl Anion

(a) Enolate anions. Deprotonation of the methyl group of acetaldehyde leads to the resonance-stabilized formylmethyl anion (48).

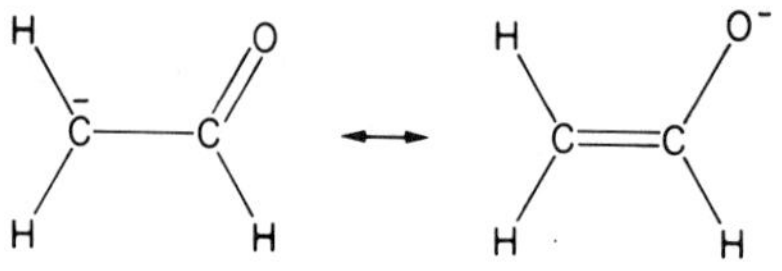

(48a) (48b)

The formylmethyl anion represents the simplest example of an enolate anion. The calculated structure[116,118,303,304] and Mulliken charge distributions[116,303,304] of (48) are fully in accord with a strongly delocalized π system. A measure of the resonance stabilization of CH_2CHO^- comes from the barrier to rotation about the CC bond. The most recent calculation[67] yielded a value of 167 kJ mol^{-1}.

The calculated proton affinity of the formylmethyl anion[67] of 1545 kJ mol^{-1} is in good agreement with an experimental value[140] of 1533±8 kJ mol^{-1}. The carbonyl hydrogen of acetaldehyde has been calculated[67,116,118] to be less acidic than the methyl hydrogens by

115–140 kJ mol^{-1}.

Configuration interaction calculations predict a Rydberg-like excited state of CH_2CHO^- of virtually the same energy as the ground state of the $CH_2CHO^{\cdot}$ radical.[304] This excited anion state is thought to be responsible for the threshold resonance observed in the photodetachment cross section of the formylmethyl anion.[304]

The electronic structure of the formylmethyl anion has been compared with that of acetaldehyde using electron density difference maps.[305] The ambident nature of the nucleophilicity of CH_2CHO^- has been investigated using reaction potential maps.[306]

Enolate anions are prominent in the ion cyclotron resonance spectra of reaction systems containing alkoxide anions. _Ab initio_ calculations have been used to study the mechanisms of formation of enolate anions by hydrogen loss from alkoxide anions[307,308] and via reaction of alkoxide-alkanol negative ions with carbonyl compounds.[309]

(b) _Methanimidamide anion_. _Ab initio_ calculations[310] indicate that the methanimidamide anion prefers conformation (49) over (50) and (51) by 15 and 51 kJ mol^{-1}, respectively.

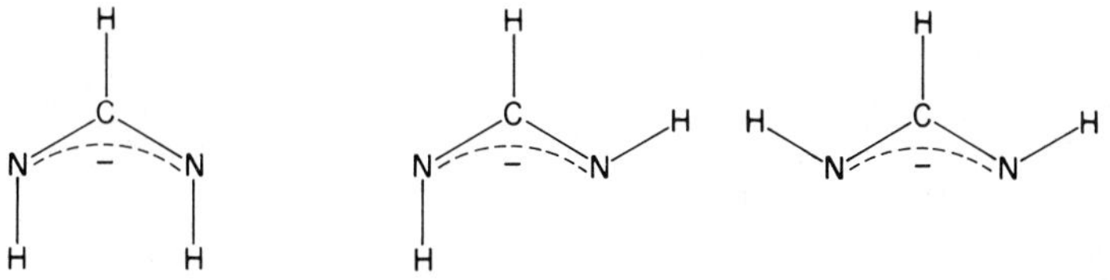

(49) (50) (51)

The barrier to rotation about one of the CN bonds in (49) is 96 kJ mol^{-1}, and the proton affinity is calculated[310] to be 1695 kJ mol^{-1}.

(3) _Substituted allyl and polyenyl anions and their analogues_. Substituted allyl anions are generally more stable in the _syn_ (52) than in the _anti_ (53) conformation (for leading references, see 311 and 312).

(52) (53)

Classical interpretations have invoked electrostatic and counter-ion (chelation) effects.[313] On the other hand, on the basis of extended Hückel calculations, it has been argued that it is a π-delocalization effect which is responsible for the observed preference for the _syn_ anions.[314] In simple terms, this interpretation holds that (52) is isoelectronic with the butadiene dianion, which in the _cis_ conformation (54) shows a small amount of aromatic stabilization because of the in-phase overlap of the terminal p_π orbitals in the highest occupied molecular orbital.

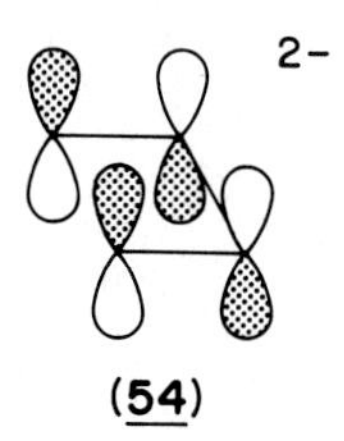

(54)

Conformational preferences in substituted allyl anions and their analogues have been studied theoretically in some detail.[213,287,311,312,315] The results are unfortunately less clear-cut than the earlier qualitative arguments had suggested. Of the semiempirical approaches, CNDO/2 generally predicts the _syn_ anions (52), including the _syn_ crotyl anion (55), to be more stable than the _anti_ isomers (53) and (56).[287] MNDO, however, predicts (55) to be _less_ stable than (56).[213]

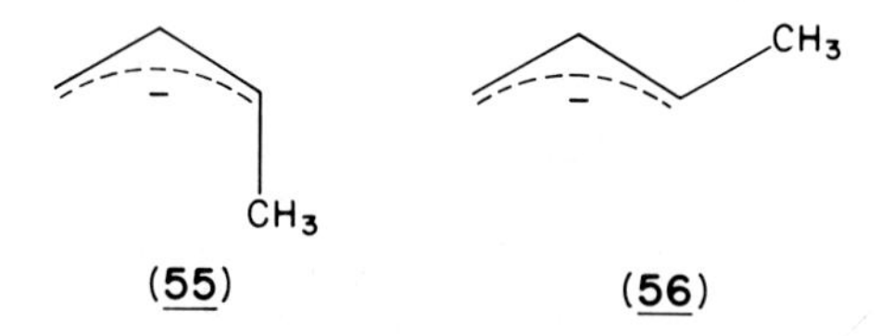

On the _ab initio_ side, minimal-basis-set calculations predict a stability order (55) < (56), while a preference for the _syn_ isomer over the _anti_ is predicted with an extended basis set.[311] Experimentally, the crotyl anion seems to prefer a _syn_ conformation in solution, but gas-phase results have been interpreted in terms of a preferred _anti_ conformation.[290] Higher-level theoretical calculations would be desirable to provide a definitive resolution of this interesting problem.

Imine anions can also exist in syn (57) and anti (58) forms.

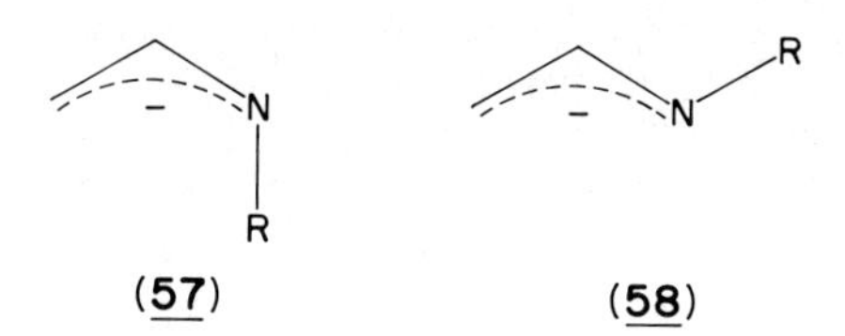

(57) (58)

Experimentally, the former appear to be thermodynamically more stable,[316] usually by 10-40 kJ mol^{-1}. Small-basis-set _ab initio_ calculations[312] reproduce this trend. Since not only substituted imine anions but also the parent system [(57), R=H] is _syn_, the π delocalization interpretation of the _syn_ preference is clearly not universally applicable, and the electrostatic explanation first put forward by Bank[313] might be more appropriate.[312] MNDO calculations suggest that the _syn_ and _anti_ forms of deprotonated acetaldehyde hydrazone, $H_2N-N=CH-CH_2^-$, are of essentially identical energies.[213]

The pentadienyl anion and its various mono-, di- and tri-methyl derivatives have been studied at the MINDO/3 and MNDO levels.[315] The W form (59) is predicted to be more stable than the S (60) or U (61) forms. This agrees qualitatively with the results of STO-3G calculations,[317] but geometrical constraints imposed in the latter study lead to a substantial overestimation of the energy of the U form (61). CNDO/2 predicts the three isomers to be essentially isoenergetic.[318] It is clear that the early prediction of Hoffmann and Olofson[314] of a preference for the U form (61) because of favourable 1,5-overlap in the highest occupied molecular orbital is not supported, presumably because of the severe steric interactions in that structure.

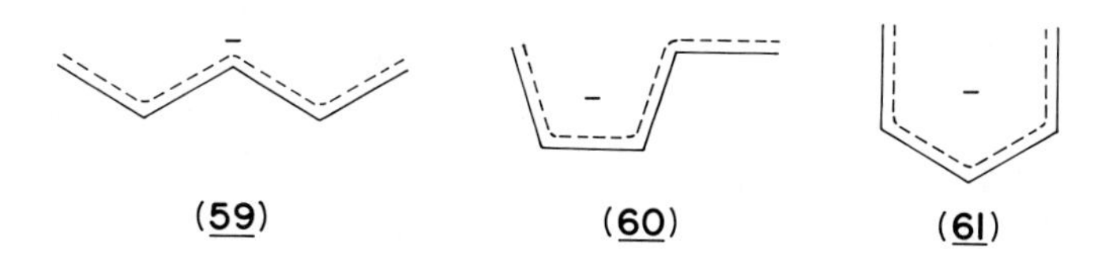
(59) (60) (61)

(4) _Cycloalkenyl anions_. Kinetic acidities in the cycloalkenyl series decrease in the order cycloheptene > cyclooctene > cyclohexene > cyclopentene. Ring strain and rehybridization effects are believed to be responsible for this trend, but model calculations[319] on the cycloalkenes and cycloalkenyl anions do not lead to a simple interpretation of the experimental data.

Model calculations are useful, however, in rationalizing

relative acidities of smaller cycloalkenes.[291] At the equilibrium structures, the allyl anion is found to be 75-85 kJ mol^{-1} more stable than the isomeric 1-methylvinyl (62) and 2-methylvinyl (63) anions.[293]

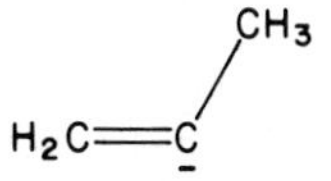

Significantly, the CCC angle in the allyl anion is larger (around 130°) than in the 1-methylvinyl anion (around 110°). Therefore, if the CCC angle is constrained at smaller values, the allyl anion is expected to become destabilized relative to the 1-methylvinyl anion, and small-ring cycloalkenes should deprotonate at the vinylic rather than at the allylic position. This qualitative argument, based on ab initio calculations on model compounds,[291] accounts for the experimental observation that the kinetic acidity of allylic protons in cyclopropene and cyclobutene is less than that of vinylic protons, but the acidity of the allylic protons of cyclopentene rivals or exceeds the acidity of the vinylic protons in this molecule.[320]

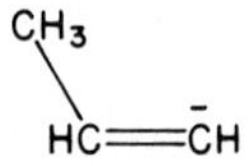

The cyclopropenyl anion is discussed in more detail in Sect. IVI1.

(5) Cyclohexadienyl and substituted cyclohexadienyl anions. Cyclohexadienyl anions are the products of the penultimate step in the Birch reduction sequence (Scheme 1).

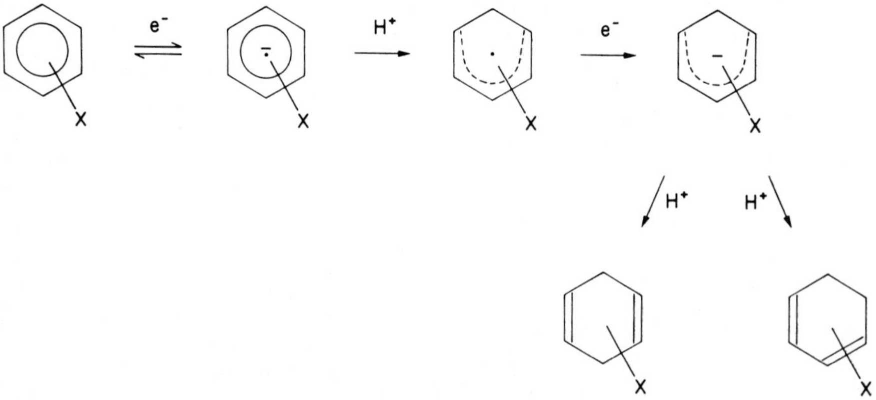

SCHEME 1

The parent anion (64) and a series of substituted derivatives (65)-(67) have been investigated using *ab initio* methods.[321]

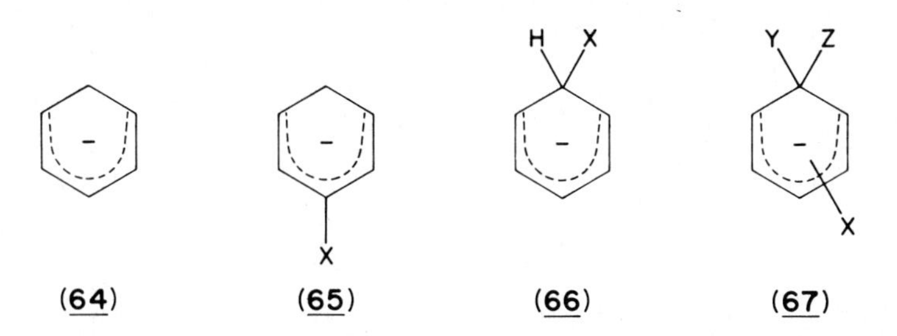

The lowest-energy structure of the cyclohexadienyl anion has a planar carbon skeleton and shows no sign of homoaromatic stabilization (cf. Sect. IVI3).[322] The excess π charge is largest at the position *para* to the site of first protonation. Electron-withdrawing substituents (CN, NO_2) prefer the *para* isomer (65), while electron-donors (F, OR, NH_2) prefer the *ipso* isomer (66). It should be noted, however, that it is the irreversible first protonation step which controls the orientation of the Birch reduction, and not the relative stability of the intermediate carbanions. The excess π charge at the *para* carbon is reflected in the electrostatic potential felt by an approaching proton, and regardless of the type and position of substituents, the second protonation step is predicted to occur *para* to the first. Experimentally, non-conjugated cyclohexadienes are formed preferentially under Birch conditions, consistent with the theoretical observations.[321]

Substituted cyclohexadienyl anions (67) are involved in nucleophilic aromatic substitutions (addition/elimination mechanism), and it has been proposed that their relative stabilities control the rate and orientation of substitution.[323,324] *Ab initio* calculations have been used to test such a proposal.[321,325,326] The calculated stabilization energies of substituents in cyclohexadienyl anions,[321] as given by the energy changes for the reaction

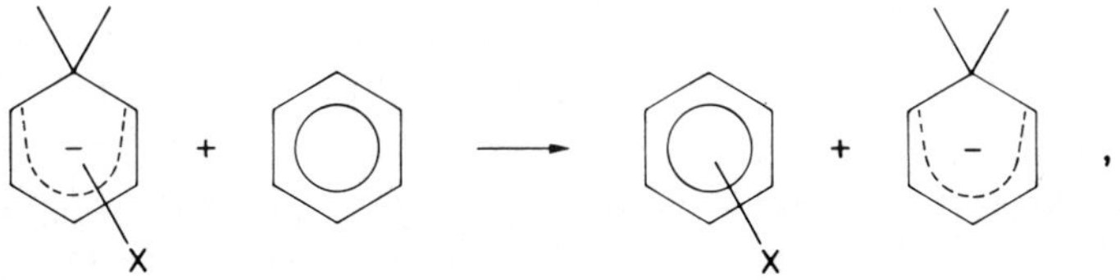

(29)

correlate well with experimental rate data for aromatic nucleophilic substitution reactions.[327] Thus, for example, strongly electron-withdrawing groups (X) such as NO_2 and CN, which greatly activate the ring toward nucleophilic attack, lead to large calculated stabilization energies. NO_2 is more stabilizing than CN, in accord with experiment. Substituent position increases the rate in the order *meta* << *ortho* < *para*, and this is again reflected in the calculated stabilization energies.[321] For π-donor substituents (NH_2, OH, F), the stabilization energies increase in the order *meta* > *ortho* > *para*,[321,325] and this is also consistent with experimental information.

(6) *Benzyl anions*. Semiempirical MNDO calculations[328] indicate that benzyl anions are trigonal planar species. The effect of substituents on the stabilities of benzyl anions and on the acidities of substituted toluenes has been examined in detail through *ab initio* STO-3G calculations.[329,330] The effect of substituents on acidity is dominated by the interactions in the anions rather than in the neutral acids. The relative gas-phase acidities of substituted toluenes are well reproduced at the minimal-basis-set level.[249,329] The benzyl anion has also been studied as a model for π-donor-substituted benzenes in electrophilic aromatic substitution.[331] The charge distribution in the benzyl anion has been compared with that in its 2-furyl and 2-thienyl analogues.[332]

(7) *The propargyl anion and substituted propargyl anions*. The propargyl anion ($HCCCH_2^-$) is of interest as the prototype for intermediates believed to be involved in the base-catalyzed equilibration of allenes and acetylenes:[333-335]

$$R\text{-}C{\equiv}C\text{-}CH_2\text{-}R' + B^- \rightleftharpoons [R\text{-}CCCH\text{-}R']^- + BH \rightleftharpoons R\text{-}CH{=}C{=}CH\text{-}R' + B^-. \quad (30)$$

Calculations[99,116,118,254,336-338] on the parent propargyl system suggest that both the acetylenic (68a) and allenic (68b) resonance structures are important. The CH_2 end is almost planar and the CH end is bent, as one would expect from structure (68b). On the other hand, there is a significant difference between the two CC bond lengths, reflecting a significant contribution from structure (68a). Isotope effects observed in the photodetachment spectra of CH_2CCH^- are more consistent with an allenic rather than an acetylenic structure.[339]

$HC{\equiv}C{-}CH_2^-$ ⟷ $H\bar{C}{=}C{=}CH_2$

(68a) (68b)

The barrier to inversion of the propargyl anion through a C_{2v} transition structure has been calculated[336-338] to be ≈30 kJ mol^{-1}, substantially less than the barrier in the vinyl anion. As in saturated carbanions, chlorine substitution increases the inversion barrier. The calculated value[337] for $HC{\equiv}C{-}CHCl^-$, ≈75 kJ mol^{-1}, compares reasonably well with an experimental estimate of 92 kJ mol^{-1} for a related species.[340]

The propargyl anion is calculated to be slightly less stable than the prop-1-ynyl anion,[118,254,338] in agreement with gas-phase experimental data.[341,342] The 1,3-hydrogen shift (68) → (69) has a high activation energy.[343]

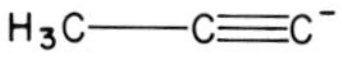

$H_3C{-}C{\equiv}C^-$

(69)

CNDO/2 calculations on the diphenyl-substituted anion (70) point to a dominance of the acetylenic structure (70a) (which in this case has a planar carbanionic centre), and are consistent with ^{13}C NMR data.[336]

$Ph{-}C{\equiv}C{-}\bar{C}H{-}Ph$ ⟷ $Ph{-}\bar{C}{=}C{=}CH{-}Ph$

(70a) (70b)

(8) Analogues of propargyl anion. Because a negative charge can be accommodated more easily on a nitrogen or oxygen atom than on carbon, the anions $HCCNH^-$ (71) and $HCCO^-$ (72) are acetylenic rather than allenic and are 65 and 213 kJ mol^{-1}, respectively, more stable than their ethynyl anion isomers (73) and (74).[254]

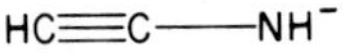

$HC{\equiv}C{-}NH^-$ (71) $HC{\equiv}C{-}O^-$ (72)

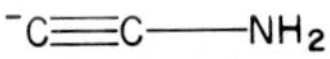

$^-C{\equiv}C{-}NH_2$ (73) $^-C{\equiv}C{-}OH$ (74)

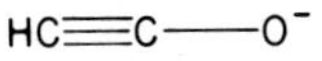

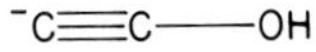

In the case of the silyl substituent, however, (75) is favoured over (76)[344,345] by 24 kJ mol^{-1}.

$^{-}C{\equiv}C{-}SiH_3$ (75) $HC{\equiv}C{-}SiH_2^{-}$ (76)

The proton affinities of these species have been calculated.[108,254,344]

(I) Aromatic, antiaromatic and homoaromatic anions

(1) Cyclopropenyl anion. According to Hückel's rule, systems with (4n+2) π electrons are aromatic and should be more stable than their acyclic counterparts. Conversely, systems with 4n π electrons are antiaromatic and should be less stable than the corresponding open-chain compounds.[346]

The cyclopropenyl anion is potentially the simplest antiaromatic system and as such has received considerable theoretical attention.[67,87,338,347-355] Although some early work[353,354] focussed on Jahn-Teller distortions of a planar singlet species, the most recent papers[67,87,338,355] leave no doubt that the singlet cyclopropenyl anion is strongly distorted from planarity, with one hydrogen bent substantially out of plane and the other two hydrogens bent out of plane in the opposite direction (77). The preferred structure is vinylic, (77) rather than allylic (78), the latter lying about 40 kJ mol^{-1} higher in energy. The barrier to inversion through the planar C_{2v} structure (79) was computed[87] to be 148 kJ mol^{-1}.

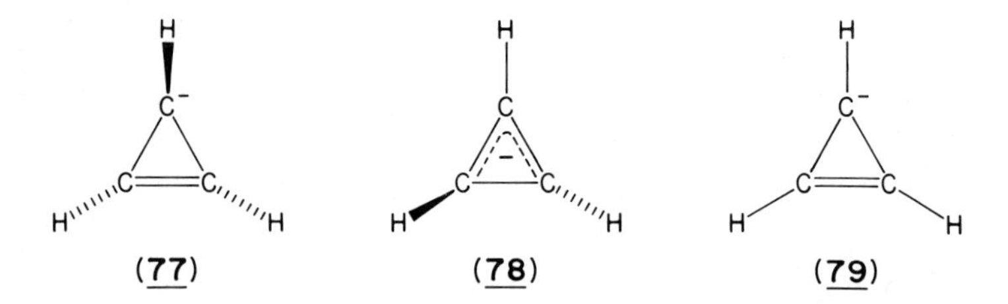

(77) (78) (79)

The triplet cyclopropenyl anion has also been examined.[87,352] Although semiempirical CNDO/2 calculations predict a planar D_{3h} triplet to lie lower in energy than the best singlet,[352] the more recent *ab initio* study[87] indicated that triplets of C_2, C_s and C_{3v} symmetries lie *higher* in energy than the ground-state singlet by at least 60 kJ mol^{-1}.

The open-chain $(CH)_3^-$ isomers (80)-(82) are comparable in energy to the singlet cyclopropenyl anion (77).[87,338] However, ring opening of (77) is a symmetry-forbidden process, and is expected to have a sizable barrier.

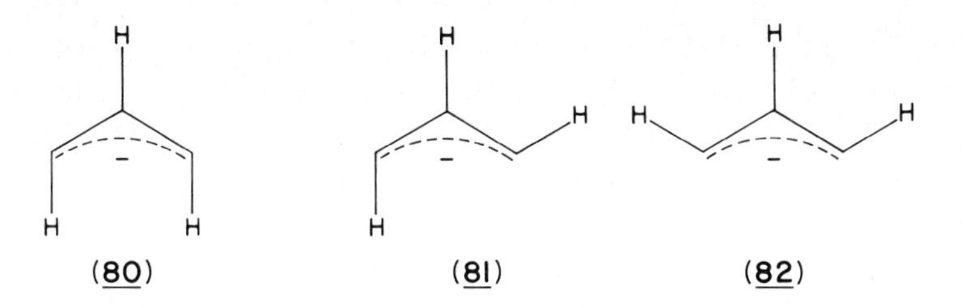

(80) (81) (82)

The electron affinity of the cyclopropenyl anion has been estimated[87] through use of the isogyric reaction

$$CH_3^{\bullet} \; + \; \longrightarrow \; + \; CH_3^{-} \quad . \qquad (31)$$

This leads to the conclusion that the electron affinity of the cyclopropenyl radical is about 75 kJ mol^{-1} less than that of the methyl radical. Since the experimental electron affinity[125] of $CH_3^{\bullet}$ is only 8 kJ mol^{-1}, this would imply that the cyclopropenyl anion is unstable in the gas phase with respect to spontaneous ejection of an electron.

The "antiaromatic" destabilization of the cyclopropenyl anion relative to the cyclopropyl anion has been estimated as 15 kJ mol^{-1}. For comparison, the cyclopropenyl radical and cation, containing three and two π electrons, are stabilized with respect to the cyclopropyl radical and cation by 63 and 293 kJ mol^{-1}.[87]

The structure of the experimentally known triphenylcyclopropenyl anion has been optimized at the MNDO level.[87] The cyclopropenyl moiety has an allylic rather than a vinylic structure due to extensive charge delocalization into the phenyl groups.

(2) 6π- and 8π-electron carbanions. The aromaticity of the 6π-electron oxetenyl anion (83) is exemplified by the calculated large positive enthalpy change[356] for the proton-transfer reaction

$$+ \; \longrightarrow \; + \qquad (32)$$

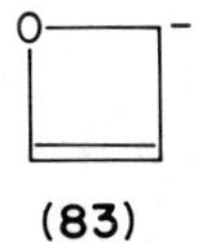

(83)

Interestingly, oxetene is nevertheless predicted to be less acidic than propene,[356] which seems to reflect the inherent low acidity of the cyclobutene allylic hydrogens,[322] as noted in Sect. IVH4.

The aromatic cyclopentadienyl anion and the antiaromatic cycloheptatrienyl anion have, as expected, singlet and triplet ground states, respectively; their description in terms of valence-bond theory has been discussed in detail.[357] Cyclopentadiene is a much stronger acid than 1,4-pentadiene, as reflected in the energy change (ΔG = 29 kJ mol^{-1})[358] for the gas-phase reaction

(33)

because of the aromatic stabilization of the cyclopentadienyl anion. This is well reproduced by semiempirical MINDO/3 calculations, but the effect is overestimated at the minimal-basis _ab initio_ level.[296] The silacyclopentadienyl anion (84) has been estimated to be only 25% as aromatic as the all-carbon analogue.[359]

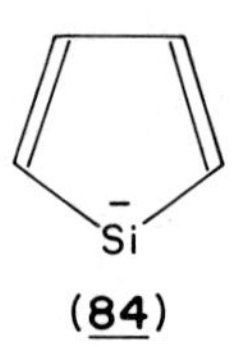

(84)

(3) Homoaromatic anions. According to Winstein,[360] stabilizing interactions in (4n+2)-π-electron systems are not destroyed completely by insertion of intervening saturated groups. Systems such as the cyclobutenium cation retain some aromatic stabilization and are termed "homoaromatic".

A potentially homoaromatic carbanion is the cyclohexadienyl anion (85).

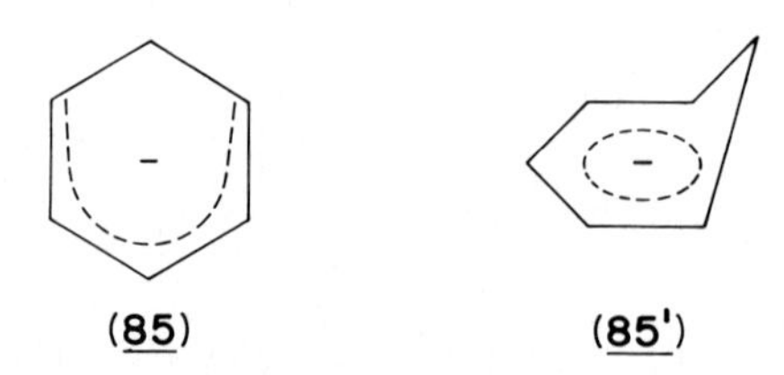
(85) (85')

If this species were indeed homoaromatic, it would be expected (by analogy with the homotropylium cation) to be nonplanar (85′), since such an equilibrium geometry would maximize the overlap of p orbitals at the termini of the unsaturated system. Experimental (^{13}C and ^{1}H NMR) and theoretical (MINDO/3) evidence suggests, however, that the lowest-energy structure of the cyclohexadienyl anion has a *planar* ring.[322] A distinct second minimum in the potential energy surface which corresponds to the homoaromatic species (85′) does exist, but is calculated to be 151 (MINDO/3[361]) or 186 (*ab initio*[321]) kJ mol^{-1} less stable than (85). Any homoaromatic stabilization of planar (85) should be reflected in an enhanced acidity of 1,3-cyclohexadiene compared with a suitable model system such as 1,3-hexadiene. However, according to MINDO/3 calculations,[322] 1,3-cyclohexadiene is actually slightly *less* acidic than 1,3-hexadiene, i.e. the cyclohexadienyl anion is not stabilized by homoaromaticity. In contrast, calculations suggest that the cyclononatetraenyl dianion (86) is a homoaromatic system.[361]

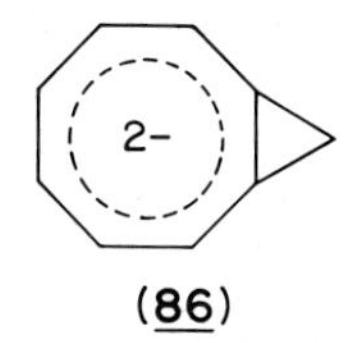

(86)

The bicyclic olefins (87)-(89) are known[362-365] to be more acidic than their monoene counterparts (90) and (91).

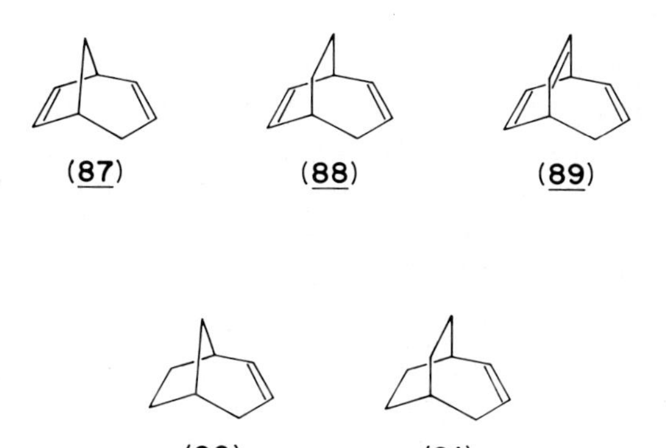

(87) (88) (89)

(90) (91)

This is usually ascribed to homoaromatic stabilization (92′)-(94′) of the corresponding anions (92)-(94).

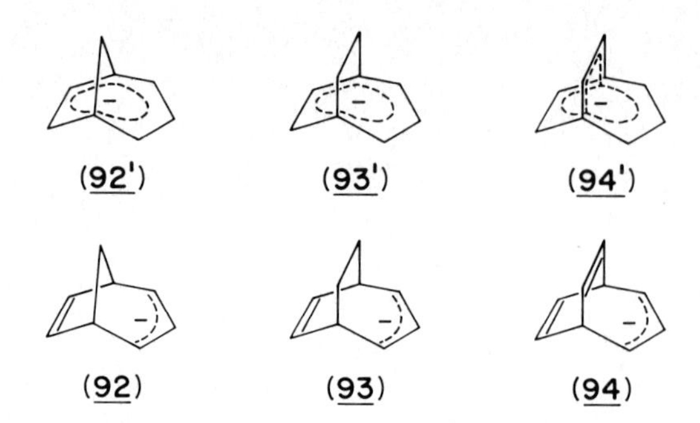

Ab initio calculations with minimal basis sets reproduce the experimental relative acidities at least qualitatively.[296,366] Any homoconjugation in these anions should be reflected in their geometries. In particular, the olefinic double bond participating in homoconjugation should be lengthened compared with that in the neutral hydrocarbon, because homoconjugation implies charge delocalization from the allylic anion part of the molecule into the antibonding olefinic π* orbital. The calculated (MINDO/3 and MNDO) geometries of (92)-(94) show no sign of such lengthening. Furthermore, the calculated acidities of (95) and (96) are similar to that of (87), even though homoconjugation is for symmetry reasons not possible in the anions (97) and (98).[366]

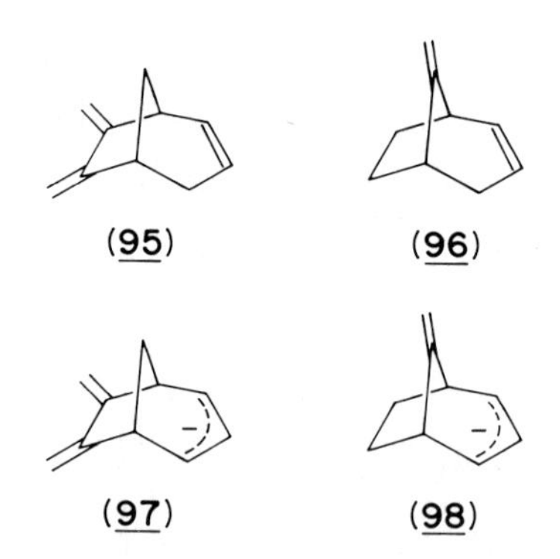

It has therefore been concluded that the enhanced acidity of (87)-(89) is not due to homoconjugation but to the inductive effect of the extra double bonds, and that homoconjugation in anions is generally unimportant.[296,366] This conclusion has been criticized on the grounds that MINDO/3 and MNDO fail to account for the enhanced acidity of (87)-(89) and therefore cannot be expected to reproduce geometry changes associated with homoconjugative stabilization.[367] An attempt has been made to model the possible homoaromatic stabilization of (87) through STO-3G calculations on the approach of an allyl anion to ethylene. These calculations provide support for the original postulate of long-range delocalization in (87).[367] Washburn[365] has also challenged the early theoretical results. By measuring the pK_a of compounds (87)-(90) in cyclohexylamine solution containing cesium cyclohexylamide, he found the anion of (87) to be more stable than the anion of (90) by 50 kJ mol^{-1}, thus providing "the first quantification of anionic homoaromatic stabilization". However, a very recent *ab initio* study,[368] employing multiconfiguration SCF calculations with minimal and split-valence basis sets, provides strong support for the unimportance of homoconjugation in (92). By analysing geometries, orbital density plots, charge distributions and the coefficients of configurations in the MCSCF expansion, it was concluded[368] that homoaromaticity is negligible. Rather, it was proposed that it is electrostatic interaction with the counterion which is responsible for the observed stability of anions such as (92).

Another potentially homoaromatic anion, (99), has been investigated theoretically; little or no homoaromatic stabilization is found.[369]

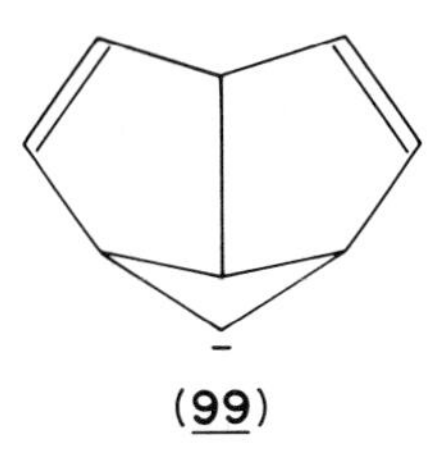

(**99**)

(J) Multiply-charged carbanions

(1) General considerations. Since there is generally little or no binding of the extra electron in singly-charged carbanions, i.e. the electron affinity of the corresponding radicals is negative or only slightly positive, multiply-charged carbanions would not be expected to be stable gas-phase species. Not surprisingly, relatively few calculations have been carried out on such species.[22,64,77,106,108,112,169,242,257,370-382] The relevance of these calculations to free carbanions is questionable. As we shall see, if sufficiently large basis sets are used, the best descriptions of many of these polyanions correspond to a neutral molecule with the additional electrons residing in very diffuse

orbitals.

(2) Methylene dianion CH_2^{2-}. The calculations on methylene dianion or dideprotonated methane (CH_2^{2-}) have all employed model geometries and used quite simple levels of theory.[112,370-372] Most of the studies[112,371,372] involve comparisons with other ten-electron hydrides. The disilyl derivative of CH_2^{2-} has also been examined.[169] A linear heavy-atom skeleton is found at the 3-21G level for $SiH_3CSiH_3^{2-}$ but better calculations (including d functions on silicon) yield a bent structure.

(3) Ethylene dianion $CH_2CH_2^{2-}$. Calculations on the ethylene dianion ($CH_2CH_2^{2-}$) show some interesting features. With small basis sets or with standard geometrical parameters, a preferred conformation is found in which the vicinal lone pair orbitals are nearly perpendicular to one another,[22,77,106,373] as expected by analogy with the isoelectronic hydrazine molecule. However, with diffuse-function-augmented basis sets and with full geometry optimization, the lowest-energy structure of ethylene dianion corresponds to a planar ethylene moiety with the two additional electrons residing in diffuse orbitals.[22,77]

(4) Acetylene dianion $HCCH^{2-}$. Acetylene dianion (or dideprotonated ethylene, $HCCH^{2-}$) exists as cis (100) and trans (101) isomers.[77] The latter lies lower in energy by 44 kJ mol^{-1}. Both isomers show bond lengths (1.376 and 1.359 Å, respectively, at 6-31+G(d)) which are slightly longer than the double bond of ethylene.

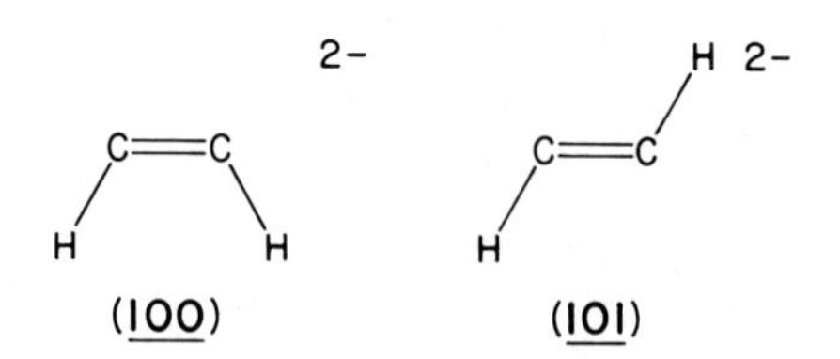

(5) Acetylide dianion CC^{2-}. Most of the emphasis in the calculations on the $C{\equiv}C^{2-}$ dianion has been placed on the equilibrium bond length.[64,77,257,374] The best calculations[77] yield 1.271 Å. Experimental values from X-ray crystal structures of carbide salts are strongly dependent on the counterion and range from 1.19 to 1.34 Å, with most structures yielding between 1.27 and 1.30 Å.[383]

(6) <u>Dideprotonated vinylacetylene</u>. Dianions arising from dideprotonation of vinylacetylene have been examined in relation to the dimetallation of vinylacetylene.[242] It is found that $H\bar{C}=CH-C\equiv\bar{C}$ is more stable than $H_2C=\bar{C}-C\equiv\bar{C}$, in contrast to $H\bar{C}=CH-C\equiv CLi$ which is less stable than $H_2C=\bar{C}-C\equiv CLi$. The latter result predicts that double metallation of vinylacetylene leads to 1,3- rather than 1,4-derivatives, in agreement with experimental evidence.

(7) <u>Y delocalization</u>. Theoretical calculations[375-377] have shown that small multiply-charged anions, in contrast to corresponding neutral or singly-charged species, favour Y delocalization compared with cyclic or extended conjugation. Thus, MNDO calculations find (<u>102</u>) to be lower in energy than (<u>103</u>) which is lower than (<u>104</u>).[375]

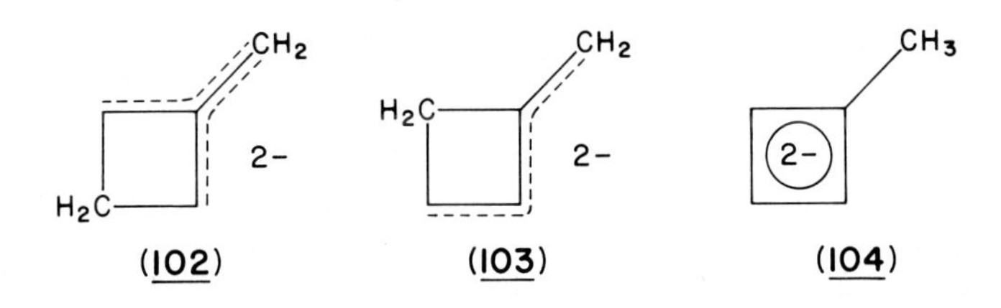

(<u>102</u>) (<u>103</u>) (<u>104</u>)

This has been attributed to electrostatic effects and becomes less important for larger ring sizes.

(8) <u>Aromatic multiply-charged anions</u>. Minimal-basis-set calculations for the dianion of cyclooctatetraene predict a planar D_{8h} structure (<u>105</u>).[378] There is a substantial energy cost associated with either distortion to a planar D_{4h} structure (<u>106</u>) or with distortion to the preferred D_{2d} structure of the neutral parent (<u>107</u>).

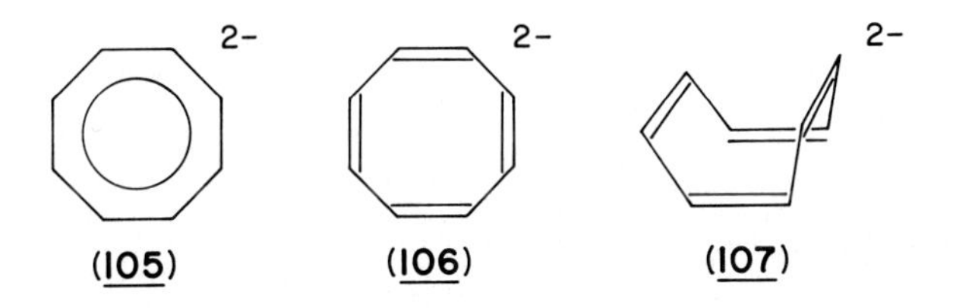

(<u>105</u>) (<u>106</u>) (<u>107</u>)

Ring opening of the bicyclo[3.3.0]octadienediyl dianion (<u>108</u>) to the cyclooctatetraene dianion (<u>105</u>) has been investigated using MNDO calculations.[379]

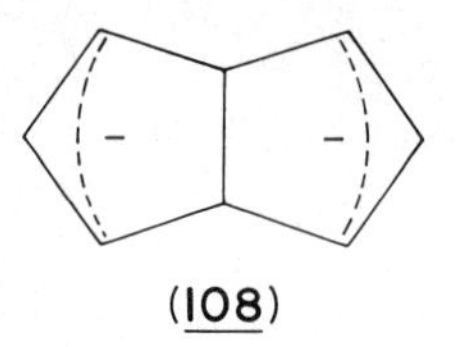

(108)

The cycloheptatrienyl trianion (109), which has been termed[384] the third member in the series of 10π aromatic anions that begins with cyclononatetraenyl anion and cyclooctatetraene dianion, has been the subject of MNDO calculations by Schleyer *et al.*[380] The best structure for (109) is planar with D_{7h} symmetry and C-C lengths of 1.475 Å. However, Schleyer *et al.* point out that this species is most unlikely to exist as a free trianion and the product observed experimentally[384] is probably the trilithio derivative, which the calculations show to be strongly puckered.

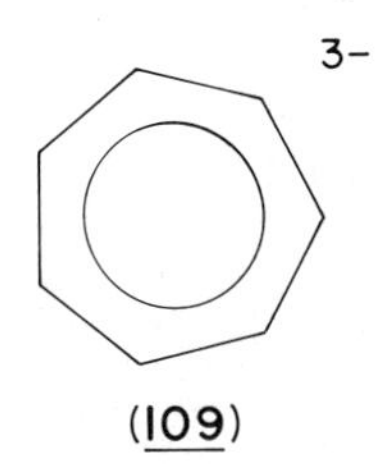

(109)

V. CONCLUDING REMARKS

Molecular orbital calculations on anions have traditionally been regarded with some caution by computational quantum chemists and as a consequence have attracted considerably less attention than calculations on neutral or positively charged species. At the conclusion of a 1977 review[9] it was pointed out, however, that "useful chemical information *can* be derived from *ab initio* single-determinant molecular orbital calculations on anions." During the past ten years, considerable further progress has been made. It can now be stated that *quantitatively* useful chemical information can be obtained for anions from calculations with sufficiently large basis sets and with adequate incorporation of electron correlation. At high levels of theory, properties such as ion structures and energies of reactions involving closed-shell species (e.g. proton affinities of anions) may be determined with an accuracy comparable to that achievable for normal neutral molecules. Theory can be used as a truly predictive tool here. Even modest levels of theory can provide useful results although reliable energy comparisons would normally require the use of diffuse-function-augmented basis sets. It remains difficult, however, to calculate accurate molecular electron affinities with methods which can at present be routinely applied to moderate-sized systems.

Carbanions are synthetically important reagents and represent a specific class of anions for which application of theory is potentially very useful. Calculations reported to date have already yielded valuable information regarding the structures and stabilities of carbanions and on the mechanisms of some of the reactions in which they are involved. There is nevertheless substantial scope for the much more extensive use of theory in the study of carbanions in the years ahead.

ACKNOWLEDGEMENTS

We thank Professors John Bowie, Bjorn Roos and Paul Schleyer for providing us with preprints of a number of papers in advance of publication.

NOTE ADDED IN PROOF

Since this manuscript was completed, a number of additional relevant papers have appeared. These include: (i) calculations[385,386] of the infrared intensities of CH_3^- (cf. Sect. IVA1); (ii) a study[387] of the cyanomethyl and isocyanomethyl anions and their lithium, sodium and magnesium derivatives (cf. Sects. IVA2c and IVA2d); (iii) a study[388] using experimental data and theoretical calculations to show that, with the exception of cyclopropyl anion, simple primary, secondary and tertiary carbanions (such as ethyl, 1-propyl and 2-propyl) are unstable with respect to electron loss and are not expected to exist as stable gas-phase species (cf. Sects. IVB1, IVC and IVD1); (iv) a study[389] of anionic hyperconjugation in fluorinated carbanions (cf. Sect. IVB2); (v) an experimental and theoretical (3-21+G) study[390] of isomeric $C_3H_5^-$ anions including allyl, 2-propenyl, 1-propenyl and cyclopropenyl anions (cf. Sects. IVD1 and IVH1); (vi) a paper[391] which argues that hyperconjugation and inductive effects rather than homoaromaticity are responsible for the stabilization of the bicyclo[3.2.1]octa-2,6-dienyl anion (<u>92</u>) in the gas phase and that interaction of the double bond at C_6 with the Li^+ gegenion in the endo geometry contributes additionally to the stabilization observed in solution; and (vii) a study[392] of the rearrangement of the fulminate anion (CNO^-) to the cyanate anion (OCN^-) with emphasis on the possible intermediacy of the oxaziranyl anion.

REFERENCES

1. Present address: CIBA-GEIGY Limited, CH-4002 Basle, Switzerland.
2. Permanent address: Department of Chemistry, The Chinese University of Hong Kong, Shatin, N.T., Hong Kong.
3. D.J. Cram, Fundamentals of Carbanion Chemistry, Academic Press, New York, 1965.
4. E. Buncel, Carbanions: Mechanistic and Isotopic Aspects, Elsevier, Amsterdam, 1975.
5. E. Buncel and T. Durst (Eds.), Comprehensive Carbanion Chemistry. Part A: Structure and Reactivity, Elsevier, New York, 1980.
6. R.B. Bates and C.A. Ogle, Carbanion Chemistry, Springer-Verlag, Berlin, 1983.
7. M.T. Bowers (Ed.), Gas Phase Ion Chemistry, Vols. 1-3, Academic Press, New York, 1979 (Vol. 1), 1979 (Vol. 2), 1984 (Vol. 3).
8. W.J. Hehre, L. Radom, P.v.R. Schleyer and J.A. Pople, Ab Initio Molecular Orbital Theory, Wiley, New York, 1986.
9. L. Radom, in H.F. Schaefer III (Ed.), Applications of Electronic Structure Theory, Plenum Press, New York, 1977, Ch. 8.
10. P.v.R. Schleyer, Pure Appl. Chem., 55 (1983) 355.
11. P.v.R. Schleyer, T. Clark, A.J. Kos, G.W. Spitznagel, C. Rohde, D. Arad, K.N. Houk and N.G. Rondan, J. Am. Chem. Soc., 106 (1984) 6467.
12. B.A. Murtagh and R.W.H. Sargent, Comput. J., 13 (1970) 185.
13. D. Poppinger, Chem. Phys. Lett., 34 (1975) 332.
14. P. Pulay, G. Fogarasi, F. Pang and J.E. Boggs, J. Am. Chem. Soc., 101 (1979) 2550.
15. H.B. Schlegel, Adv. Chem. Phys., in press.
16. A. Banerjee, N. Adams, J. Simons and R. Shepard, J. Phys. Chem., 89 (1985) 52.
17. J.W. McIver, Jr and A. Komornicki, J. Am. Chem. Soc., 94 (1972) 2625.
18. D. Poppinger, Chem. Phys. Lett., 35 (1975) 550.
19. A. Komornicki, K. Ishida, K. Morokuma, R. Ditchfield and M. Conrad, Chem. Phys. Lett., 45 (1977) 595.
20. C.J. Cerjan and W.H. Miller, J. Chem. Phys., 75 (1981) 2800.
21. S. Bell and J.S. Crighton, J. Chem. Phys., 80 (1984) 2464.
22. A.J. Kos, P.v.R. Schleyer, C.H. Heathcock, S.L. Graham, Y. Apeloig and K.N. Houk, to be published.
23. W.J. Hehre, R.F. Stewart and J.A. Pople, J. Chem. Phys., 51

(1969) 2657.
24. W.J. Hehre, R. Ditchfield, R.F. Stewart and J.A. Pople, J. Chem. Phys., 52 (1970) 2769.
25. W.J. Pietro, B.A. Levi, W.J. Hehre and R.F. Stewart, Inorg. Chem., 19 (1980) 2225.
26. W.J. Pietro, E.S. Blurock, R.F. Hout, Jr, W.J. Hehre, D.J. DeFrees and R.F. Stewart, Inorg. Chem., 20 (1981) 3650.
27. W.J. Pietro and W.J. Hehre, J. Comput. Chem., 4 (1983) 241.
28. J.S. Binkley, J.A. Pople and W.J. Hehre, J. Am. Chem. Soc., 102 (1980) 939.
29. M.S. Gordon, J.S. Binkley, J.A. Pople, W.J. Pietro and W.J. Hehre, J. Am. Chem. Soc., 104 (1982) 2797.
30. R. Ditchfield, W.J. Hehre and J.A. Pople, J. Chem. Phys., 54 (1971) 724.
31. W.J. Hehre and J.A. Pople, J. Chem. Phys., 56 (1972) 4233.
32. W.J. Hehre and W.A. Lathan, J. Chem. Phys., 56 (1972) 5255.
33. J.D. Dill and J.A. Pople, J. Chem. Phys., 62 (1975) 2921.
34. W.J. Hehre, R. Ditchfield and J.A. Pople, J. Chem. Phys., 56 (1972) 2257.
35. J.S. Binkley and J.A. Pople, J. Chem. Phys., 66 (1977) 879.
36. M.M. Francl, W.J. Pietro, W.J. Hehre, J.S. Binkley, M.S. Gordon, D.J. DeFrees and J.A. Pople, J. Chem. Phys., 77 (1982) 3654.
37. R. Krishnan, J.S. Binkley, R. Seeger and J.A. Pople, J. Chem. Phys., 72 (1980) 650.
38. P.C. Hariharan and J.A. Pople, Theor. Chim. Acta, 28 (1973) 213.
39. A.J. Duke and R.F.W. Bader, Chem. Phys. Lett., 10 (1971) 631.
40. T. Clark, J. Chandrasekhar, G.W. Spitznagel and P.v.R. Schleyer, J. Comput. Chem., 4 (1983) 294.
41. G.W. Spitznagel, T. Clark, J. Chandrasekhar and P.v.R. Schleyer, J. Comput. Chem., 3 (1982) 363.
42. I. Shavitt, in C.E. Dykstra (Ed.), Advanced Theories and Computational Approaches to the Electronic Structure of Molecules, D. Reidel, Dordrecht, 1984, p. 185.
43. I. Shavitt, in H.F. Schaefer III (Ed.), Methods of Electronic Structure Theory, Plenum Press, New York, 1977, Ch. 6.
44. C. Møller and M.S. Plesset, Phys. Rev., 46 (1934) 618.
45. J.A. Pople, J.S. Binkley and R. Seeger, Intern. J. Quantum Chem. Symp., 10 (1976) 1.
46. R. Krishnan and J.A. Pople, Intern. J. Quantum Chem., 14

(1978) 91.
47. R. Krishnan, M.J. Frisch and J.A. Pople, J. Chem. Phys., 72 (1980) 4244.
48. W. Kutzelnigg, in H.F. Schaefer III (Ed.), Methods of Electronic Structure Theory, Plenum Press, New York, 1977, Ch. 5.
49. J.A. Pople and D.L. Beveridge, Approximate Molecular Orbital Theory, McGraw-Hill, New York, 1970.
50. M.J.S. Dewar, J. Mol. Struct., 100 (1983) 41.
51. M.J.S. Dewar, E.G. Zoebisch, E.F. Healy and J.J.P. Stewart, J. Am. Chem. Soc., 107 (1985) 3902.
52. R. Hoffmann, J. Chem. Phys., 39 (1963) 1397.
53. M.J.S. Dewar and W. Thiel, J. Am. Chem. Soc., 99 (1977) 4907.
54. D.J. DeFrees, B.A. Levi, S.K. Pollack, W.J. Hehre, J.S. Binkley and J.A. Pople, J. Am. Chem. Soc., 101 (1979) 4085.
55. D.J. DeFrees, K. Raghavachari, H.B. Schlegel and J.A. Pople, J. Am. Chem. Soc., 104 (1982) 5576.
56. L.M. Branscomb, Phys. Rev., 148 (1966) 11.
57. P.T. Ford and R.E. Richards, Disc. Faraday Soc., 19 (1955) 230.
58. J.A. Ibers, J. Chem. Phys., 40 (1964) 402.
59. J.M. Rowe, D.G. Hinks, D.L. Price, S. Susman and J.J. Rush, J. Chem. Phys., 58 (1973) 2039.
60. G.E. Pringle and D.E. Noakes, Acta Crystallogr., 24B (1968) 262.
61. S.F. Mason, J. Phys. Chem., 61 (1957) 384.
62. G.B. Carpenter, Acta Crystallogr., 8 (1955) 852.
63. M.J.S. Dewar and H.S. Rzepa, J. Am. Chem. Soc., 100 (1978) 784.
64. L. Radom, Aust. J. Chem., 29 (1976) 1635.
65. R.A. Whiteside, M.J. Frisch and J.A. Pople (Eds.), The Carnegie-Mellon Quantum Chemistry Archive, Third Edition, Carnegie-Mellon University, Pittsburgh, 1983.
66. T.J. Lee and H.F. Schaefer III, J. Chem. Phys., 83 (1985) 1784.
67. J. Chandrasekhar, J.G. Andrade and P.v.R. Schleyer, J. Am. Chem. Soc., 103 (1981) 5609.
68. D.J. DeFrees and A.D. McLean, J. Comput. Chem., 7 (1986) 321.
69. C. Glidewell, J. Mol. Struct., 65 (1980) 231.
70. A.J. Duke, Chem. Phys. Lett., 21 (1973) 275.
71. D.S. Marynick and D.A. Dixon, Proc. Natl Acad. Sci. USA, 74

(1977) 410.
72. W.R. Rodwell and L. Radom, J. Am. Chem. Soc., 103 (1981) 2865.
73. D. Papoušek, J. Mol. Struct., 100 (1983) 179.
74. C.E. Dykstra, M. Hereld, R.R. Lucchese, H.F. Schaefer III and W. Meyer, J. Chem. Phys., 67 (1977) 4071.
75. R. Ahlrichs, F. Driessler, H. Lischka, V. Staemmler and W. Kutzelnigg, J. Chem. Phys., 62 (1975) 1235.
76. J. Kalcher and R. Janoschek, Chem. Phys., 104 (1986) 251.
77. W.-K. Li, R.H. Nobes and L. Radom, J. Mol. Struct., Theochem, in press.
78. D. Feller and E.R. Davidson, J. Chem. Phys., 82 (1985) 4135.
79. K. Raghavachari, J. Chem. Phys., 82 (1985) 4142.
80. G. Frenking and W. Koch, J. Chem. Phys., 84 (1986) 3224.
81. J. Baker, R.H. Nobes and L. Radom, J. Comput. Chem., 7 (1986) 349.
82. J.J. Novoa and F. Mota, Chem. Phys. Lett., 119 (1985) 135.
83. P.S. Drzaic, J. Marks and J.I. Brauman, in M.T. Bowers (Ed.), Gas Phase Ion Chemistry, Vol. 3, Academic Press, New York, 1984, Ch. 21.
84. L.S. Cederbaum and W. Domcke, Adv. Chem. Phys., 36 (1977) 205.
85. J. Simons, Ann. Rev. Phys. Chem., 28 (1977) 15.
86. M. Dupuis and B. Liu, J. Chem. Phys., 73 (1980) 337.
87. G. Winkelhofer, R. Janoschek, F. Fratev, G.W. Spitznagel, J. Chandrasekhar and P.v.R. Schleyer, J. Am. Chem. Soc., 107 (1985) 332.
88. C.S. Gudeman and R.J. Saykally, Ann. Rev. Phys. Chem., 35 (1984) 387.
89. T. Oka, Phys. Rev. Lett., 45 (1980) 531.
90. H.-J. Werner, P. Rosmus and E.-A. Reinsch, J. Chem. Phys., 79 (1983) 905.
91. N.H. Rosenbaum, J.C. Owrutsky, L.M. Tack and R.J. Saykally, J. Chem. Phys., 84 (1986) 5308.
92. P. Botschwina, Chem. Phys. Lett., 114 (1985) 58.
93. J.A. Pople, H.B. Schlegel, R. Krishnan, D.J. DeFrees, J.S. Binkley, M.J. Frisch, R.A. Whiteside, R.F. Hout and W.J. Hehre, Int. J. Quantum Chem. Symp., 15 (1981) 269.
94. R.E. Kari and I.G. Csizmadia, J. Chem. Phys., 46 (1967) 4585.
95. M.S. Gordon and H. Fischer, J. Am. Chem. Soc., 90 (1968) 2471.
96. Ph. Millié and G. Berthier, Int. J. Quantum Chem. Symp., 2

(1968) 67.
97. M.J.S. Dewar and M. Shanshal, J. Am. Chem. Soc., 91 (1969) 3654.
98. T.P. Lewis, Tetrahedron, 25 (1969) 4117.
99. W. Gründler, Tetrahedron, 26 (1970) 2291.
100. P.H. Owens, R.A. Wolf and A. Streitwieser, Jr, Tetrahedron Lett., (1970) 3385.
101. G. Szeimies, Tetrahedron Lett., (1970) 1949.
102. P.Th. van Duijnen and D.B. Cook, Mol. Phys., 21 (1971) 475.
103. P.H. Owens and A. Streitwieser, Jr, Tetrahedron, 27 (1971) 4471.
104. A. Rauk, J.D. Andose, W.G. Frick, R. Tang and K. Mislow, J. Am. Chem. Soc., 93 (1971) 6507.
105. R.E. Kari and I.G. Csizmadia, J. Chem. Phys., 56 (1972) 4337.
106. S. Wolfe, L.M. Tel, J.H. Liang and I.G. Csizmadia, J. Am. Chem. Soc., 94 (1972) 1361.
107. F. Driessler, R. Ahlrichs, V. Staemmler and W. Kutzelnigg, Theor. Chim. Acta, 30 (1973) 315.
108. A.C. Hopkinson, J. Chem. Soc., Perkin Trans. 2, (1973) 795.
109. H. Nakatsuji, J. Am. Chem. Soc., 96 (1974) 30.
110. P.E. Stevenson and D.L. Burkey, J. Am. Chem. Soc., 96 (1974) 3061.
111. J.E. Williams, Jr and A. Streitwieser, Jr, J. Am. Chem. Soc., 97 (1975) 2634.
112. R. Daudel, E. Kapuy, C. Kozmutza, J.D. Goddard and I.G. Csizmadia, Chem. Phys. Lett., 44 (1976) 197.
113. E.D. Jemmis, V. Buss, P.v.R. Schleyer and L.C. Allen, J. Am. Chem. Soc., 98 (1976) 6483.
114. G.T. Surratt and W.A. Goddard III, Chem. Phys., 23 (1977) 39.
115. J. Burdon, D.W. Davies and G. del Conde, J. Chem. Soc., Perkin Trans. 2, (1979) 1205.
116. A.C. Hopkinson and M.H. Lien, Int. J. Quantum Chem., 18 (1980) 1371.
117. K. Ishida, S. Kadowaki and T. Yonezawa, Bull. Chem. Soc. Japan, 54 (1981) 967.
118. A. Pross, D.J. DeFrees, B.A. Levi, S.K. Pollack, L. Radom and W.J. Hehre, J. Org. Chem., 46 (1981) 1693.
119. N.G. Rondan, K.N. Houk, P. Beak, W.J. Zajdel, J. Chandrasekhar and P.v.R. Schleyer, J. Org. Chem., 46 (1981) 4108.
120. J.B. Moffat, Int. J. Quantum Chem., 22 (1982) 299.

121. J.B. Moffat, J. Am. Chem. Soc., 104 (1982) 3949.
122. K.E. Edgecombe and R.J. Boyd, Can. J. Chem., 61 (1983) 45.
123. R.H. Nobes, D. Poppinger, W.-K. Li, and L. Radom, present work.
124. L. Andrews, J. Chem. Phys., 47 (1967) 4834.
125. G.B. Ellison, P.C. Engelking and W.C. Lineberger, J. Am. Chem. Soc., 100 (1978) 2556.
126. T. Clark, H. Körner and P.v.R. Schleyer, Tetrahedron Lett., (1980) 743.
127. S. Wolfe, Can. J. Chem., 62 (1984) 1465.
128. F. Bernardi, A. Mangini, G. Tonachini and P. Vivarelli, J. Chem. Soc., Perkin Trans. 2, (1985) 111.
129. M.H. Lien, A.C. Hopkinson and M.A. McKinney, J. Mol. Struct., Theochem, 105 (1983) 37.
130. H. Kollmar, J. Am. Chem. Soc., 100 (1978) 2665.
131. W. Cherry and N. Epiotis, J. Am. Chem. Soc., 98 (1976) 1135.
132. H. Yamabe, H. Kato and T. Yonezawa, Bull. Chem. Soc. Japan, 43 (1970) 2702.
133. J.N. Murrell, B. Vidal and M.F. Guest, J. Chem. Soc., Faraday Trans. 2, 71 (1976) 1577.
134. P.G. Mezey, A.J. Kresge and I.G. Csizmadia, Can. J. Chem., 54 (1976) 2526.
135. J.R. Murdoch, A. Streitwieser, Jr and S. Gabriel, J. Am. Chem. Soc., 100 (1978) 6338.
136. B. Bigot, D. Roux and L. Salem, J. Am. Chem. Soc., 103 (1981) 5271.
137. O.H. Crawford, Mol. Phys., 20 (1971) 585.
138. W.R. Garrett, Chem. Phys. Lett., 62 (1979) 325.
139. S. Tsuda, A. Yokohata and M. Kawai, Bull. Chem. Soc. Japan, 42 (1969) 614.
140. J.E. Bartmess, J.A. Scott and R.T. McIver, Jr, J. Am. Chem. Soc., 101 (1979) 6046.
141. K.E. Edgecombe and R.J. Boyd, Can. J. Chem., 62 (1984) 2887.
142. A.H. Zimmerman and J.I. Brauman, J. Am. Chem. Soc., 99 (1977) 3565.
143. I.N. Juchnovski, J.S. Dimitrova, I.G. Binev and J. Kaneti, Tetrahedron, 34 (1978) 779.
144. P.G. Mezey, M.A. Robb, K. Yates and I.G. Csizmadia, Theor. Chim. Acta, 49 (1978) 277.
145. A. Loupy, J.-M. Lefour, D. Deschamps and J. Seyden-Penne, Nouv. J. Chim., 4 (1980) 121.
146. D.J. Swanton, G.B. Bacskay, G.D. Willett and N.S. Hush, J. Mol. Struct., Theochem, 91 (1983) 313.

147. F.G. Bordwell, G.E. Drucker and H.E. Fried, J. Org. Chem., 46 (1981) 632.
148. J.B. Cumming and P. Kebarle, Can. J. Chem., 56 (1978) 1.
149. F. Delbecq, J. Org. Chem., 49 (1984) 4838.
150. G. Albagnac, B. Brun, B. Calas and L. Giral, Bull. Soc. Chim. France, (1974) 1469.
151. S. Durmaz and H. Kollmar, J. Am. Chem. Soc., 102 (1980) 6942.
152. B.A. Levi, E.S. Blurock and W.J. Hehre, J. Am. Chem. Soc., 101 (1979) 5537 and unpublished data.
153. A.D. Walsh, Trans. Faraday Soc., 45 (1949) 179.
154. E. Heilbronner, Tetrahedron Lett., (1964) 1923.
155. W.-D. Stohrer and R. Hoffmann, J. Am. Chem. Soc., 94 (1972) 1661.
156. T. Clark, P.v.R. Schleyer, K.N. Houk and N.G. Rondan, J. Chem. Soc., Chem. Commun., (1981) 579.
157. D.K. Bohme and G.I. Mackay, cited in ref. 116.
158. S.P. So, Chem. Phys. Lett., 67 (1979) 516.
159. D.S. Marynick, J. Mol. Struct., Theochem, 87 (1982) 161.
160. F. Bernardi, I.G. Csizmadia, A. Mangini, H.B. Schlegel, M.-H. Whangbo and S. Wolfe, J. Am. Chem. Soc., 97 (1975) 2209.
161. A. Streitwieser, Jr and J.E. Williams, Jr, J. Am. Chem. Soc., 97 (1975) 191.
162. N.D. Epiotis, R.L. Yates, F. Bernardi and S. Wolfe, J. Am. Chem. Soc., 98 (1976) 5435.
163. J.-M. Lehn and G. Wipff, J. Am. Chem. Soc., 98 (1976) 7498.
164. W.T. Borden, E.R. Davidson, N.H. Andersen, A.D. Denniston and N.D. Epiotis, J. Am. Chem. Soc., 100 (1978) 1604.
165. S. Wolfe, L.A. LaJohn, F. Bernardi, A. Mangini and G. Tonachini, Tetrahedron Lett., 24 (1983) 3789.
166. S. Wolfe, A. Stolow and L.A. LaJohn, Tetrahedron Lett., 24 (1983) 4071.
167. C. Eaborn, R. Eidenschink, P.M. Jackson and D.R.M. Walton, J. Organomet. Chem., 101 (1975) C40.
168. A.C. Hopkinson and M.H. Lien, J. Org. Chem., 46 (1981) 998.
169. C. Glidewell and C. Thomson, J. Comput. Chem., 3 (1982) 495.
170. S.M. Adams and S. Bank, J. Comput. Chem., 4 (1983) 470.
171. R.W. Strozier, P. Caramella and K.N. Houk, J. Am. Chem. Soc., 101 (1979) 1340.
172. D.M. Golden and S.W. Benson, Chem. Rev., 69 (1969) 125.
173. A. Streitwieser, Jr, P.H. Owens, R.A. Wolf and J.E. Williams, Jr, J. Am. Chem. Soc., 96 (1974) 5448.

174. D.J. DeFrees, W.J. Hehre and D.E. Sunko, J. Am. Chem. Soc., 101 (1979) 2323.
175. C. Van Alsenoy, J.N. Scarsdale, J.O. Williams and L. Schäfer, J. Mol. Struct., Theochem, 86 (1982) 365.
176. A.C. Hopkinson and M.H. Lien, J. Chem. Soc., Chem. Commun., (1980) 107.
177. D.R. Kelsey and R.G. Bergman, J. Am. Chem. Soc., 93 (1971) 1953.
178. D.A. Hunter, J.B. Stothers and E.W. Warnhoff, in P. de Mayo (Ed.), Rearrangements in Ground and Excited States, Academic Press, New York, 1980, Ch. 6.
179. R.S. Mulliken, J. Chem. Phys., 7 (1939) 339.
180. R.S. Mulliken, C.A. Rieke and W.G. Brown, J. Am. Chem. Soc., 63 (1941) 41.
181. J.W. Baker, Hyperconjugation, Oxford University Press, 1952.
182. M.J.S. Dewar, Hyperconjugation, Ronald Press, New York, 1962.
183. D. Holtz, Chem. Rev., 71 (1971) 139.
184. D. Holtz, Prog. Phys. Org. Chem., 8 (1971) 1.
185. L. Radom, Prog. Theor. Org. Chem., 3 (1982) 1.
186. J.D. Roberts, R.L. Webb and E.A. McElhill, J. Am. Chem. Soc., 72 (1950) 408.
187. D. Holtz, A. Streitwieser, Jr and R.G. Jesaitis, Tetrahedron Lett., (1969) 4529.
188. A. Streitwieser, Jr, C.M. Berke, G.W. Schriver, D. Grier and J.B. Collins, Tetrahedron, 37 (1981) Suppl. 1, p. 345.
189. R. Hoffmann, L. Radom, J.A. Pople, P.v.R. Schleyer, W.J. Hehre and L. Salem, J. Am. Chem. Soc., 94 (1972) 6221.
190. R.C. Bingham, J. Am. Chem. Soc., 97 (1975) 6743.
191. Y. Apeloig and Z. Rappoport, J. Am. Chem. Soc., 101 (1979) 5095.
192. R.D. Bach, R.C. Badger and T.J. Lang, J. Am. Chem. Soc., 101 (1979) 2845.
193. A. Pross and L. Radom, Aust. J. Chem., 33 (1980) 241.
194. J.G. Stamper and R. Taylor, J. Chem. Res. (S), (1980) 128.
195. Y. Apeloig, J. Chem. Soc., Chem. Commun., (1981) 396.
196. Y. Apeloig, M. Karni and Z. Rappoport, J. Am. Chem. Soc., 105 (1983) 2784.
197. P.v.R. Schleyer and A.J. Kos, Tetrahedron, 39 (1983) 1141.
198. D.S. Friedman, M.M. Francl and L.C. Allen, Tetrahedron, 41 (1985) 499.
199. M. Roy and T.B. McMahon, Can. J. Chem., 63 (1985) 708.
200. L. Libit and R. Hoffmann, J. Am. Chem. Soc., 96 (1974) 1370.

201. D.J. McLennan, Quart. Rev. Chem. Soc., 21 (1967) 490.
202. Z. Rappoport and S. Patai, in S. Patai (Ed.), The Chemistry of Alkenes, Wiley, New York, 1964.
203. Z. Rappoport, Acc. Chem. Res., 14 (1981) 7.
204. S.I. Miller, Tetrahedron, 33 (1977) 1211.
205. J.F. Kincaid and F.C. Henriques, Jr, J. Am. Chem. Soc., 62 (1940) 1474.
206. A.C. Hopkinson, M.A. McKinney and M.H. Lien, J. Comput. Chem., 4 (1983) 513.
207. D.E. Applequist and A.H. Peterson, J. Am. Chem. Soc., 83 (1961) 862.
208. J.B. Pierce and H.M. Walborsky, J. Org. Chem., 33 (1968) 1962.
209. H.M. Walborsky, L.E. Allen, H.-J. Traenckner and E.J. Powers, J. Org. Chem., 36 (1971) 2937.
210. D.T. Clark and D.R. Armstrong, Theor. Chim. Acta, 14 (1969) 370.
211. D.T. Clark and D.R. Armstrong, J. Chem. Soc., Chem. Commun., (1969) 850.
212. J. Tyrrell, V.M. Kolb and C.Y. Meyers, J. Am. Chem. Soc., 101 (1979) 3497.
213. M.J.S. Dewar and D.J. Nelson, J. Org. Chem., 47 (1982) 2614.
214. H.M. Walborsky and F.M. Hornyak, J. Am. Chem. Soc., 77 (1955) 6026.
215. H.-U. Wagner and G. Boche, Helv. Chim. Acta, 66 (1983) 842.
216. D.T. Clark and G. Smale, Tetrahedron Lett., (1968) 3673.
217. M.J.S. Dewar and S. Kirschner, J. Am. Chem. Soc., 93 (1971) 4290.
218. M.J.S. Dewar, Fortschr. Chem. Forsch., 23 (1971) 1.
219. P. Merlet, S.D. Peyerimhoff, R.J. Buenker and S. Shih, J. Am. Chem. Soc., 96 (1974) 959.
220. G. Boche, D. Martens and H.-U. Wagner, J. Am. Chem. Soc., 98 (1976) 2668.
221. G. Boche, K. Buckl, D. Martens, D.R. Schneider and H.-U. Wagner, Chem. Ber., 112 (1979) 2961.
222. T. Jonvik and J.E. Boggs, J. Mol. Struct., Theochem, 85 (1981) 293.
223. R.G. Jesaitis and A. Streitwieser, Jr, Theor. Chim. Acta, 17 (1970) 165.
224. R.B. Bates, R.S. Cutler and R.M. Freeman, J. Org. Chem., 42 (1977) 4162.
225. J.L. Dickstein and S.I. Miller, in S. Patai (Ed.), The Chemistry of the Carbon-Carbon Triple Bond, Wiley, New York,

1978, Ch. 19.
226. J.M. Lehn, B. Munsch and Ph. Millié, Theor. Chim. Acta, 16 (1970) 351.
227. C.E. Dykstra, A.J. Arduengo and T. Fukunaga, J. Am. Chem. Soc., 100 (1978) 6007.
228. L.L. Lohr and S.H. Ponas, J. Phys. Chem., 88 (1984) 2992.
229. S.I. Miller and W.G. Lee, J. Am. Chem. Soc., 81 (1959) 6313.
230. J.E. Bartmess and R.T. McIver, Jr, in M.T. Bowers (Ed.), Gas Phase Ion Chemistry, Vol. 2, Academic Press, New York, 1979, Ch. 11.
231. C.H. DePuy, V.M. Bierbaum and R. Damrauer, J. Am. Chem. Soc., 106 (1984) 4051.
232. D.K. Bohme and L.B. Young, J. Am. Chem. Soc., 92 (1970) 3301.
233. W. Lindinger, D.L. Albritton, F.C. Fehsenfeld and E.E. Ferguson, J. Chem. Phys., 63 (1975) 3238.
234. O. Eisenstein, G. Procter and J.D. Dunitz, Helv. Chim. Acta, 61 (1978) 2538.
235. J.E. Baldwin, J. Chem. Soc., Chem. Commun., (1976) 734.
236. P. Caramella and K.N. Houk, Tetrahedron Lett., (1981) 819.
237. J. Weber, M. Yoshimine and A.D. McLean, J. Chem. Phys., 64 (1976) 4159.
238. S.J. Gould and B.D. Remillard, Tetrahedron Lett., (1978) 4353.
239. A.R. Rossi, B.D. Remillard and S.J. Gould, Tetrahedron Lett., (1978) 4357.
240. C.D. Broaddus, J. Org. Chem., 29 (1964) 2689.
241. T.F. Crimmins and E.M. Rather, J. Org. Chem., 43 (1978) 2170.
242. L. Brandsma, H. Hommes, H.D. Verkruijsse, A.J. Kos, W. Neugebauer, W. Baumgärtner and P.v.R. Schleyer, to be published.
243. N.D. Epiotis, R.L. Yates, J.R. Larson, C.R. Kirmaier and F. Bernardi, J. Am. Chem. Soc., 99 (1977) 8379.
244. G. Seconi, C. Eaborn and J.G. Stamper, J. Organomet. Chem., 204 (1981) 153.
245. R.R. Schmidt, J. Talbiersky and P. Russegger, Tetrahedron Lett., (1979) 4273.
246. J. Chandrasekhar, J.G. Andrade and P.v.R. Schleyer, J. Am. Chem. Soc., 103 (1981) 5612.
247. D.K. Bohme, G.I. Mackay and S.D. Tanner, J. Am. Chem. Soc., 102 (1980) 407.
248. A.C. Hopkinson and I.G. Csizmadia, J. Chem. Soc., Chem.

Commun., (1971) 1291.
249. L. Radom, J. Chem. Soc., Chem. Commun., (1974) 403.
250. L. Radom, Aust. J. Chem., 28 (1975) 1.
251. N.M. Vitkovskaya, N.N. Murav'eva and Yu A. Frolov, J. Struct. Chem., 16 (1975) 309.
252. P. Čársky, R. Zahradník and I. Kozák, Chem. Phys. Lett., 41 (1976) 165.
253. A. Hinchliffe,, J. Mol. Struct., 37 (1977) 145.
254. A.C. Hopkinson, M.H. Lien, K. Yates, P.G. Mezey and I.G. Csizmadia, J. Chem. Phys., 67 (1977) 517.
255. H. Lischka, P. Čársky and R. Zahradník, Chem. Phys., 25 (1977) 19.
256. J. Pacansky and G. Orr, J. Chem. Phys., 67 (1977) 5952.
257. N.L. Summers and J. Tyrrell, J. Am. Chem. Soc., 99 (1977) 3960.
258. A. Pross and L. Radom, J. Am. Chem. Soc., 100 (1978) 6572.
259. K. Vasudevan and F. Grein, J. Chem. Phys., 68 (1978) 1418.
260. B.K. Janousek, J.I. Brauman and J. Simons, J. Chem. Phys., 71 (1979) 2057.
261. G. Caldwell, M.D. Rozeboom, J.P. Kiplinger and J.E. Bartmess, J. Am. Chem. Soc., 106 (1984) 4660.
262. M.F. Powell, M.R. Peterson and I.G. Csizmadia, J. Mol. Struct., Theochem, 92 (1983) 323.
263. R. Bonaccorsi, C. Petrongolo, E. Scrocco and J. Tomasi, J. Chem. Phys., 48 (1968) 1500.
264. R. Bonaccorsi, C. Petrongolo, E. Scrocco and J. Tomasi, Chem. Phys. Lett., 3 (1969) 473.
265. E. Clementi and D. Klint, J. Chem. Phys., 50 (1969) 4899.
266. J.B. Moffat and H.E. Popkie, J. Mol. Struct., 6 (1970) 155.
267. G. Doggett and A. McKendrick, J. Chem. Soc. (A), (1970) 825.
268. M. Dixon and G. Doggett, J. Chem. Soc. (A) (1973) 298.
269. K.M. Griffing and J. Simons, J. Chem. Phys., 64 (1976) 3610.
270. B. Liu, J. Chem. Phys., 67 (1977) 373.
271. J. Pacansky and B. Liu, J. Chem. Phys., 66 (1977) 4818.
272. J.E. Gready, G.B. Bacskay and N.S. Hush, Chem. Phys., 31 (1978) 467.
273. P.R. Taylor, G.B. Bacskay, N.S. Hush and A.C. Hurley, J. Chem. Phys., 70 (1979) 4481.
274. T.-K. Ha and G. Zumofen, Mol. Phys., 40 (1980) 445.
275. J.V. Ortiz, R. Basu and Y. Öhrn, Chem. Phys. Lett., 103 (1983) 29.
276. L.L. Lohr, J. Phys. Chem., 89 (1985) 3465.
277. J. Berkowitz, W.A. Chupka and T.A. Walter, J. Chem. Phys.,

50 (1969) 1497.
278. D.S. Marynick, L. Throckmorton and R. Bacquet, J. Am. Chem. Soc., 104 (1982) 1.
279. R.L. DeKock and D.S. Caswell, J. Phys. Chem., 85 (1981) 2639.
280. J.I. Brauman and L.K. Blair, J. Am. Chem. Soc., 93 (1971) 4315.
281. J.I. Brauman and L.K. Blair, J. Am. Chem. Soc., 90 (1968) 6561.
282. W.J. Hehre and J.A. Pople, Tetrahedron Lett., (1970) 2959.
283. D.J. DeFrees, J.E. Bartmess, J.K. Kim, R.T. McIver, Jr and W.J. Hehre, J. Am. Chem. Soc., 99 (1977) 6451.
284. S. Ikuta, J. Comput. Chem., 5 (1984) 374.
285. S.D. Peyerimhoff and R.J. Buenker, J. Chem. Phys., 51 (1969) 2528.
286. M. Shanshal, Z. Naturforsch. B, 25 (1970) 1065.
287. J.R. Grunwell and J.F. Sebastian, Tetrahedron, 27 (1971) 4387.
288. A.Y. Meyer and M. Chrinovitzky, J. Mol. Struct., 12 (1972) 157.
289. E.R. Tidwell and B.R. Russell, J. Organomet. Chem., 80 (1974) 175.
290. J.E. Bartmess, W.J. Hehre, R.T. McIver, Jr and L.E. Overman, J. Am. Chem. Soc., 99 (1977) 1976.
291. D.W. Boerth and A. Streitwieser, Jr, J. Am. Chem. Soc., 100 (1978) 750.
292. T. Clark, E.D. Jemmis, P.v.R. Schleyer, J.S. Binkley and J.A. Pople, J. Organomet. Chem., 150 (1978) 1.
293. G.I. Mackay, M.H. Lien, A.C. Hopkinson and D.K. Bohme, Can. J. Chem., 56 (1978) 131.
294. V. Barone, F. Lelj and N. Russo, Int. J. Quantum Chem., 19 (1981) 1197.
295. P. Cremaschi, G. Morosi and M. Simonetta, J. Mol. Struct., Theochem, 85 (1981) 397.
296. J.B. Grutzner and W.L. Jorgensen, J. Am. Chem. Soc., 103 (1981) 1372.
297. R.J. Elliott and W.G. Richards, J. Mol. Struct., Theochem, 87 (1982) 211.
298. R. González-Luque, I. Nebot-Gil and F. Tomás, Chem. Phys. Lett., 104 (1984) 203.
299. P.v.R. Schleyer, J. Am. Chem. Soc., 107 (1985) 4793.
300. R. González-Luque, I. Nebot-Gil, M. Merchán and F. Tomás, Theor. Chim. Acta, 69 (1986) 101.

301. J.M. Oakes and G.B. Ellison, J. Am. Chem. Soc., 106 (1984) 7734.
302. T.B. Thompson and W.T. Ford, J. Am. Chem. Soc., 101 (1979) 5459.
303. S. Wolfe, H.B. Schlegel, I.G. Csizmadia and F. Bernardi, Can. J. Chem., 53 (1975) 3365.
304. R.W. Wetmore, H.F. Schaefer III, P.C. Hiberty and J.I. Brauman, J. Am. Chem. Soc., 102 (1980) 5470.
305. D.L. Grier and A. Streitwieser, Jr, J. Am. Chem. Soc., 104 (1982) 3556.
306. H. Moriishi, O. Kikuchi, K. Suzuki and G. Klopman, Theor. Chim. Acta, 64 (1984) 319.
307. R.N. Hayes, J.C. Sheldon, J.H. Bowie and D.E. Lewis, J. Chem. Soc., Chem. Commun., (1984) 1431.
308. R.N. Hayes, J.C. Sheldon, J.H. Bowie and D.E. Lewis, Aust. J. Chem., 38 (1985) 1197.
309. G. Klass, J.C. Sheldon and J.H. Bowie, Aust. J. Chem., 35 (1982) 2471.
310. T.J. Zielinski, M.R. Peterson, I.G. Csizmadia and R. Rein, J. Comput. Chem., 3 (1982) 62.
311. P.v.R. Schleyer, J.D. Dill, J.A. Pople and W.J. Hehre, Tetrahedron, 33 (1977) 2497.
312. K.N. Houk, R.W. Strozier, N.G. Rondan, R.R. Fraser and N. Chuaqui-Offermanns, J. Am. Chem. Soc., 102 (1980) 1426.
313. S. Bank, J. Am. Chem. Soc., 87 (1965) 3245.
314. R. Hoffmann and R.A. Olofson, J. Am. Chem. Soc., 88 (1966) 943.
315. M.J.S. Dewar, M.A. Fox and D.J. Nelson, J. Organomet. Chem., 185 (1980) 157.
316. R.R. Fraser, N. Chuaqui-Offermanns, K.N. Houk and N.G. Rondan, J. Organomet. Chem., 206 (1981) 131.
317. A. Bongini, G. Cainelli, G. Cardillo, P. Palmieri and A. Umani-Ronchi, J. Organomet. Chem., 92 (1975) C1.
318. R.J. Bushby and A.S. Patterson, J. Organomet. Chem., 132 (1977) 163.
319. A. Streitwieser, Jr and D.W. Boerth, J. Am. Chem. Soc., 100 (1978) 755.
320. G. Schröder, Chem. Ber., 96 (1963) 3178.
321. A.J. Birch, A.L. Hinde and L. Radom, J. Am. Chem. Soc., 102 (1980) 6430.
322. G.A. Olah, G. Asensio, H. Mayr and P.v.R. Schleyer, J. Am. Chem. Soc., 100 (1978) 4347.
323. J. Burdon, Tetrahedron, 21 (1965) 3373.

324. E. Buncel, M.R. Crampton, M.J. Strauss and F. Terrier, Electron Deficient Aromatic- and Heteroaromatic-Base Interactions, Elsevier, Amsterdam, 1984.
325. J. Burdon, I.W. Parsons and E.J. Avramides, J. Chem. Soc., Perkin Trans. 2, (1979) 1201.
326. J. Burdon, I.W. Parsons and E.J. Avramides, J. Chem. Soc., Perkin Trans. 1, (1979) 1268.
327. J. Miller, Aromatic Nucleophilic Substitution, Elsevier, Amsterdam, 1968.
328. K.B. Lipkowitz, C. Uhegbu, A.M. Naylor and R. Vance, J. Comput. Chem., 6 (1985) 662.
329. G. Kemister, A. Pross, L. Radom and R.W. Taft, J. Org. Chem., 45 (1980) 1056.
330. A. Pross and L. Radom, Prog. Phys. Org. Chem., 13 (1981) 1.
331. R.J. Elliott, V. Sackwild and W.G. Richards, J. Mol. Struct., Theochem, 86 (1982) 301.
332. M. Fiorenza, A. Ricci, G. Sbrana, G. Pirazzini, C. Eaborn and J.G. Stamper, J. Chem. Soc., Perkin Trans. 2, (1978) 1232.
333. R.J. Bushby and G.H. Whitham, J. Chem. Soc. (B), (1969) 67.
334. J.H. Wotiz, in H.G. Viehe (Ed.), Chemistry of Acetylenes, Marcel Dekker, New York, 1969, p. 371.
335. J. Klein and S. Brenner, Tetrahedron, 26 (1970) 5807.
336. R.J. Bushby, A.S. Patterson, G.J. Ferber, A.J. Duke and G.H. Whitham, J. Chem. Soc., Perkin Trans. 2, (1978) 807.
337. J.K. Wilmshurst and C.E. Dykstra, J. Am. Chem. Soc., 102 (1980) 4668.
338. W.-K. Li, R.H. Nobes and L. Radom, to be published.
339. J.M. Oakes and G.B. Ellison, J. Am. Chem. Soc., 105 (1983) 2969.
340. W.J. le Noble, D-.M. Chiou and Y. Okaya, J. Am. Chem. Soc., 100 (1978) 7743.
341. R.J. Schmitt, V.M. Bierbaum and C.H. DePuy, J. Am. Chem. Soc., 101 (1979) 6443.
342. C.H. DePuy, V.M. Bierbaum, L.A. Flippin, J.J. Grabowski, G.K. King, R.J. Schmitt and S.A. Sullivan, J. Am. Chem. Soc., 102 (1980) 5012.
343. A.C. Hopkinson, Prog. Theor. Org. Chem., 2 (1977) 194.
344. A.C. Hopkinson and M.H. Lien, J. Organomet. Chem., 206 (1981) 287.
345. A.C. Hopkinson and M.H. Lien, J. Mol. Struct., Theochem, 104 (1983) 303.
346. R. Breslow, Angew. Chem. Int. Ed. Eng. 7 (1968) 565.

347. D.T. Clark, J. Chem. Soc., Chem. Commun., (1969) 637.
348. N.C. Baird, Tetrahedron, 28 (1972) 2355.
349. M.J.S. Dewar and C.A. Ramsden, J. Chem. Soc., Chem. Commun., (1973) 688.
350. T.-K. Ha, F. Graf and Hs.H. Günthard, J. Mol. Struct., 15 (1973) 335.
351. C.U. Pittman, Jr, A. Kress, T.B. Patterson, P. Walton and L.D. Kispert, J. Org. Chem., 39 (1974) 373.
352. J. Pancíř and R. Zahradník, Tetrahedron, 32 (1976) 2257.
353. E.R. Davidson and W.T. Borden, J. Chem. Phys., 67 (1977) 2191.
354. W.T. Borden and E.R. Davidson, Acc. Chem. Res., 14 (1981) 69.
355. B.A. Hess, Jr, L.J. Schaad and P. Čársky, Tetrahedron Lett., (1984) 4721.
356. L.E. Friedrich and P.Y.-S. Lam, Tetrahedron Lett., (1980) 1807.
357. G.A. Gallup and J.M. Norbeck, J. Am. Chem. Soc., 97 (1975) 970.
358. P. Kebarle, Ann. Rev. Phys. Chem., 28 (1977) 445.
359. M.S. Gordon, P. Boudjouk and F. Anwari, J. Am. Chem. Soc., 105 (1983) 4972.
360. S. Winstein, in G.A. Olah and P.v.R. Schleyer (Eds.), Carbonium Ions, Vol. 3, Wiley-Interscience, New York, 1972, Ch. 22.
361. R.C. Haddon, J. Org. Chem., 44 (1979) 3608.
362. J.M. Brown and J.L. Occolowitz, J. Chem. Soc., Chem. Commun., (1965) 376.
363. M.J. Goldstein and S. Natowsky, J. Am. Chem. Soc., 95 (1973) 6451.
364. M.V. Moncur and J.B. Grutzner, J. Am. Chem. Soc., 95 (1973) 6449.
365. W.N. Washburn, J. Org. Chem., 48 (1983) 4287.
366. E. Kaufmann, H. Mayr, J. Chandrasekhar and P.v.R. Schleyer, J. Am. Chem. Soc., 103 (1981) 1375.
367. J.M. Brown, R.J. Elliot and W.G. Richards, J. Chem. Soc., Perkin Trans. 2, (1982) 485.
368. R. Lindh, B.O. Roos, G. Jonsäll and P. Ahlberg, J. Am. Chem. Soc., 108 (1986) 6554.
369. L.A. Paquette, H.C. Berk, C.R. Degenhardt and G.D. Ewing, J. Am. Chem. Soc., 99 (1977) 4764.
370. S. Wolfe, L.M. Tel and I.G. Csizmadia, Can. J. Chem., 51

(1973) 2423.

371. R. Daudel, M.E. Stephens, I.G. Csizmadia, C. Kozmutza, E. Kapuy and J.D. Goddard, Int. J. Quantum Chem., 11 (1977) 665.

372. C. Kozmutza, E. Kapuy, M.A. Robb, R. Daudel and I.G. Csizmadia, J. Comput. Chem., 3 (1982) 14.

373. R.B. Davidson and M.L. Hudak, J. Am. Chem. Soc., 99 (1977) 3918.

374. D. Poppinger and L. Radom, Chem. Phys., 30 (1978) 415.

375. T. Clark, D. Wilhelm and P.v.R. Schleyer, Tetrahedron Lett., 23 (1982) 3547.

376. S. Inagaki, H. Kawata and Y. Hirabayashi, J. Org. Chem., 48 (1983) 2928.

377. D. Wilhelm, T. Clark and P.v.R. Schleyer, J. Chem. Soc., Perkin Trans. 2, (1984) 915.

378. G. Wipff, U. Wahlgren, E. Kochanski and J.M. Lehn, Chem. Phys. Lett., 11 (1971) 350.

379. D. Wilhelm, T. Clark, P.v.R. Schleyer and A.G. Davies, J. Chem. Soc., Chem. Commun., (1984) 558.

380. P.v.R. Schleyer, D. Wilhelm and T. Clark, J. Organomet. Chem., 281 (1985) C17.

381. S. Wolfe, A. Rauk, L.M. Tel and I.G. Csizmadia, J. Chem. Soc., Chem. Commun., (1970) 96.

382. D. Wilhelm, T. Clark, P.v.R. Schleyer, J.L. Courtneidge and A.G. Davies, J. Am. Chem. Soc., 106 (1984) 361.

383. M. Atoji, J. Chem. Phys., 35 (1961) 1950.

384. J.J. Bahl, R.B. Bates, W.A. Beavers and C.R. Launer, J. Am. Chem. Soc., 99 (1977) 6126.

385. Y. Yamaguchi, M.J. Frisch, T.J. Lee, H.F. Schaefer III and J.S. Binkley, Theor. Chim. Acta, 69 (1986) 337.

386. D.J. Swanton, G.B. Bacskay and N.S. Hush, Chem. Phys., 107 (1986) 9.

387. J. Kaneti, P.v.R. Schleyer, T. Clark, A.J. Kos, G.W. Spitznagel, J.G. Andrade and J.B. Moffatt, J. Am. Chem. Soc., 108 (1986) 1481.

388. P.v.R. Schleyer, G.W. Spitznagel and J. Chandrasekhar, Tetrahedron Lett., 27 (1986) 4411.

389. D.A. Dixon, T. Fukunaga and B.E. Smart, J. Am. Chem. Soc., 108 (1986) 4027.

390. S.W. Froelicher, B.S. Freiser and R.R. Squires, J. Am. Chem. Soc., 108 (1986) 2853.

391. P.v.R. Schleyer, E. Kaufmann, A.J. Kos, H. Mayr and J.

Chandrasekhar, J. Chem. Soc., Chem. Commun., in press.
392. W.-K. Li, J. Baker and L. Radom, Aust. J. Chem., 39 (1986) 913.

CHAPTER 2

ELECTROCHEMISTRY OF CARBANIONS

MARYE ANNE FOX
Department of Chemistry, University of Texas, Austin, TX 78712

CONTENTS

I. INTRODUCTION 94
A. APPROACH 94
B. THE ELECTROCHEMICAL EXPERIMENT 95
II. ELECTROCHEMICAL CHARACTERIZATION OF CARBANIONS 99
A. MOLECULAR ORBITAL THEORY 99
B. pKa DETERMINATION 107
C. CONFORMATIONAL ANALYSIS 117
III. SYNTHETIC IMPLICATIONS OF OXIDATION OF CARBANIONS 119
A. RADICAL COUPLING 119
B. KOLBE ELECTROLYSES 122
C. ISOELECTRONIC HETEROATOMIC SYSTEMS 127
D. PHOTOELECTROCHEMISTRY 128
IV. ELECTROGENERATED CARBANIONS 130
A. ALKYL HALIDE REDUCTION 130
B. ARYL HALIDE REDUCTION 135
C. ELECTROGENERATED BASES 138
D. ELECTROGENERATED NUCLEOPHILES 141
V. REDUCTIONS VIA CARBANIONIC INTERMEDIATES 142
A. HYDROCARBON REDUCTIONS 143
B. CARBONYL REDUCTIONS 146
C. REDUCTION OF NITRO COMPOUNDS 151
D. REDUCTION OF ACID DERIVATIVES 154
E. REDUCTIVE ELIMINATIONS 156
F. REDUCTION OF ORGANOMETALLIC COMPOUNDS 159
VI. INDIRECT ELECTROREDUCTIONS 161
VII. SUMMARY 163
VIII. ACKNOWLEDGEMENT 164

INTRODUCTION

(A) Approach

Electrochemistry has been considered for years to be a useful and significant technique of sufficient fundamental importance to be taught to students at the earliest stages of understanding chemical principles. Organic electrochemical transformations also have a secure position historically, with an extensive collection of such reactions having been compiled as early as 1942 in Fichter's classic text.[1] Nonetheless, the field has blossomed recently, largely because of the development of instrumentation that can allow the non-specialist to explore mechanistic and analytical consequences of electron exchange-initiated reactions. Not only can the identity of products now be determined but also the route by which these products are formed can be easily monitored.

In particular, the advent of easy access to controlled electrochemical techniques has also permitted detailed studies of reactive intermediates generated in secondary chemical steps following a primary electron transfer. It is easy to imagine the conversion of carbocations to free radicals to carbanions, eqn 1, by the

$$R^{\oplus} \underset{}{\overset{+e^-}{\rightleftharpoons}} R\cdot \underset{}{\overset{+e^-}{\rightleftharpoons}} R^{\ominus} \qquad (1)$$

sequential addition of two electrons. The reverse of this sequence can thus provide a route for inverting at will the electron demand of a particular reagent, and in fact, this inversion of polarity can be referred to as "redox umpolung".[2]

The addition or removal of a single electron (reduction or oxidation, respectively) to or from a neutral substrate similarly produces a charge doublet, the respective radical cation or radical anion, eqn 2. Charge dispersal in such intermediates often follows

$$\begin{array}{ccccc} RX^{\dot{-}} & \rightleftharpoons & RX & \rightleftharpoons & RX^{\dot{+}} \\ \downarrow & & & & \downarrow \\ R\cdot + X^- \ (X=\text{halide}) & & & & R\cdot + X^+ \ (X=H) \end{array} \qquad (2)$$

a course involving bond cleavage, by loss of a cation (often a proton) or a stable anion (e.g., halide). The radical thus formed may possibly participate in further electrochemical steps as we described earlier. Electrochemical reactions thus offer unique environments for the interconversion of the reactive intermediates of organic chemistry.

In this chapter is found an overview of recent work directed toward electrochemical characterization of carbanions. Included are sections which address fundamental questions concerning the electron distribution or charge localization in the occupied molecular orbitals of these electron-rich species and which describe the *in situ* electrochemical formation of carbanions. Examples of the chemical consequences of carbanion oxidation are presented. A comparison is made of chemical reactivity of electrogenerated bases and nucleophiles with parallel reactions observed under more standard conditions. A brief survey of bond-forming and -breaking reactions from radical anionic intermediates is then presented, and recent developments on indirect, often catalytic, routes to anions are summarized.

The approach taken here is illustrative rather than comprehensive. Recent texts by Baizer and Lund[3] and Mann and Barnes,[4] as well as the organic sections of the "Encyclopedia of Electrochemistry of the Elements," edited by Bard and Lund,[5] offer extensive tables of electrochemical data, arranged conveniently by functional group. This chapter is not intended to rival these or other excellent review sources as compendia for valuable data. Rather, its intent is to introduce the non-specialist to the mechanistic and structural insight afforded by electrochemical methods for the description of carbanionic intermediates.

(B) The Electrochemical Experiment

The common distinguishing feature of the transformations discussed in this chapter is that they all occur after an electron has been transferred across the interface formed when an electrode, usually an inert metal but sometimes a semiconductor, is dipped into an electrolyte. This is a solution comprised of an oxidizable or reducible substrate dissolved in a polar solvent containing a substantial concentration of ions which render the medium highly conductive. In order for charge to flow, a cell must be constructed to allow current to flow to and from the electrodes, from one electrode toward the other through the electrolyte and in the reverse direction through a direct electrical connection. The electrode which is rendered electron-rich by application of an external potential (and hence where reductions occur) is called the cathode; that rendered electron-deficient (and hence where oxidations take place) is called the anode. All electrochemical cells must have both a cathode and an anode, but since often organic chemists are more concerned with either the oxidative or reductive transformation of a given substrate, one of the electrodes is arbitrarily called the working electrode, while the other is called the auxiliary or counterelectrode. If both half-reactions are of interest, both electrodes are called working.

Chemical control of a given electrochemical transformation mandates that the potential difference between the electrodes be controlled. A potentiostat thus adjusts the external source of electrical power by maintaining a desired potential difference between the working electrode and a third (reference) electrode. The potential difference is then measured with a voltmeter with a high internal resistance. An ammeter in circuit between the anode and cathode allows for simultaneous measurement of the current flowing through the electrolyte. A complete electrochemical cell is picture in Figure 1.

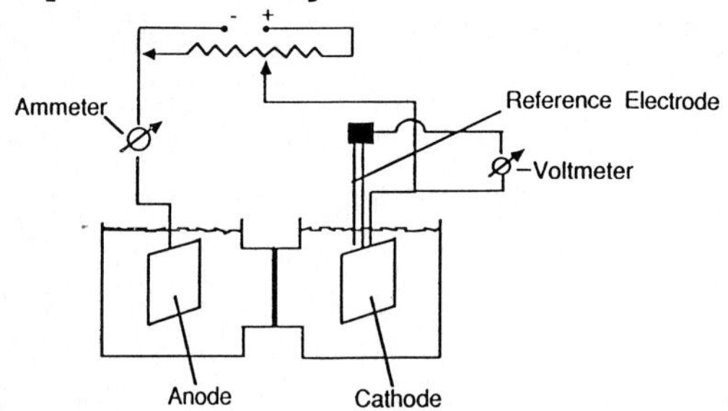

Fig. 1. A typical electrochemical cell.

As a potential difference is applied between the cathode and anode, an electrical field is formed at the interface in which polar molecules become highly oriented. This region, called the electrical double layer, may thus be distinguished from the normal bulk solution. At a certain potential difference, current begins to flow as the substrate contained within the double layer transfers electrons across the interface.

Current flow is plotted as a current-potential curve. If the potential difference continues to increase, the magnitude of the current flow will continue to rise until the electroactive substrate has been depleted in the region near the interface. Although mass transfer delivers more of the active substrate from the bulk to the depleted region, its rate is slow compared with usual rates for electron transfer across the interface. A peak in the current-potential plot is thus observed.

Suppose now that the potential difference between the two

electrodes is decreased. Since the region near the interface is rich in the reduced form of the redox couple, back electron transfer will occur and current will flow in the opposite direction from that initially observed. If the reduced intermediate is ideally stable, its concentration at the electrode surface will be equal to that lost by its redox partner on the forward scan, and the reverse current-potential curve will be exactly symmetrical to the forward curve. If the reduced intermediate is unstable with respect to a competing chemical reaction, its concentration in the double layer will be lower than expected and a dminished reverse current will be obtained.

In cyclic voltammetry, the current-potential curve obtained upon such forward and reverse potential scans is plotted. Thus, the cyclic voltammetric curve shape of an ideally reversible couple will resemble that shown in Figure 2. The half-wave potential

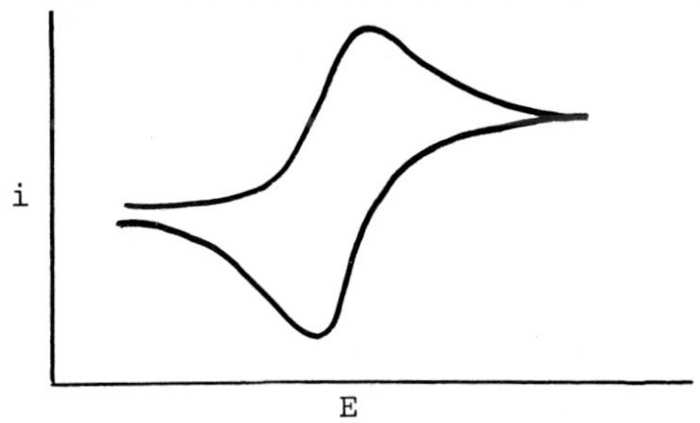

Fig. 2. Cyclic voltammetric scan of an ideally reversible redox couple.

between the oxidative and reductive peaks will have thermodynamic significance and will provide valuable information regarding the electronic properties of the molecular in question.

The cyclic voltammogram of an irreversible couple will show a peak potential in one direction only. Since its exact position will depend critically on variables in the cell (e.g., electrode material, adsorption effects, electrolyte concentration, etc.) the peak potential lacks firm thermodynamic significance, although shifts in peak potential within a series will be indicative of trends of redox reactivity. The cyclic voltammogram of such an irreversible couple is shown as Figure 3.

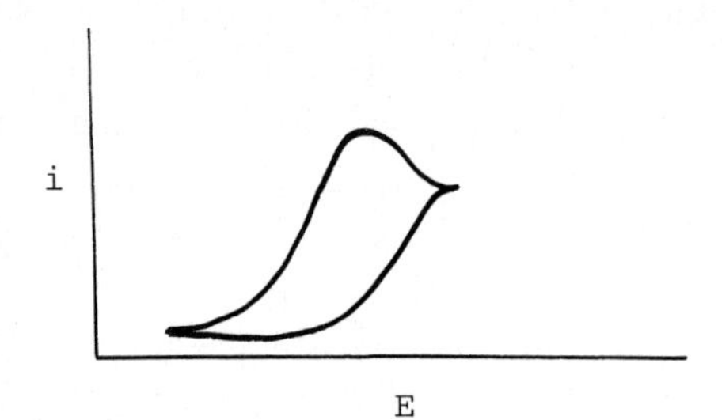

Fig. 3. Cyclic voltammetric scan of an irreversible redox couple.

Between the extremes of ideal reversibility and complete irreversibility lies, of course, a middle ground, the quasi-reversible redox couple. Here the observation and size of an inverted wave will depend on the scan rate, and rough kinetic ordering for secondary chemical reactions can often be attained by observing the wave shape as a function of scan rate. A typical series of such plots is shown in Figure 4.

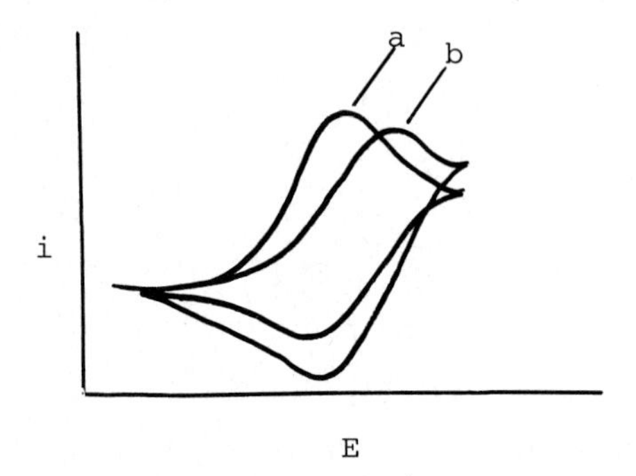

Fig 4. Cyclic voltammetric scan of a quasireversible redox couple. scan rate (a) 50 mV/s (b) 1000 mV/s

These current-potential plots can address questions of mechanistic significance to the physical organic chemist, giving a direct measure of the necessary driving force for an electron transfer and demonstrating quickly whether an electrogenerated intermediate is stable or reactive. Although a detailed discussion of the analysis of curve shapes and electrode kinetics is beyond the scope of this chapter, the interested reader can refer to an excellent text by Bard and Faulkner[6] for a more quantitative treatment of these ideas.

A complete mechanistic treatment of an organic electrochemical reaction would reflect the unique features of the electrochemical experiment. Thus, not only would there be established the relative reactivity of various substrates, the identity of reactive intermediates, and the structure of relevant transition states, but also fundamental questions would be answered concerning the nature of the electron transfer process and the movement of the reactant from the bulk solution to the electrode.[7] Adsorption effects, and consequently structure and orientation of substrates and subsequent intermediates, and the nature of operative mass transfer (convection, diffusion, or charge-induced migration) may therefore become mechanistically determinative variables.[8,9] Firm characterization of all these variables is difficult, however, and very often we are forced to infer mechanism, as in conventional chemical reactions, from incomplete information.

II. ELECTROCHEMICAL CHARACTERIZATION OF ANIONS

(A) Molecular Orbital Theory

We being by considering the nature of the orbitals involved in a primary electron exchange. We assume that electron transfer to the lowest unoccupied molecular orbital (LUMO) or from the highest occupied molecular orbital (HOMO) is accomplished at the electrode surface. Transfer to or from other orbitals would be more energetically demanding and, even if it were to occur, electron redistribution within the resulting radical ion would rapidly occur, placing the extra electron from a reduction into the LUMO and locating the electrogenerated hole from the oxidation in the HOMO. Thus, the half-wave potential for a reversible electroreduction defines the relative position of the LUMO of a molecule, while the oxidation potential orders the HOMO energy.

These potentials are commonly expressed as potential differences (in volts) from a reference, usually the normal hydrogen electrode (NHE), the saturated calomel electrode (SCE), or the Ag/AgCl reference or Ag wire pseudoreference electrodes. The electrode's Fermi level (the energy level occupied by free electrons in a medium of choice) becomes more negative as electrons are supplied by an external power source, and more positive as they are removed. Thus, in general, cathodic reductions occur at negative potentials (compared with a standard reference potential), while anodic oxidations occur at positive potentials.

Since we are interested in anions, we focus on the cathodic process, eqn 3. When the electrode Fermi level lies less negative

$$ArH \xrightarrow{+e^-} ArH^{\dot{-}} \xrightarrow{+e^-} ArH^{=} \quad (3)$$

than the LUMO of the molecule of interest, electron transfer cannot take place, Figure 5a. More precisely, at potentials positive

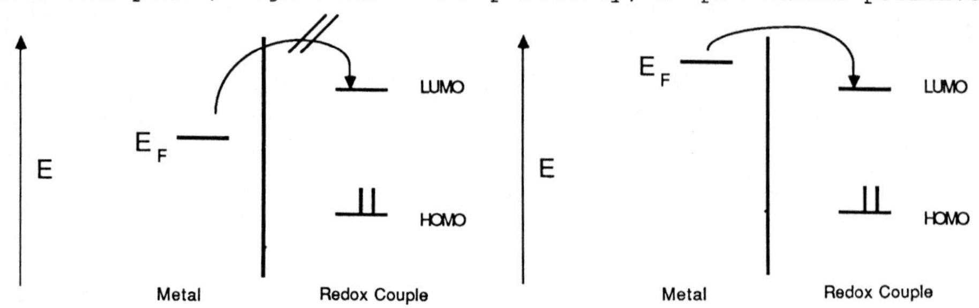

Fig. 5. Effect of an applied potential on electron transfer at a metal electrode-electrolyte interface.

of the reduction potential, the rate of electron transfer is so slow as to be negligible.

As the Fermi level of the electrode is adjusted, an energy level matching obtains an electroreduction of the acceptor present at the interface will ensue. That is, as the reduction potential is reached, Figure 5b, quantum mechanical tunnelling of the electron from the electrode Fermi level will generate the radical anion of the acceptor. Since the Fermi level is precisely defined (+ kT at ordinary temperatures), it is possible to describe the energy of the LUMO to within a few millivolts. If the electrode potential is further increased (i.e., the Fermi level becomes more negative), current will continue to flow despite energetic mismatching, because molecular distortion described by a Morse curve will permit energy matching with a higher energy state.

The well-recognized success of Hückel molecular orbital theory in predicting spectroscopic properties of conjugated hydrocarbons is also borne out for electrochemical measurements. Hoytink has developed a coherent scheme for cathodic reductions of aromatic hydrocarbons based on the correlation between experimentally observed reduction potentials and calculated molecular orbital energies.[10] Thus, excellent empirical correlations exist between solution phase reduction half-wave potentials and gas phase electron affinities, in parallel to the linear relationship between electrochemical oxidation potentials and gas phase ionization potentials. Since the ease of achieving an electrochemical reduction is governed both by the electron affinity of the substrate and by the solvation energy of the ions formed, this correlation implies either similar molecular distortions within the series (and hence constant experimental errors) or a fortuitous compensation of errors caused by differences in hybridization, substituents, ring strain, nonplanarity, etc.

Although benzene is reduced at potentials outside the accessible electrochemical window of most common solvents, the electrochemical reduction of more highly annelated aromatic hydrocarbons can be easily monitored, Table 1. Current-potential curves for

TABLE 1
Potentials and Rate Constants for Electroreduction[a] of Some Polycyclic and Substituted Aromatic Hydrocarbons[8]

Compound	E_o (V vs SCE)	k^o (cm s^{-1})
Naphthalene	-2.49	1.0
Anthracene	-1.95	>4
$Anthracene^-$	-2.55	0.009
Tetracene	-1.58	1.64
Perylene	-1.67	>4
$Perylene^-$	-2.26	0.009
t-Stilbene	-2.15	1.22
c-Stilbene	-2.18	0.42
Cyclooctatetraene[b]	-1.62	0.002
$Cyclooctatetraene^-$[b]	-1.86	0.12
Benzophenone	-1.73	>4
m-Dinitrobenzene	-0.80	5.0

a) Measured in DMF containing 0.1 M tetrabutylammonium iodide at Hg.
b) Electrolyte was tetrabutylammonium perchlorate.

many such substrates show two discrete, reversible steps in media of low proton availability,[10,11] producing sequentially a radical anion and dianion. Since protonation of the intermediate radical anion can be facile, it is often mandatory in seeking reversible potentials for the hydrocarbon - radical anion couple that such measurements be conducted in the presence of suspended neutral alumina, which scavenges traces of water or other electrophilic impurities.

That the potentials for formation of the radical anion and the dianion are separated by 500-700 mV (Table 2) implies that the

TABLE 2

Disproportionation Equilibrium Constants from Electrochemical[a] Measurements[11]

Compound	ΔE^o (mV)[b]	K_{disp}
Anthracene	740	1.0×10^{-13}
9,10-Diphenylanthracene	675	1.5×10^{-12}
1,2-Benzanthracene	665	2.0×10^{-12}
Chrysene	505	1.5×10^{-9}
Coronene	640	6.0×10^{-12}
Perylene	615	1.5×10^{-11}

a) Measured in DMF containing 0.1 M DMF at 11°C.
b) Peak separation between first and second reduction peak potenials.

radical anion is a sufficiently stable intermediate to exist for periods long enough to allow for diffusion away from the electrode surface. In several common electrolytic solvents, slow sweep cyclic voltammetry[11-13] and potentiometric titrations[14] indicate radical anion disproportionation equilibrium constants, eqn 4,

$$ArH^{\dot{-}} \rightleftharpoons ArH^{=} + ArH \quad (4)$$

to be exceedingly small (Table 2).

The radical anion formed electrochemically is associated with a cation, usually that supplied by the inert electrolyte. As with closed shell cation-anion pairs, the possibility exists of differential ion-pairing in electrogenerated cation-radical anion pairs. That is, the ion pairing equilibria shown in eqn 5 can contribute

$$\underset{\text{contact ion pair}}{ArH^{\dot{-}}M^{+}} \rightleftharpoons \underset{\text{solvent separated ion pair}}{ArH^{\dot{-}}|M^{+}} \rightleftharpoons \underset{\text{free ions}}{ArH^{\dot{-}} + M^{+}} \quad (5)$$

to the relative stability of an electrogenerated radical anion.

Whether the structure of the electrogenerated ion pairs can

be probed electrochemically remains an open question, for small effects on the disproportionation equilibrium constant as a function of the solvent and the counterion have been observed. For example, although the reduction potential for forming the perylene radical anion was independent of the supporting electrolyte cation for NMe_4^+, NEt_4^+, or $N(n\text{-}Pr)_4^+$, the separation between the two reduction peaks (forming the radical anion and dianion, respectively) did depend on the identity of the counterion. For the largest ion, a peak separation of 595 mV was observed while only a 545 mV separation was observed with NEt_4^+. With either Na^+ or with NMe_4^+, a larger separation (610 mV) was observed. Parallel shifts were also seen in the reduction of anthracene.[11]

These cation effects were rationalized as involving a solvent-separated ion pair for sodium and the small tetramethylammonium ion and progressively more weakly associated contact ion pairs as the size of the ammonium ion increased. The size of these shifts, however, is sufficiently small as to seem to make an unambiguous ion-pairing characterization risky.

Nonetheless, this explanation is parallel to previous electrochemical investigations on the nature of ion-pairing effects. The effect of tetraalkylammonium ions on the reversibility of alkali metal reduction has been attributed to the size of the ions.[15] That reductions were found to be irreversible for small counterions and reversible for larger ones was rationalized as preferential adsorption of the smaller ions on the electrode surface. Similarly, cyclic voltammetric studies of anthracene in hexamethylphosphoramide (HMPA) revealed smaller potential differences between the waves for radical anion and dianion formation upon changing the counterion from Li^+ to Bu_4N^+ to Et_4N^+, an order interpreted as indicative of increasing ion pair strength.[16]

Given these observations, it is somewhat surprising that the disproportionation equilibrium constants varied so little with solvent: less than a 35-fold variation in equilibrium constant for the perylene radical anion disproportionation was observed within the solvent series dimethylformamide (DMF), HMPA, pyridine, acetonitrile (MeCN), and tetrahydrofuran (THF). On the other hand, the disproportionation constants varied more importantly with supporting electrolyte concentration: for example, a hundred-fold increase in salt concentration brought about a 600-fold decrease in the anthracene radical anion disproportionation equilibrium constant.[11]

The reversibility for electrochemical anion and dianion formation which could be attained under the above aprotic conditions implies that dianions of extensively conjugated pi systems may also have sufficiently long lifetimes for characterization. Thus, the traditional assumption that protonation of electrochemically formed dianions by solvent is so rapid that other chemical reactions cannot compete is probably based on an erroneous premise (that purified solvents are dry). In truly aprotic solvents which lack electrophilic impurities, dianions of hydrocarbons may in fact be observed electrochemically. Their persistence will depend on the pKa of the medium, and, in some cases, almost heroic efforts may be required to provide a medium sufficiently basic (and impurity-free) to allow their survival. Unlike the dianions of acidic hydroquinones which can be readily observed in aprotic media,[17] these hydrocarbon dianions may often be more basic than typical electrochemical solvents.

The ease of protonation of electrogenerated anion radicals and dianions should depend markedly on structure. Some generalizations emerge[18]: protonation occurs more readily for hydrocarbons reduced at more negative potentials; easily accessible sites (terminal methylenes, sp-hybridized atoms, etc.) are easily protonated; hydrocarbons bearing acidic sites can act as internal protonation sources for their daughter radical anion or dianion.

The mechanism for protonation of radical anions or dianions can be extremely complex. A kinetic study of the protonation of the anion radical of 9,10-diphenylanthracene by phenol, for example, shows second order dependence on both the radical anion and phenol.[19] As we will see later, this complexity can sometimes make discrimination between mono- and dianions as critical reaction intermediates difficult.

Several routes exist for forming closed shell anions, eqn 6.

$$R \xrightarrow{+e^-} R^{\dot{-}} \begin{cases} \xrightarrow[(a)]{+H^+} \dot{R}H \begin{cases} \xrightarrow{+e^-} \overset{\ominus}{R}H \\ \xrightarrow[(b)]{R^{\dot{-}}} \overset{\ominus}{R}H + R \end{cases} \\ \xrightarrow{+e^-} R^{=} \xrightarrow[(c)]{+H^+} \overset{\ominus}{R}H \end{cases} \qquad (6)$$

In path a, the radical anion is protonated, removing one atom from conjugating and producing an odd alternant pi radical which bears the odd electron in a nonbonding orbital. The resulting radical is therefore more readily reduced than was the original hydrocarbon, and at the potential necessary to produce the radical anion will be immediately reduced further to the closed shell anion. This sequence is called an ECE process, since an electrochemical step is followed by a rapid chemical reaction to produce a species which is also electroactive.

Path b indicates that RH^- could also have been formed by electron transfer from a second electrogenerated radical anion, producing the anion and neutral precursor. Path c forms this same anion by an EEC route, i.e., by sequential formation of the radical anion and dianion which is then protonated.

Since the anion formed by any of these routes has one atom removed from conjugation, the charge will be borne in a nonbonding molecular orbital. As would be consistent with this Hückel-based prediction, the oxidation potentials of the pi anions obtained by two electron reduction and monoprotonation of a series of polycyclic aromatic hydrocarbons were nearly identical, despite significant differences in their precursors' reduction potentials.[20]

An alternate method of gauging the occurrence of protonation is to determine rates of electron transfer, either from voltammetric peak shapes or by double step chronoamperometry.[12] For example, Parker has found that electron transfer rates for the electroreduction of a series of aromatic hydrocarbons increased by as much as a factor of 10^8 in the presence of acid. Table 1 lists relative rates of heterogeneous reduction of polycyclic aromatic hydrocarbons and their anions. A persistent trend is the significant retardation of the rate of the radical anion to dianion conversion compared to that of the neutral to radical anion transfer. This electrostatic repulsion can, of course, be greatly reduced by protonation, and anion production by routes a or b in eqn 6 can thus be kinetically accelerated.

Hückel theory also describes charge dispersal in molecules and derived ions, and electrochemical methods have been used to investigate such effects. By examining the temperature dependence of the reduction potentials of aromatic hydrocarbons by phase selective second harmonic voltammetric techniques, Svaan and Parker have differentiated charge-disperse radical anion like that of triphenylene from charge-intensive radical anions like anthracene.[21] Since charge is concentrated in the 9- and 10-positions of the latter ion, much more temperature-sensitive electrochemical profiles are observed for anthracene than for triphenylene. Analogous trends were also reported for heteroatomic compounds,[22] nitro compounds,[23] and ketones.[24]

Hückel charge density calculations for electrogenerated radical ions and dianions also have consequence on the regiochemistry of protonation. For example, in alcoholic solvents the cathodic reduction of naphthalene gives 1,4-dihydronaphthalene as the major product, eqn 7, path a,[25] and that of cyclooctatetraene

(a) $+e^-$, $MeO(CH_2)_2OH$
(b) $+e^-$, DMF

(7)

gives 1,3,6-cyclooctatriene, eqn 8, path a.[26] Since these

(a) $+e^-$, EtOH
(b) $+e^-$, DMF

(8)

1,4-dihydro derivatives are less stable than their more highly conjugated 1,2-dihydro isomers, their dominant formation implies a kinetic preference for the observed protonation regiochemistry. Although steric factors may sometimes be important, ambident reactivity is determined by the relative activation energies for forming the transition states. Since protonations (or reactions with electrophiles) are generally fast, the transition state will closely resemble the anion in structure, and hence the charge density predicted from Hückel theory may be used as a reactivity index. Indeed, the orientation of the electroreduction of many hydrocarbons does follow these predictions, although the correlation with the relative rates of protonation of radical anions is somewhat less successful.[27]

Under slow protonation conditions, regiochemical preference can sometimes be reversed. For example, the reduction of cyclooctatetraene in anhydrous DMF gives 1,3,5-cyclooctatriene, the

1,2-protonation product, almost exclusively, eqn 8, path b.[26] 1,2-Protonation presumably also occurs when naphthalene is reduced in anhydrous acetonitrile containing tetrabutylammonium salts, eqn 7, path b, for tetralin becomes the major isolable product. This tetrahydro derivative forms because the 1,2-dihydro derivative is electroactive and is further reduced to the observed tetrahydro product.[28] Whether protic impurities, solvent, or the tetraalkylammonium ion acts as proton donor is unclear. Isolation of the more stable product can reflect either a shift in the site of the original protonation or the occurrence of base-catalyzed isomerization of the dihydro product.

Thus, we conclude that electrochemical methods are exceedingly useful for specifying molecular orbital energies and charge distributions. They provide an experimental test of the predictions of Hückel molecular orbital theory and a preliminary basis for describing solvent and ion pairing effects.

(B) pKa Determination

Since carbanions are stable in many common aprotic electrochemical solvents, electrochemical methods can be used directly to determine some of their fundamental characteristics. One of the most important properties of an anion is the equilibrium constant which describes its coexistence with its conjugate acid. The pKa of a molecule of interest is critical to the physical organic chemist's description of polar reactions, and whether the anion in question is generated electrochemically or by deprotonation of a hydrogenated precursor, an electrochemical description of its physical properties provides an important method for quantitative comparison within a series of related structures.

Electrochemical data can provide thermodynamic estimates of the basicities of carbanions if the following cyclie is constructed, eqn 9.[29] If the bond dissociation energy is available from

$$\mathrm{RH} \underset{}{\overset{\mathrm{BDE}}{\rightleftharpoons}} \mathrm{R\cdot} + \mathrm{H\cdot} \underset{}{\overset{-e^-}{\rightleftharpoons}} \mathrm{H^+} \qquad (9)$$

$$\mathrm{R\cdot} \overset{+e^-}{\rightleftharpoons} \mathrm{R^{\ominus}}$$

calorimetric measurements, the potential for the R· to $R^{\ominus}$ interconversion can be established for a series of compounds, for the energy of the H· to H^+ is common to the series. The potential difference for the anion oxidation can be readily converted to a pKa when compared with the known pKa of a model compound, assuming that other thermodynamic changes (e.g., from gas phase bond dissociation energies to solution phase electrochemical events) are comparable in both compounds.

A parallel cycle involving carbon - halogen homolysis, eqn 10,

$$RX \rightleftharpoons R\cdot + X\cdot \rightleftharpoons X^{\ominus} \quad (10)$$
$$R\cdot \underset{}{\overset{+e^-}{\rightleftharpoons}} R^{\ominus}$$

can be analyzed in the same way.

Electrochemical techniques thus allow for experimental estimation of pKa values for compounds whose pKa are much higher than those of common organic solvents, and for which therefore standard equilibrative methods for measuring pKa fail. For example, by employing second harmonic ac voltammetry techniques, Breslow and Goodin were able to determine the pKa of isobutane to be 71.[30] This value was obtained by observing that the peak potential for the second reductive wave (and thus presumably the conversion of the t-butyl radical to the anion) for t-butyl bromide in DMSO (-2.46 V vs. SCE) was nearly equivalent to that seen in t-butyl iodide in DME (-2.49 V vs. SCE).[31] Furthermore, the peak position showed only minor shifts with the reaction medium[32] or when the electrode material was changed from mercury (-2.43 V) to gold (-2.56 V) to platinum (-2.69 V). The same two reduction waves were also observed on vitreous carbon.[33] This invariance implied that organometallic bond formation is not an appreciable perturbation of the measured potential. From the difference between the average values for the reduction potentials and bond dissociation energies, the pKa difference between isobutane and triphenylmethane was calculated to be about 40 pKa units. Since the latter is known to have a pKa of 31.5,[34] the pKa of isobutane can thus be determined.

Note that this cycle will provide valid data only if the second reductive wave can be distinguished from the precursor alkyl halide reduction and if adsorption of neither the anion precursor nor the electrogenerated radical or anion leads to intermediate formation of organometallic species, thereby altering the identity of the bond cleaved in the dissociation step and fundamentally altering the calculation. Although these requirements appeared to be met for the isobutane calculation,[33] they were not met for other alkyl halides whose anion oxidation potentials were more positive by virtue of lower pKas. For example, in the first reported parallel studies of benzyl iodide, allyl iodide, and propargyl iodide, two reduction waves were observed, the second of which was assigned to reduction of the corresponding radical to the anion. Unfortunately, more careful investigation showed that the "first" wave was actually due to adsorption of the iodide on the metal electrodes, for the wave is absent on vitreous carbon.[34] Thus, the "second" wave was not correctly attributable to the reduction of the radical, and the pKas derived from it should be considered upper limits since the true reduction potential is presumably anodic of the potential required to reduce the iodide.

Similar complications with alkyl halide reductions have been observed in coulometric studies which showed that decyl iodide reduction proceeded via radical intermediates at potentials positive of the reduction peak potential and by carbanionic intermediates at potentials negative of this value.[35] Peters and his group also found that in preparative experiments the number of electrons taken up in the electroreduction of alkyl iodides varied from one to two depending on the availability of a ready proton source in anhydrous solvents.[35] Furthermore, Saveant and coworkers have clearly shown that the relative peak potential and the true thermodynamic potential for ECE-type reductions are almost always influenced by the rate of dimerization of the electrogenerated radical.[36]

In principle, this complication with alkyl halides could be obviated if the anion itself could be examined directly, and Breslow and coworkers have attempted pKa determinations derived from electrochemical oxidation potentials of the corresponding organolithiums.[33] Since many organolithiums with delocalized carbanion components do not have covalent carbon-lithium bonds but exist, rather, as carbanion-lithium ion pairs in solvents such as THF or HMPA, parallel behavior of the organolithium and the free carbanion might be expected. Accordingly, peak potentials from cyclic voltammetry and second harmonic ac voltammetry were obtained for benzyllithium, allyllithium, t-butylpropargyllithium, diphenylmethyllithium, cyclopentadienyllithium, and triphenylmethyllithium in THF containing HMPA, Table 3. Also listed are the pKas calculated from known bond dissociation energies by comparison with a known pKa for triphenylmethane. Close agreement with accepted values are found in each case.[33]

TABLE 3

pKas of Hydrocarbons Derived from Electrooxidation of Their Corresponding Organolithiums[33]

Compound	Ep[a]	BDE for CH	pKa
ϕ_3CH	-1.12	75	31.5
ϕ_2CH_2-H	-1.16	76.7-78.3	33.4
ϕCH_2-H	-1.39	88	44.4-45.2
$CH_2=CH-CH_2$-H	-1.61	86.6	47.1-48.0
$(CH_3)_3C-C\equiv C$-H	-1.12	94	44.2-45.4
cyclopentadiene C-H	-0.37	81.2	22.2-23.4

a) Oxidative peak potentials vs. SCE.

The situation became more complex, however, when phenyllithium, vinyllithium, methyllithium, and n-butyllithium were so examined, since they give broad irreversible waves and extensive electrode fouling. The poor electrochemical behavior of these lithio derivatives of less stable anions may be caused by high chemical reactivity of the relevant intermediates. Since such species often exist as clusters, even the extent to which they can be considered models of free carbanions is questionable. In any case, data of sufficient precision for assigning pKa values could not be attained.[33]

Other attempts at electrochemical characterization of organometallic compounds as a method to establish pKa were likewise fraught with problems. Psarras and Dessy, for example, found the same oxidation potential for both dimethylmagnesium and dibenzylmagnesium in DME,[37] and the pKa reported for methane (58.7) based on a photoelectrochemical route[38] remains the most reliable experimental value to date, despite rather severe assumptions made in the experimental design.

Formation of more stable anions should allow for more reliable pKa measurements, and several electrochemical studies[39-45] have shown that enolates and other extensively conjugated anions show electrochemical behavior of sufficient reliability for accurate pKa compilations. A selection of the oxidation potentials attained for several stable anions are shown in Table 4. Clearly,

TABLE 4.
Oxidation Peak Potentials for Several Stable Carbanions[105]

Carbanions	Solvent	Potential (V)	Ref
Me_2Mg	DME	-1.2 (vs Ag/AgCl)	37
$(\phi CH_2)_2Mg$	DME	-1.2 "	37
ϕ_3CLi	THF	-1.33 (vs SCE[a])	33
ϕ_2CHLi	THF	-1.37 "	33
ϕCH_2Li	THF	-1.45 "	33
(allyl)Li	THF	-1.40 "	33
MeCH(CHO)(CN) /Base	DMSO	-0.06 (Fc/Fc[+b])	41
2-cyanocyclohexanone /Base	DMSO	-0.23 "	41
$\phi C(O)CH_2CN$ /Base	DMSO	+0.06 "	41
$CH_3COCH_2COCH_3$ /Base	DMSO	+0.49 "	41
$CH_3COCH_2CO_2Et$ /Base	DMSO	+0.38 "	41
$CH_2(CO_2Et)_2$	DMSO	+0.39 "	41

as the anion in question becomes more stable, its oxidation potential shifts anodic, and relative pKas can be predicted with confidence by this simple measurement. Kern and Federlin, moreover, showed that an excellent correlation exists between electrochemically measured pKa values and those obtained by standard equilibration methods.[39]

This correlation allows for quantitative study of more subtle electronic effects. Banks and coworkers, for example, have shown the effect of methyl[46,47] and trimethylsilyl[48] substitution on the oxidation peak potential of diphenylmethyl and triphenylmethyl anions. Excellent linearity was observed in plots of half-wave oxidation potentials against changes in pi energy from SCF calculations or against pKa. Similar studies by Bordwell and coworkers[49,50] on substituted fluorenide or 9-phenylfluorenide anions showed good predictive ability of electrochemical peak potentials and basicity. Since basicity was also found to correlate with the ability to transfer electrons to an electrophile, a preliminary prediction of whether a given anion would react by single electron transfer or by a normal S_N2 route could be made.[49] These predictions are thus in accord with those made earlier by Fox and Singletary in describing photoactivated electron transfer from the cyclooctadienyl anion.[45]

Bordwell and Bausch[50] also found that electrochemically measured oxidation potentials could be used for estimating relative bond dissociation energies and radical stabilities and for predicting electron transfer rate constants. For example, the oxidation potentials of twenty-one 2-substituted fluorenide ions were found to plot linearly with the corresponding pKa vlues in DMSO. Substituents which deviated from the line were found to preferentially stabilize the fluorenyl radical by 0.4 - 1.5 kcal/mol. Similar results with alpha-cyanobenzyl anions and with 9-substituted fluorenyl anions revealed preferential radical stabilizing effects with Me_2N, MeO, Me, MeS, Ph, H_2NCO, MeOCO, and CN, with these substituents contributing in the optimal case (where the substituent is directly attached to a carbon atom bearing a relatively high spin density) as much as 2 - 11 kcal/mol. Similar linear plots of $E_{1/2}$ for the $(Ph_2C{=}CH)_2C(C_6H_4G)^-$ family against pKa produced a Bronsted beta of 1.1, while those of rate data (k_{obs} for electron transfer to an acceptor) again E_{ox} gave an improved plot that covered a rate range of greater than 10^5 and a potential range of 7 kcal/mol to yield a Bronsted beta of 1.04. Thus, electrochemical data can obviously be usefully employed to predict relative rates of electron transfer reactions.

Precise potential values in these anion oxidations sometimes showed dependence on the identity of the associated cation, and Banks concluded in a study of methylated diphenylmethyl anions that the intrinsic donor effect and that of ion pairing were of comparable importance in determining anion physical properties.[51] Presumably, the rates of electron transfer for cathodic reduction of alkyl halides will also depend on ion pairing, for the cathodic reduction of benzonitrile in DMF showed appreciable dependence on the identity of the alkyl group in the tetraalkylammonium electrolyte.[52] These effects were sufficiently large to cause the authors to propose different sites within the double layer for the requisite electron exchange, since a simple Frumkin double layer correction could not be applied directly.

Valuable information can also be attained by electrochemical oxidations of stabilized dianions. Two electron oxidation of the dianion shown in eqn 11 gave rise to the corresponding quinodi-

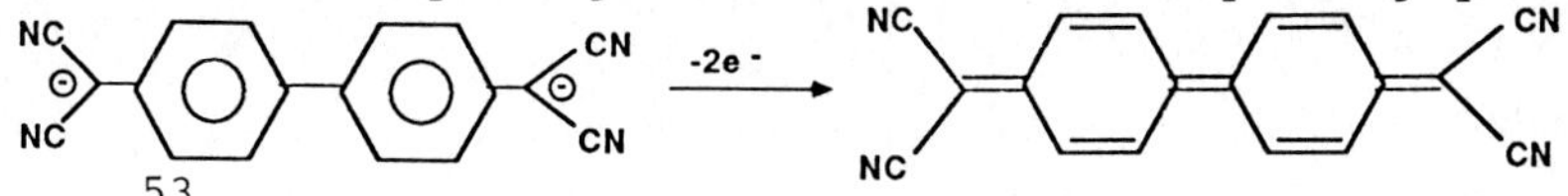

(11)

methane.[53] Electrochemical characterization of the analogous pyrenyl derivative, eqn 12, revealed a reversible one electron

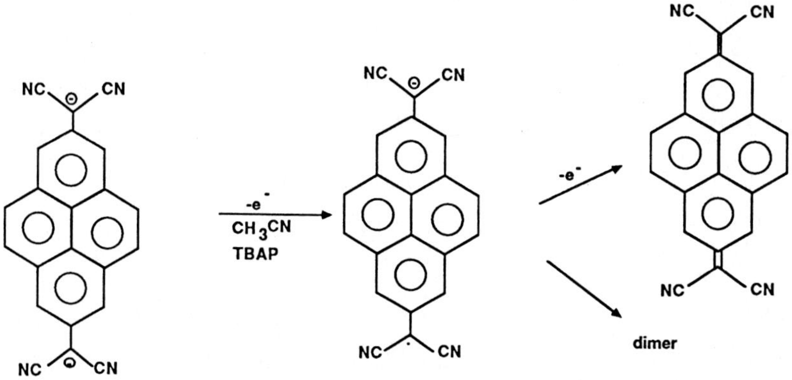

(12)

oxidation to form a dimerizable radical anion accompanied by a second irreversible one electron oxidation to form the quinoid neutral.[54]

That electrooxidation of dianions is chemically important can be seen in the recent observation of accompanying styrene radical anion formation in the deuterium exchange and coupling shown as eqn 13.[55] Such electrooxidation is almost certainly

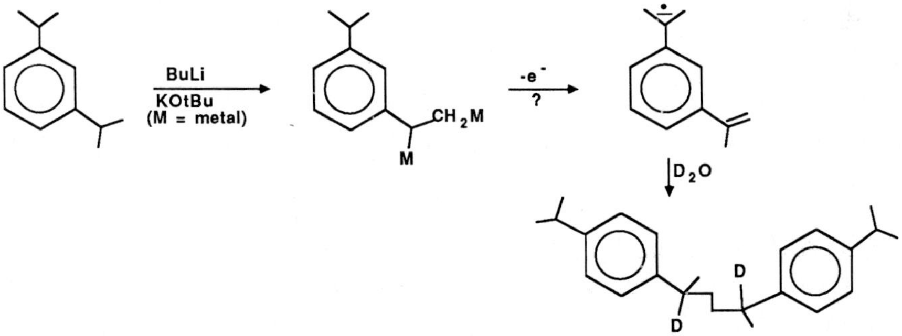

(13)

involved as well in the quenching of ketone dianions, for parallel reactivity is observed upon quenching with iodine or alkyl dihalides.[56,57]

The electrochemical potentials for one and two electron oxidation of fused hydroquinone dianions have also been used as an indication of resistance to bond localization in cyclobutadienoid derivatives, and hence as a test for antiaromaticity.[29] Effecting a two electron oxidation as shown in eqn 14, for example, can

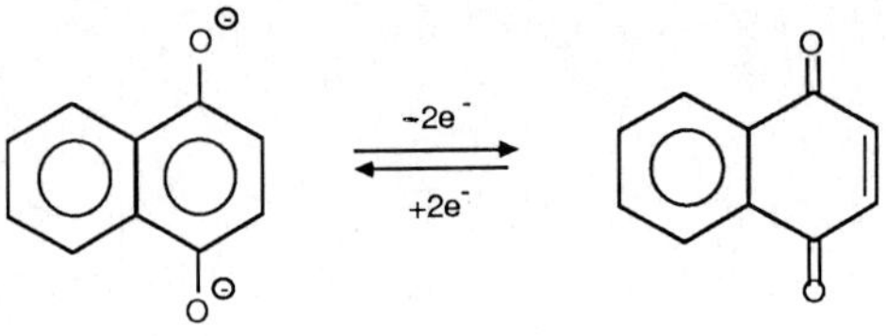

$$\text{naphthalene-1,4-diolate dianion} \underset{+2e^-}{\overset{-2e^-}{\rightleftharpoons}} \text{1,4-naphthoquinone} \qquad (14)$$

assess the energetic consequence of the changing bond order in the unsaturated fused ring.[58,59] Thus, as shown in Table 5, shifts in

TABLE 5

Oxidation potentials for Fused Hydroquinone Dianions[58,59]

Compound	Reduction Potentials (V vs SCE)
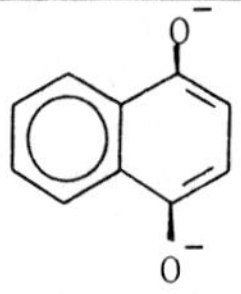 1,4-naphthalenediolate dianion	-0.68, -1.50
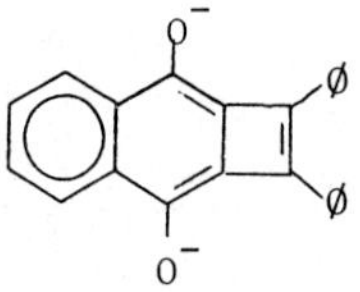 cyclobuta-fused naphthalenediolate dianion (Ø, Ø)	-0.25, -0.90
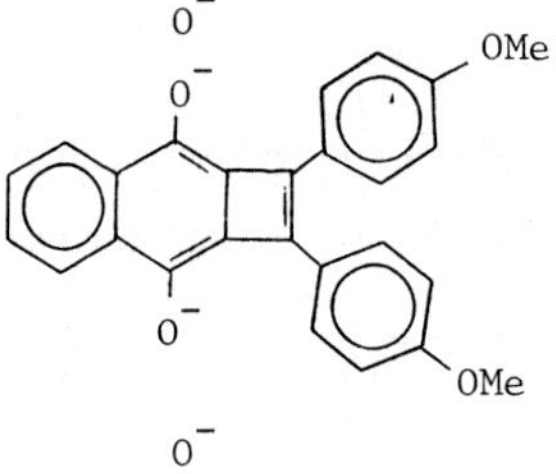 cyclobuta-fused naphthalenediolate dianion (p-OMe-C$_6$H$_4$, p-OMe-C$_6$H$_4$)	-0.38, -1.00
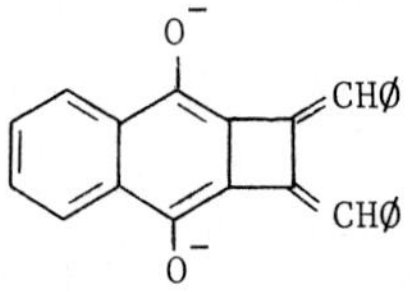 cyclobuta-fused naphthalenediolate dianion (=CHØ, =CHØ)	-0.45, -1.22
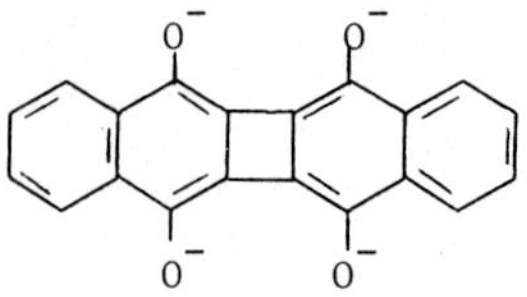 bis(naphthalenediolate)-fused cyclobutadiene tetraanion	-0.36, -1.17
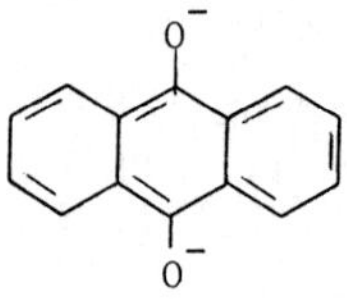 9,10-anthracenediolate dianion	-0.96, -1.67

the potential upon 2,3-ring fusion indicate a resistance to delocalization in the cyclobutadiene-like quinone.

The search for electrochemical evidence for antiaromaticity also led to studies of electroreduction as a route to potentially antiaromatic anions. Examples are known in which aromatic anions are easily formed by electroreduction of a nonconjugated neutral precursor. For example, electroreductive routes to aromatic cyclopentadienide anions are known.[60] The radical anion formed by electrochemical reduction of the neutral hydrocarbon loses a hydrogen atom to form the conjugated anion, a route shown here for tetraphenylcyclopentadiene, eqn 15.[44] The irreversible peak

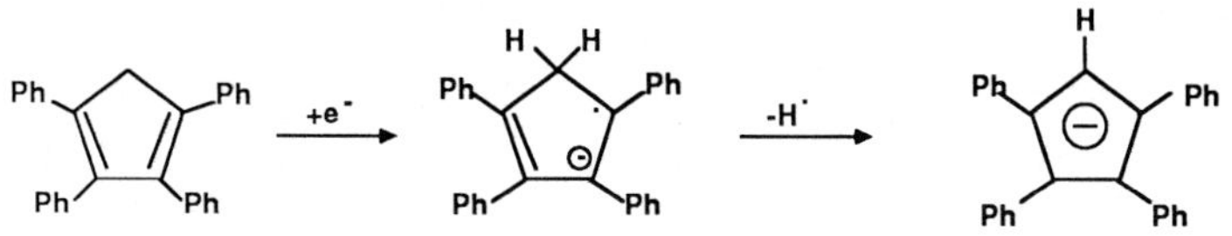

(15)

potential can thus be used as a rough calibration for the ease (or difficulty) in forming the corresponding cyclic anions. In the electroreduction of hydrocarbon precursors of antiaromatic anions, e.g., cyclopropene[61] and cycloheptatriene,[61,62] the observed peak potentials (-1.78 and -1.63 V, respectively) were more cathodic than might have been expected from model systems, and antiaromatic contributions to the electron distribution in the cleavage of the CH bond in the radical anion were inferred.

The electrooxidative formation of radicals from anions can be thought of as complementary to the reductive formation of the radical attained by adding electrons to the corresponding trivalent cation. Thus the following cycle, eqn 16, can approach the anion

$$RH \longrightarrow ROH \xrightarrow{H^{\oplus}} R^{\oplus} + H_2O \qquad (16)$$

$$R^{\oplus} \xrightarrow{+e^-} R\cdot \xrightarrow{+e^-} R^{\ominus}$$

oxidation potential determination from the reverse direction.[29] If the heats of formation of the hydrocarbon and alcohol are known and if the heat of dehydration of the alcohol can be measured, the potential of the second reduction wave will define the energy of the reduction of the free radical, and hence, that of the oxidation of the carbanion. In this way, the triphenylcyclopropenyl cation could be doubly reduced, eqn 17,[63] and its pKa assigned as greater than 51, a value consistent with destabilizing cyclic delocalization

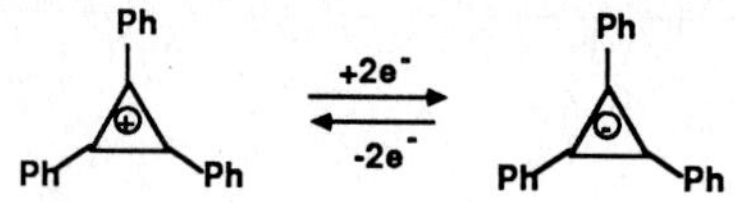

(17)

of charge. That this route is reliable was established by the demonstration that close correspondence was observed between pK_R^+ values obtained by this electrochemical method and by standard equilibration techniques for fluorenyl, triphenylmethyl, benzo-fused tropylium, and other stable cations.[64-66] That is, predictable potentials were observed for the first and sometimes second reduction of the cation.[67]

With sufficiently stabilized pi systems, all three oxidation levels of trivalent intermediates are occasionally stable and reversible potentials for their interconversion can be observed. Two such amphielectronic radicals are shown here, as eqns 18[50,68,69] and 19,[70] respectively.

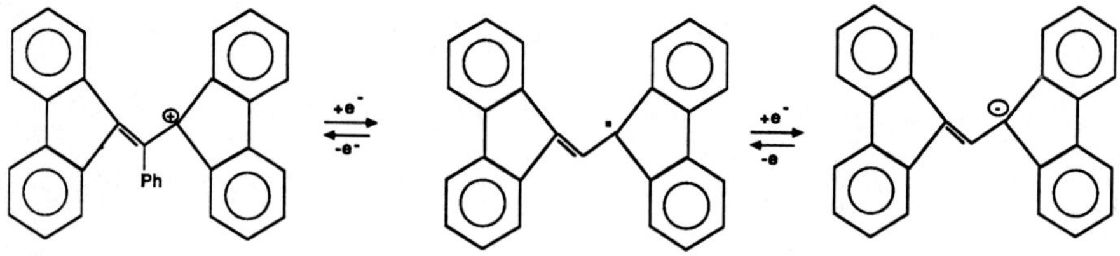

(18)

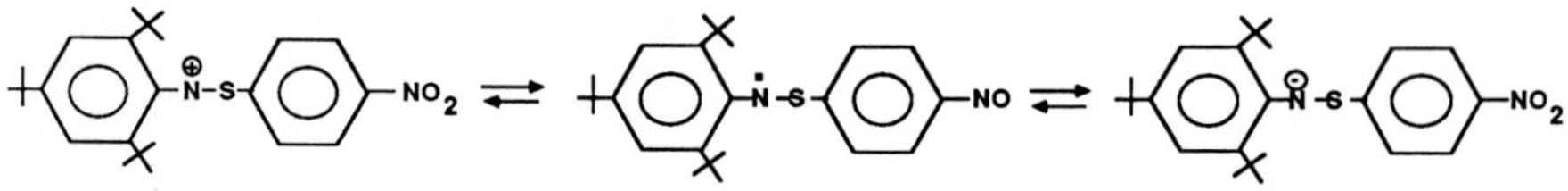

(19)

This same kind of electrochemical measurement has recently been employed by Arnold and coworkers to determine pKas of radical cations.[71] Using the thermodynamic cycle shown in eqn 20,

$$RH_{sol} \xrightarrow{-e^-} RH^{+\cdot}_{sol} \longrightarrow R\cdot_{sol} + H^+_{sol} \longrightarrow H^+_{aq} \qquad (20)$$

$$RH^{+\cdot}_{sol} \xrightarrow{B^-_{sol}} R\cdot_{sol} + BH_{sol} \longrightarrow RH_{sol} + B\cdot_{sol}$$

$$R\cdot_{sol} \longrightarrow R\cdot(g) \qquad H^+_{aq} \longrightarrow 1/2\ H_2(g)$$

eletrochemical methods have again yielded physical properties of a series of reactive intermediates which are difficult to attain in other ways.

(C) Conformation Analysis

Understanding the electronic properties of isomers and conformers is important in designing selective reactions both in homogeneous solution and at the surface of electrodes. Bianthrone is an example of a thermochromic, conformationally mobile molecule which has been intensely studied by electrochemical methods.[72-80] It is a highly hindered aromatic species which relieves strain either by twisting or bending about the ethene bond, giving rise to two stable interconverting conformations, eqn 21. Cyclic

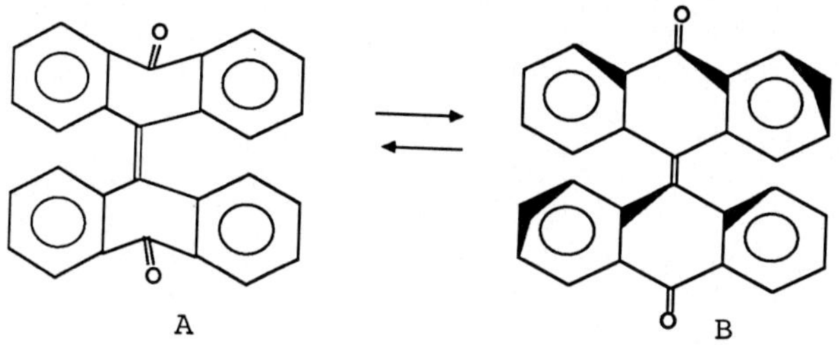

A ⇌ B (21)

voltammetric study indicated that the two forms interconvert with low activation barrier as anion radicals. Anion radical intermediates were also implied in a pulse radiolytic study of the conformation equilibration.[81] When electrogenerated, the anion radicals are produced either directly by a one electron reduction or by rapid disproportionation of the electrogenerated dianion formed in a two electron process, eqn 22. The conformational

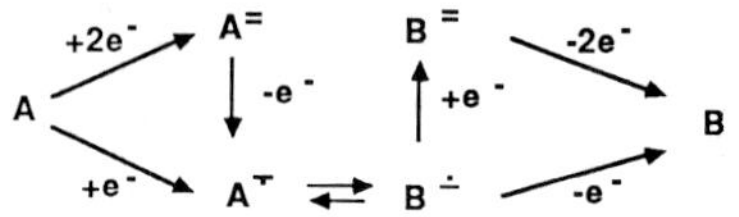

$$A \xrightarrow{+2e^-} A^{=} \xrightarrow{-e^-} A^{\cdot-} \rightleftarrows B^{\cdot-} \xrightarrow{-e^-} B \qquad (22)$$
$$A \xrightarrow{+e^-} A^{\cdot-}$$
$$B^{\cdot-} \xrightarrow{+e^-} B^{=}$$
$$B^{=} \xrightarrow{-2e^-} B$$

interconversion is often accompanied by a diagnostic curve crossing on the return cyclic voltammetric scan. The faster equilibration at the radical anion oxidation level has been attributed to a narrowed HOMO-LUMO gap which occurs upon twisting.[80]

Conformational equilibration can also be reflected in variant redox potentials for the contributing isomers. Dixanthylene[80] and trans-1,2-dibromocyclohexane, eqn 23,[82] for example, exist as an

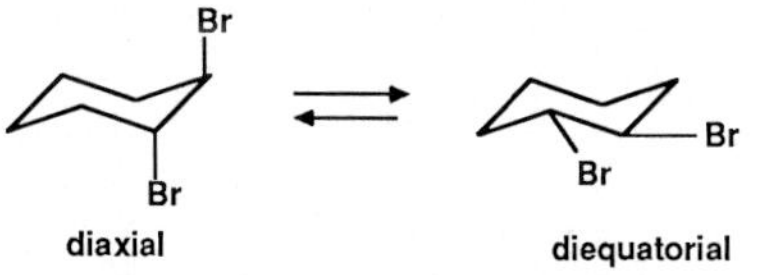

Br, Br diaxial ⇌ Br, Br diequatorial (23)

equilibrating mixture of conformers, each of which reacts at a different potential. In the latter compound, the diequatorial isomer is reduced at more negative potentials and with a lower absolute rate constant than its diaxial conformational isomer since the bromines are not located stereoelectronically in an optimum orientation for C-Br stretching, i.e., the bromines are not anti as would be required for a trans-periplanar elimination.

Radical anions themselves can fold and undergo conformationally specific reduction. Such an explanation has been advanced, for example, in describing the second electroreductive wave of anthraquinone.[83]

Perhaps the most dramatic electrochemical probe of contributing conformations of anions, though, is found in electrochemical studies of cyclooctatetraene and its derivatives. Here the conformational change required in moving from the tub-shaped neutral to the planar monoanion, eqn 24,[26] is accompanied by a

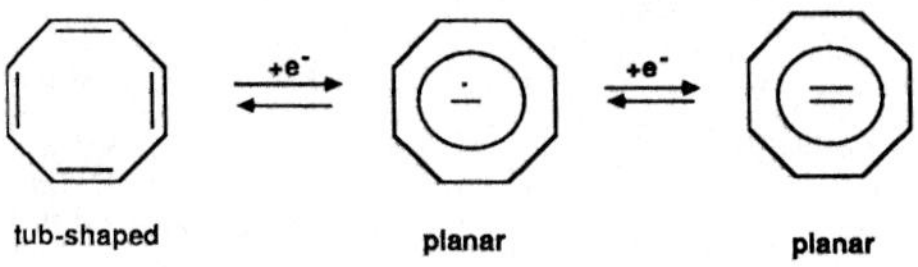

tub-shaped $\underset{}{\overset{+e^-}{\rightleftharpoons}}$ planar $\underset{}{\overset{+e^-}{\rightleftharpoons}}$ planar (24)

substantial kinetic barrier. Depending on the solvent and electrolyte counterion, a one or two electron reduction may be observed, because formation of the aromatic dianion from the monoanion encounters no such conformational barrier.[84-86] Recent evidence that the barrier for reduction is related to the conformational approach to planarity has been supplied by a counter-example, a derivatized cyclooctatetraene in which the neutral parent is planar, eqn 25.[87] Here no such barrier to reduction

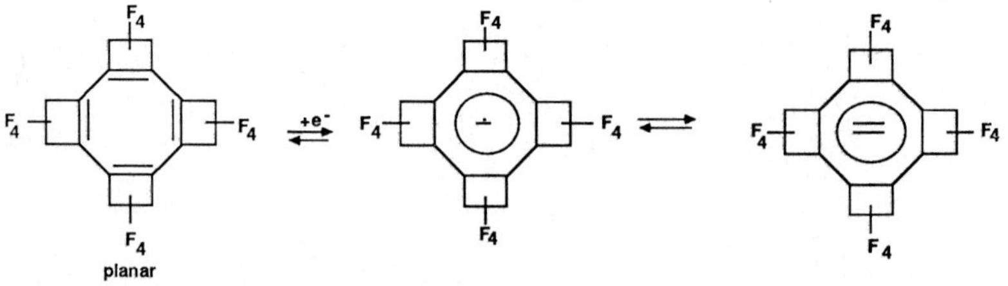

F_4 F_4 F_4 F_4 $\underset{}{\overset{+e^-}{\rightleftharpoons}}$ F_4 F_4 F_4 F_4 $\rightleftharpoons$ F_4 F_4 F_4 F_4 (25)

planar

(one or two electrons) is found and nearly ideal reversible behavior is observed. These conformational ideas apparently prevail in the presence of heteroatoms, for methoxyazocines, eqn 26,[88] exhibit analogous behavior.

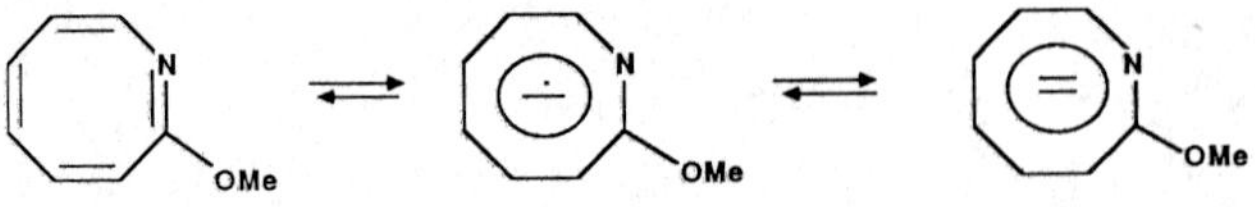

N OMe $\rightleftharpoons$ N OMe $\rightleftharpoons$ N OMe (26)

The generation of conformationally labile radical anions or dianions implies that, in some cases, reversible electrochemical transformations can effect geometric isomerization. Thus, for example, since the cis-stilbene radical ion isomerizes to the trans-isomer, electroreduction of cis-stilbene, followed by reoxidation, will produce the trans olefin, eqn 27.[89] Flash

$$\text{cis-PhCH=CHPh} \underset{}{\overset{+e^-}{\rightleftharpoons}} \text{Ph}\dot{\text{C}}\text{H}-\overset{\ominus}{\text{C}}\text{HPh} \rightleftharpoons \text{PhCH}_2\text{-}\overset{\ominus}{\text{C}}\text{HPh (skew)} \overset{-e^-}{\rightleftharpoons} \text{trans-PhCH=CHPh} \quad (27)$$

photolysis studies[90] indicate that isomerism occurs by disproportionation of the radical anion to give a dianion. This species exists in a skew conformation, and collapses to a planar form, with donation of an electron to an acceptor, to produce the isomerized alkene. Cis-azobenzene also isomerizes in like fashion.[91] Electron deficient vinyl ethers similarly undergo rapid geometrical isomerization, mediated by radical anions participating in chain electron exchange.[92]

Thus, electrochemical methods can be used to probe conformations of both neutral precursors and of stable anion radicals and dianions and can effect their facile interconversion.

III. SYNTHETIC IMPLICATIONS OF OXIDATION OF CARBANIONS

(A) Radical Couplings

In addition to its role in defining fundamental properties of carbanions, anodic oxidation of anions has synthetic importance. As we have seen, stable anions are easily oxidized electrochemically, usually at potentials below 1 V. That such electrooxidations can produce coupling by combination of radicals, eqn 28,[93] has been

$$^{\ominus}\text{CH(CO}_2\text{Et)}_2 \xrightarrow{-e^{\ominus}} \cdot\text{CH(CO}_2\text{Et)}_2 \longrightarrow (\text{EtO}_2\text{C})_2\text{CH-CH(CO}_2\text{Et)}_2 \quad (28)$$

known since before the turn of the century. Although the yields of reactions involving radical-radical combination are often low, higher yields of dimer (50%) were obtained with triethyl orthoacetate by this electrochemical route than by chemical methods.[94] The anions of alkylated malonic esters can be coupled to dimers in acceptable yields (20-55%), eqn 29,[95] although the solvent can become involved and specific electrolytic conditions will often

$$Et\text{-}\overset{\ominus}{C}(CO_2Et)_2 \xrightarrow[EtONa,\ CH_3CN]{-e^-} (EtO_2C)_2C\text{—}C(Et)(CO_2Et)_2 \text{ (with Et on first C)} \qquad (29)$$

dictate the course of a coupling. In ethanol, for example, the sodium salt of diethylmalonate gives not only simple coupling product, but also a product of further coupling and one of condensation with acetaldehyde (solvent oxidation) subsequent dehydration followed by a Michael addition reaction, eqn 30.[96]

$$\ominus CHE_2 \xrightarrow[EtOH]{-e^-} H\text{-}CE_2\text{-}CE_2\text{-}H + E_2HC\text{-}C(CH_3)(H)\text{-}CHE_2 + E_2CH\text{-}CE_2\text{-}CHE_2 \qquad (30)$$

$E=CO_2Me$

Two mechanisms have been proposed for formation of the hexaester. In the first, the tetraester formed by simple coupling is deprotonated by the anionic starting material, producing a radical which cross-couples with the malonate radical, eqn 31, path a.[96]

$$\ominus CE_2\text{-}CE_2\text{-}H \xrightarrow{(a)} \cdot CE_2\text{-}CE_2\text{-}H \xrightarrow{\cdot CHE_2} E_2CH\text{-}CE_2\text{-}CE_2\text{-}H$$

$$\ominus CE_2\text{-}CE_2\text{-}H \xrightarrow[-e^-,\ -H^\cdot]{(b)} E_2C{=}CE_2 \xrightarrow{\ominus CHE_2} Et_2CH\text{-}CE_2\text{-}\overset{\ominus}{C}E_2 \xrightarrow{H^+\text{-}CHEt_2} E_2CH\text{-}CE_2\text{-}CE_2\text{-}H \qquad (31)$$

$E = CO_2Me$

In the second mechanism, eqn 31, path b, oxidation of the tetraester anion produces the conjugated species, which is attacked by malonate in Michael addition to afford the observed product.[97]

Non-enolate anions can also be dimerized, as can be seen with the anion of 2-nitrobutane, eqn 32,[98] with the phenylacetylide

$$Et\text{-}\overset{\ominus}{C}(NO_2)\text{-}CH_3 \xrightarrow[NaOH]{-e^-,\ 25\%} Me\text{-}C(NO_2)(Et)\text{—}C(NO_2)(Et)\text{-}Me + Et\text{-}C(NO_2)_2\text{-}Me \qquad (32)$$

anion, eqn 33,[99] and with alkyl and aryl Grignard reagents,

$$Ph\text{-}C{\equiv}C^{\ominus} \xrightarrow[THF]{-e^-} Ph\text{-}C{\equiv}C\text{-}C{\equiv}C\text{-}Ph \qquad (33)$$

eqn 34.[100]

$$RMgBr \xrightarrow{-e^-} R\text{-}R \qquad (34)$$

$R = C_5H_{11},\ C_{18}H_{37},\ Ph$

If the electrochemical oxidation is performed in the presence of activated olefins, cross coupling products can be isolated in excellent yields. Sodium diethylmalonate, for example, give the adducts shown in eqn 35.[101] Reduction coupling product is

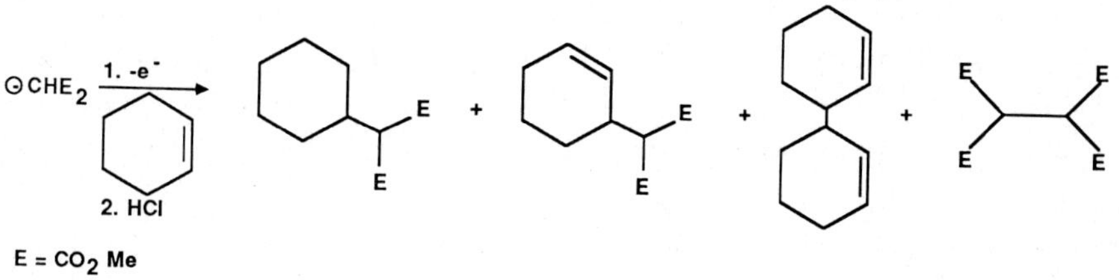

(35)

E = CO_2 Me

accompanied by products formed by radical recombination. The intermediate dimethyl malonate radical and cyclohexenyl radical (presumably formed via hydrogen abstraction) thus couple with thermselves or each other. Similar adducts are also observed with vinyl ethers, arylethylenes, and dienes. In a parallel cross coupling with an enamine, eqn 36, no simple dimers were observed,

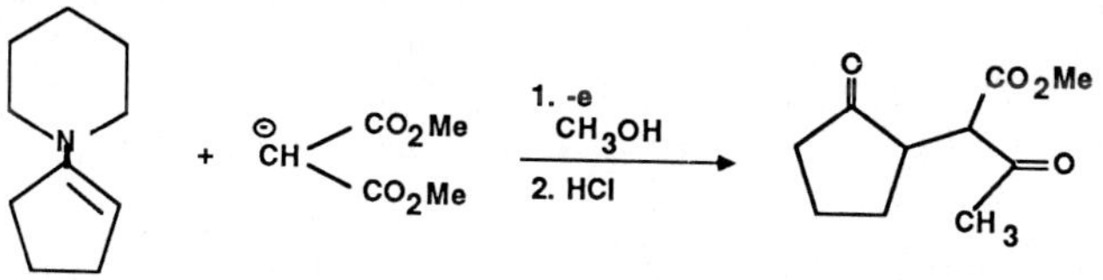

(36)

although the mechanism (i.e., the identity of the primary electro-oxidation) remains obscure.[102]

The products ultimately observed in such cross coupoings can often depend on the potential applied to initiate the reaction. Electrooxidation of the anion of 2-nitropropane in the presence of styrene, eqn 37, gave dimeric coupled radicals at potentials

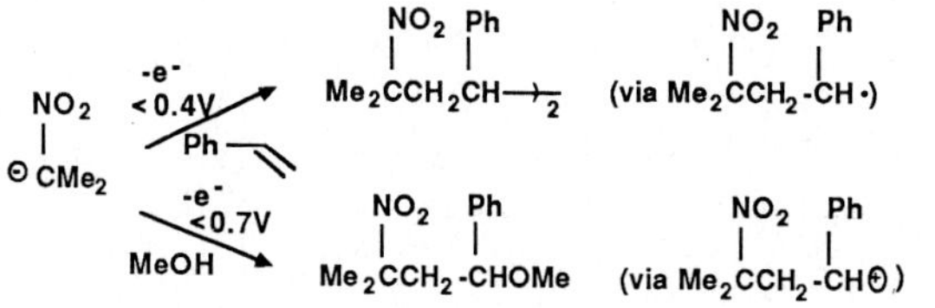

(37)

cathodic of 0.4 V, but products characteristic of cation trapping at potentials anodic of 0.7 V.[103]

1,3-Diketones add to enol ethers, giving rise to adducts important as synthetic intermediates, eqn 38,[104]

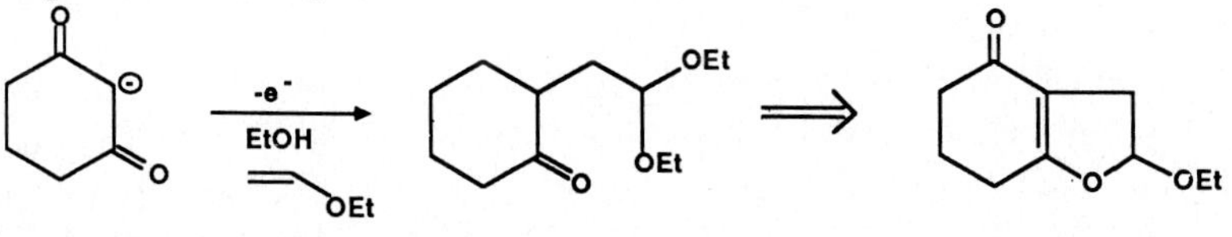

(38)

and occasionally such cyclizations occur spontaneously if geometrically allowed, eqn 39.[101]

1. $-e^-$, OEt; 2. HCl → OEt (39)

Because of the growing importance of radical couplings and cyclizations in organic synthesis, such electrooxidations are likely to occupy a place of increasing importance as novel methodologies.[105]

(B) Kolbe Electrolyses

The Kolbe reaction is one of the oldest and most widely used electrochemical syntheses. In its simplest form, it involves the preparation of dimeric product generated by radical coupling of free radicals formed via decarboxylation of electrogenerated carboxyl radicals, eqn 40.[106] The reaction can be complex, though,

$$RCO_2^{\ominus} \xrightarrow{-e^-} RCO_2\cdot \longrightarrow R\cdot + CO_2 \qquad (40)$$

for further oxidation of the radical can produce a cation which is trapped by nucleophiles, eqn 41. If water is used as the

$$RCO_2^{\ominus} \xrightarrow[-CO_2]{-e^-} R\cdot \xrightarrow{-e^-} R^+ \xrightarrow{Nuc^{\ominus}} RNuc \qquad (41)$$

nucleophile, alcohols are obtained, and the transformation is called the Hofer-Moest reaction. Esters can also be formed if the cation is trapped by carboxylate. The course of the reaction is significantly influenced by a number of electrochemical variables: electrode material, potential, solvent, current density, electrolyte, temperature, and/or the identity of the carboxylate substrate.

Kolbe dimerization via electrogenerated radicals is most efficient at high potentials under conditions in which oxygen evolution is suppressed. Thus, high concentration of acid, high current density, low pH, and inert electrolyte counterions and electrode materials often lead to maximal radical combination. For example, on platinum anodes, discharge of carboxylates often give rise to dimers, for example, in the high yield (93%) of ethane obtained from acetate, eqn 42.[107] Varying substitution

$$CH_3CO_2^{\ominus} \xrightarrow{-e^-} CH_2CH_3 \qquad (42)$$

patterns, particularly at the alpha position of acetates, strongly influence the isolable yield of coupling product.[108] Trialkylacetates, diphenyl acetates, or donor-substituted acetates often give low or no yields of radical-derived products, presumably because the more stable radical can be easily oxidized further to the cation. Such an effect is certainly reasonable, since substituents will affect significantly the electrogenerated radical's further oxidizability.

Mixed Kolbe reactions, involving mixtures of carboxylates, usually give rise to three products, those formed by cross-coupling and self-coupling, respectively. For example, as shown in eqn 43,

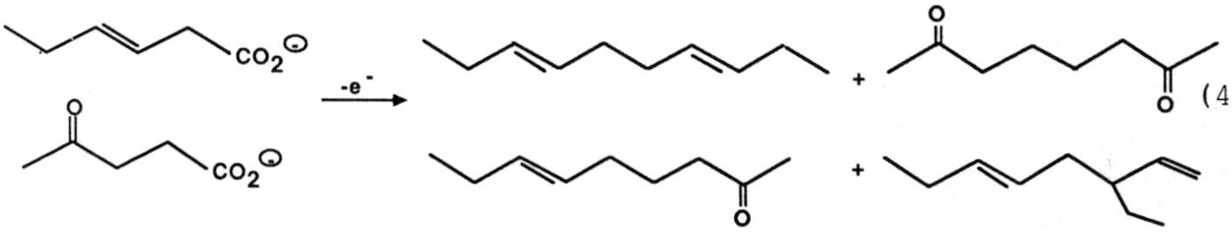

$CO_2^{\ominus}$ $CO_2^{\ominus}$ $\xrightarrow{-e^-}$ + + (43)

all possible coupling products were obtained in a mixed Kolbe coupling. One isomer (trans-6-nonen-2-one) was later converted to the pheromone brevicomin.[109] Although mixtures of products generally complicate synthetic schemes, the maintenance of geometrical and stereochemical integrity of centers remote from the site of bond formation make this reaction synthetically useful.

If the electrode material is graphite or porous carbon, however, the products usually obtained are those derived from cationic intermediates, characterized by capture with nucleophiles and by rearrangement. For example, oxidation of cyclobutanecarboxylate affords a mixture of cyclopropylcarbinol, cyclobutanol, and allylcarbinol, eqn 44, essentially identical to that obtained by

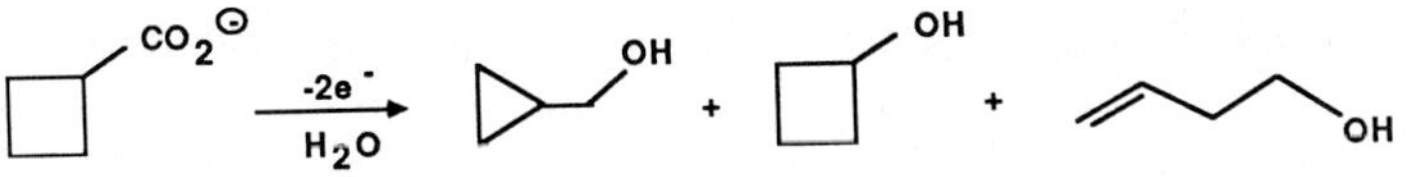

$CO_2^{\ominus}$ $\xrightarrow[H_2O]{-2e^-}$ OH + OH + OH (44)

deamination of cyclobutylamine.[110] The whole array of products expected from the cyclooctyl cation are observed when cyclooctylcarboxylate is electrolyzed, eqn 45.[111] Intramolecular trapping

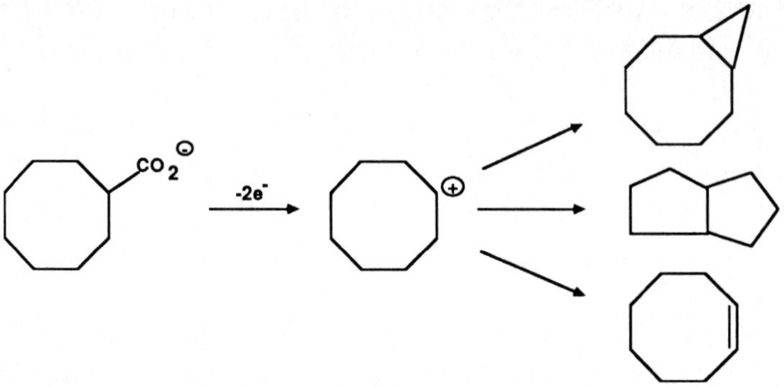

(45)

of the cation can be sometimes observed, as can be seen in eqn 46.[112] Analogous intermolecular trapping by added nucleophiles

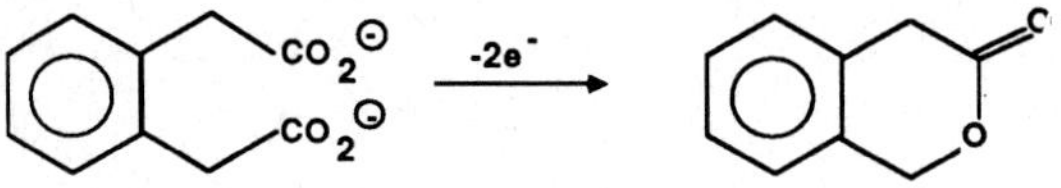

(46)

is similarly well-documented, e.g., eqn 47.[113]

$$Ph_2CHCO_2^{\ominus} \xrightarrow[MeOH]{-2e} Ph_2CHOMe \qquad (47)$$

Ring expansion, e.g., eqn 48,[114] and Wagner-Meerwein

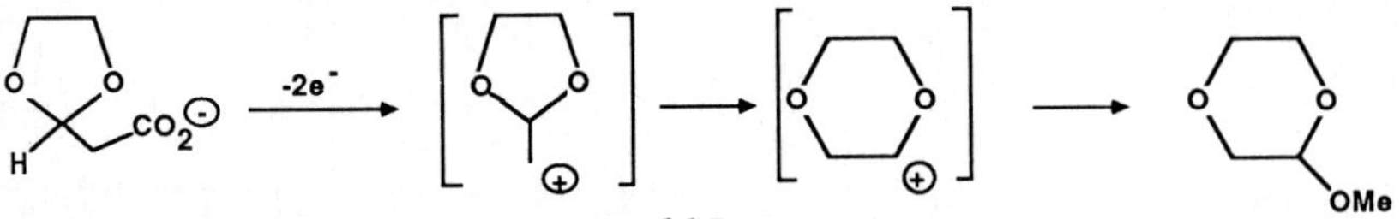

(48)

rearrangement, e.g., eqn 49,[115] can also be induced in this way.

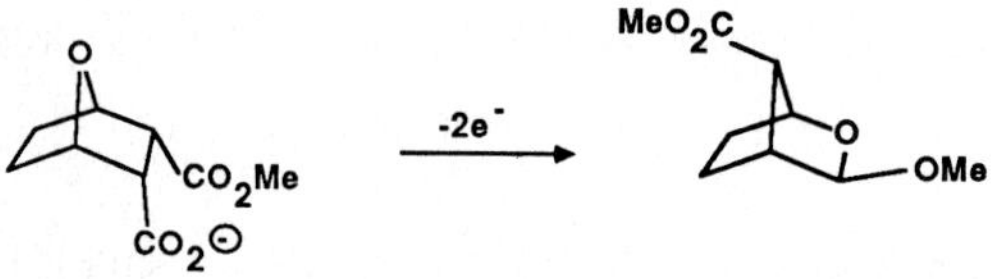

(49)

Migratory aptitudes in electrochemically induced pinacol-pinacolone rearrangements parallel those observed in conventional cation solvolysis. An electrochemically induced ring expansion has been exploited, for example, in the synthesis of muscone, eqn 50.[116]

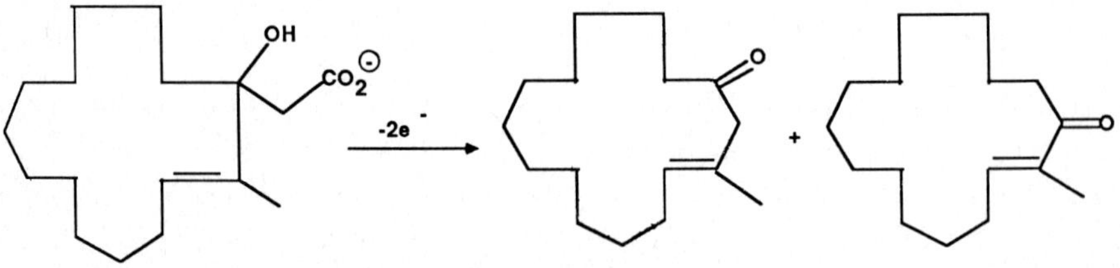

(50)

Two-electron oxidations are also implicated in analogous

bis-decarboxylations. 1,2-Dicarboxylic acids can be electrolyzed to produce alkenes, eqn 51,[117] and 1,3-diacids have been used as

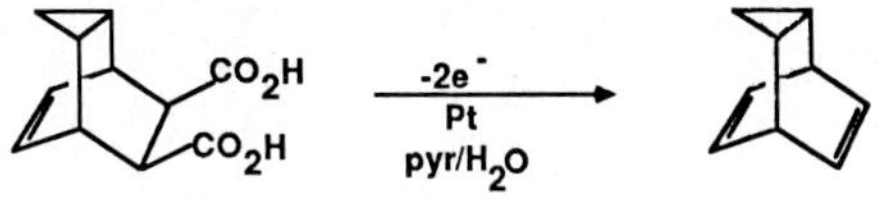

(51)

substrates to form strained rings, eqn 52.[118] Although yields are

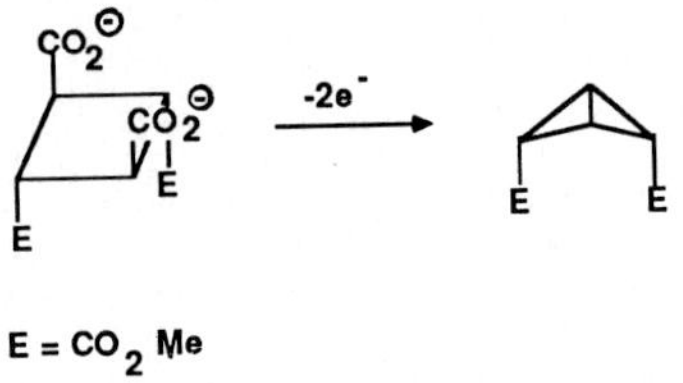

(52)

E = CO_2Me

not high, this method is usually preferable to the alternative chemical oxidation with lead tetraacetate. Mechanistic details are sparse, but the reactions appear to proceed through cationic intermediates, for no sterespecificity could be observed in the bis-decarboxylation of meso- and d,l-2,3-diphenylbutanedioic acid, eqn 53.[119] As with other radicals formed by direct electrooxidation, those formed by electrooxidative decarboxylation can also be trapped by oxygen[120] or by active olefins, eqn 54.[121]

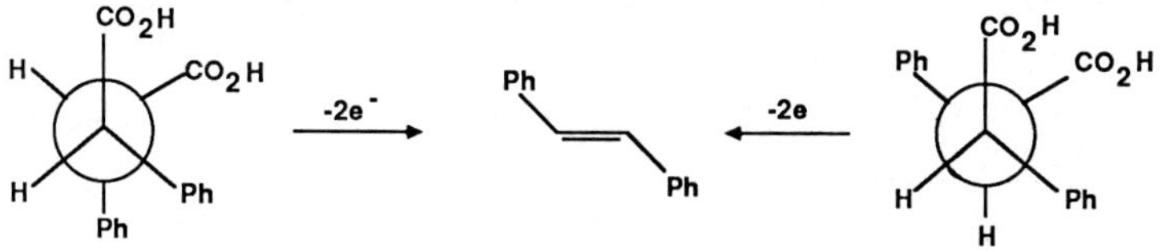

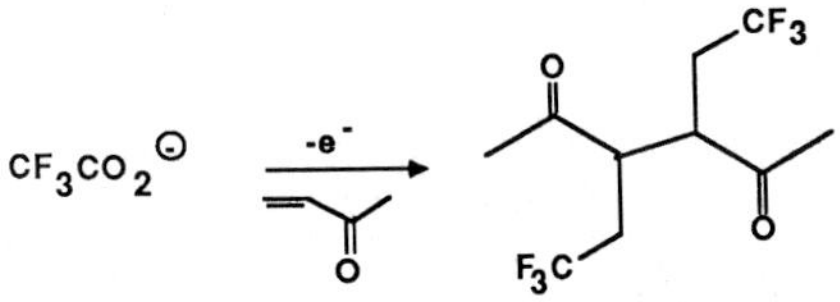

(53)

(54)

Since these radical and cationic intermediates are adsorbed on the surface of the electrode as they are formed in electrochemical routes, the possibility that the intermediates might behave as if they were chiral seemed reasonable. This proposition seemed to derive support from the observation that electrooxidatively generated cations often seemed "hotter" than those obtained by solvolysis routes: i.e., they rearranged more rapidly and were less sensitive to environmental effects than were analogous ions formed under solvolytic conditions. Unfortunately, the Kolbe decarboxylation of a chiral acetate, eqn 55,[122] gave complete

$$\overset{*}{\mathrm{EtCH}}(\mathrm{Me})\mathrm{CO_2^{\ominus}} + \mathrm{MeO_2CCH_2CO_2^{\ominus}} \xrightarrow{-e^-} \mathrm{MeO_2CCH_2}\overset{*}{\mathrm{CH}}(\mathrm{Me})(\mathrm{Et}) \quad \text{racemic} \tag{55}$$

racemization and fully random coupling was observed from absorptively biased carboxylates, eqn 56.[123] Thus, no evidence could be

garnered for the synthetic utility of these anion oxidations for asymmetric formation of carbon-carbon bonds.

Although we ordinarily think of the carboxylate anion as the primary oxidation site in such Kolbe transformations, careful mechanistic investigation has shown that if other oxidizable groups are present they may participate in the primary electron transfer. Because of effects of current density, electrode potential, and the identity of the counterion on product distributions, Utley and coworkers concluded that in para-methoxyphenylacetate electron transfer occurred initially from the aromatic nucleus.[124]

Access to the full array of electrochemical apparatus is not necessary to effect these electrooxidations, for Bard and coworkers have shown that irradiated semiconductor powders (platinized TiO_2) can also induce Kolbe oxidation.[125] Back electron transfer to the surface-adsorbed radical apparently is important for simple reduction products are often observed in addition to or instead of the usual dimer, eqn 57. A wide variety of functionalized

$$\mathrm{CH_3CO_2^{\ominus}} \xrightarrow[\mathrm{HOAc}]{h\nu,\ \mathrm{Pt/TiO_2}} \underset{\text{major}}{\mathrm{CH_4}} + \mathrm{CH_3CH_3} \tag{57}$$

carboxylates can be employed[126] and the principles governing such photoinduced electron transfers are applicable to other organic oxidations as well.[127] The course of the reaction can be at least partially controlled by the choice of the semiconductor, eqn 58,[128]

$$\mathrm{CH_3CH(OH)CO_2H} \xrightarrow{h\nu,\ \mathrm{Pt/TiO_2}} \mathrm{H_2} + \mathrm{CO_2} + \mathrm{CH_3CHO}$$
$$\mathrm{CH_3CH(OH)CO_2H} \xrightarrow{h\nu,\ \mathrm{Pt/CdS}} \mathrm{CH_3}\overset{\mathrm{O}}{\overset{\|}{\mathrm{C}}}\mathrm{CO_2H} + \mathrm{H_2} \tag{58}$$

for lactic acid produces different oxidation products on CdS and on TiO_2. This selectivity may be attributed either to the shift in valence band positions of the two semiconductors or to differential adsorption on the oxide and sulfide surfaces.

(C) Isoelectronic Heteroatomic Systems

A trivalent carbanionic center is isoelectronic with the lone pair of a trivalent nitrogen atom, and one might reasonably expect parallel electroreactivity from pi systems bearing like electron densities. Electrooxidation of carbazole in the presence of pyridine, eqn 59,[129] for example, produces a dimer structurally

-e⁻ pyr → ; -H⁺ → ; ↓ dimer (59)

analogous to those observed in the radical-radical coupling of substituted cyclopentadienyl anions, eqn 60.[44] Oxidation of

-e⁻ → ; → dimer (60)

pyrrole itself causes polymerization and electrode coating, but the resulting polymer can be further electrooxidized to produce a protective electroconductive layer. Although this polymerization limits the synthetic potential of pyrrole oxidations, this reaction has engendered an exciting new route for chemical modification of electrodes.[130] In the presence of benzaldehyde, tetraphenylporphyrins are formed as pyrrole is oxidized,[131] presumably through radical cation trapping. If the N-H proton is substituted by an alkyl group, ring oxidation can ensue to produce monomeric product. In methanol a tetramethoxypyrroline is formed from N-methylpyrrole, eqn 61.[132] This latter reaction finds analogy in the electro-

-2e MeOH → ; -2e⁻ MeOH → (61)

oxidation of furan, eqn 62,[133] in which a 2,5-dihydro-2,5-dimeth-

-2e → (62)

oxyfuran is formed in a Clauson-Kaas reaction sequence.

If a heteroatom center is present in a molecule, its ready oxidizability may enhance the initial electron exchange, but the final site of the electrooxidative transformation may be an another position in the molecule. With 1-azabicyclo(2.2.2)octane-2-carboxylic acid, for example, the oxidation potential is significantly lower than a typical acetic acid derivative, and yet the Kolbe decarboxylation product is obtained, eqn 63.[134]

(63)

Such reaction are merely representative of a wide array of electrooxidation of heteroatomic systems, and the interested reader should consult the several excellent reviews of these topics for more detail.[135-137]

(D) Photoelectrochemistry of Anions

Carbanions by their nature are electron rich and are well-known to act as effective electron donors. This tendency can be even further amplified by photoexcitation: promotion of electrons to a high lying orbital reduces the energetic requirement for electron ejection.[138] Furthermore, the rate of recapture of the ejected electron is lower than that typically encountered for photoinduced electron transfers from neutral precursors.[139] This observation follows from the absence of strong electrostatic attraction between the ejected electron and the neutral free radical derived from a carbanionic excited state, eqn 64, compared

$$M^{\ominus} \underset{\Delta}{\overset{h\nu}{\rightleftharpoons}} M\cdot + e^- \qquad (64)$$

with that encountered in the electron - radical cation pair obtained from a neutral precursor, eqn 65.

$$M \underset{\Delta}{\overset{h\nu}{\rightleftharpoons}} M^{\dot{+}} + e^- \qquad (65)$$

The chemical consequences of such electron transfers are manifold, and they are covered in more detail in another chapter of this text. Here we consider only those excited state anion oxidations which can be monitored by electrochemical methods. We have already referred to a semiconductor-mediated carboxylate oxidation (eqn 57). In this reaction, wavelength selection ensured excitation of the light-responsive, heterogeneous photocatalyst (in this case TiO_2), thus creating an electron-hole pair, eqn 66. The hole

$$TiO_2 \xrightarrow{h\nu} h^+ + e^-$$
$$h^+ + RCO_2^- \longrightarrow RCO_2\cdot \longrightarrow R\cdot + CO_2 \quad (66)$$

is poised at the band edge of the semiconductor valence band and the electron at that of the conduction band. The hole is scavenged by the carboxylate, leading to formation of the same carboxyl radical produced by conventional one electron oxidation. Further oxidation of the radical to the corresponding cation is less of a problem at low light flux on photosensitive semiconductors, since the irradiated semiconductor becomes an effective oxidation catalyst only when irradiated. A variety of other functional groups can also be oxidized under these conditions.[127]

An alternate mode of photoelectrochemical activation employs the semiconductor as an electron trap. Thus, the excited anion transfers an electron to the semiconductor's conduction band. Band bending at the semiconductor-electrolyte interface in n-type semiconductors causes the injected electron to move from the interface toward the bulk. If the semiconductor is formulated as the anode of a working electrochemical cell, an anodic photocurrent is produced, Figure 6. Since carbanions are often highly colored,

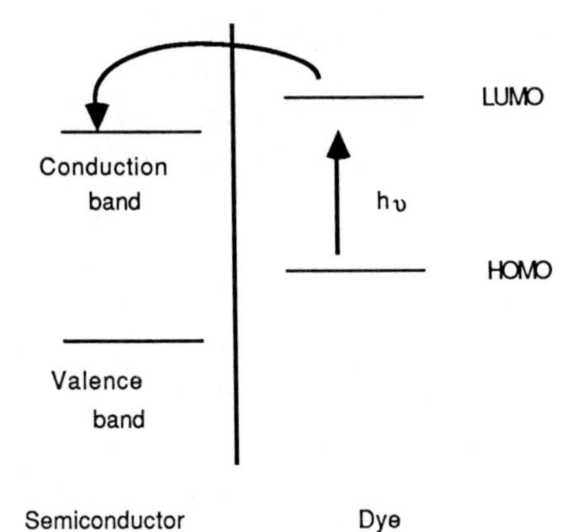

Fig. 6. Sensitization of Anodic Photocurrent

this sequence represents a method for the conversion of visible light to electrochemical potential.

An example of this process is found in the photolysis of a mixture of cyclooctatetraene/cyclooctatetrene dianion in an electrochemical cell. Upon visible light irradiation, electron ejection occurs, producing the monoanion, eqn 67. The stability of the monoanion with respect to disporoportionation is dependent

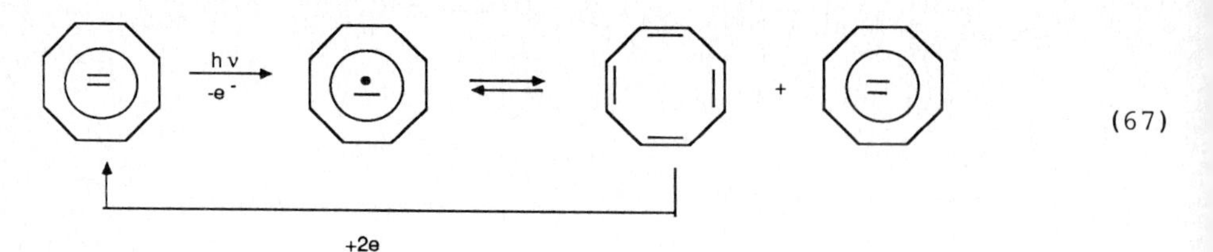

(67)

on solvent and the identity of the associated cation, and under appropriate conditions it is possible for the semiconductor electrode to scavenge the ejected electron as disproportionation inhibits the back electron transfer. The oxidized hydrocarbon can then migrate to the counterelectrode, completing a reversible cycle.[140] Covalent attachment of cyclooctatetranene derivatives, when coupled with a mobile electron relay, improves the current efficiency by an order of magnitude.[141] Other organic anionic dyes can also be used as effective excited state sensitizers.[142-144]

Radical anions can also be photoexcited at semiconductor electrode surfaces as a means of inducing potential gradients in electrochemical cells. Pyrene and perylene radical anions showed enhanced currents with respect to chlorobenzene electroreduction upon photolysis.[145] The enhanced peak current was interpreted as establishing the excited radical anion as a more electron reactive donor than the parent hydrocarbon. However, a quantitative separation of the relative importance of excitation and photochemically produced thermal gradients near the electrode surface proved to be difficult.[146]

IV. ELECTROGENERATED CARBANIONS

Not only can the properties of carbanions be studied by these techniques, but they can themselves be formed with electrochemical methods. This section summarizes several ways in which carbanions can be produced and their subsequent reactions defined by electrogeneration of carbanionic intermediates.

(A) Alkyl Halide Reductions

The electroreductive activation of a carbon-halogen bond has proven to be a versatile electrochemical reaction. It has been known for several decades that with appropriate homogeneous reducing agents or at the surface of a poised electrode, alkyl halides are reduced in a single two electron wave to give carbanionic intermediates.[147] The observation of one irreversible wave implies that the anion radical generated upon electron capture is rapidly dissociated, producing a radical which is instantaneously reduced at the first potential much more rapidly than it can diffuse from the electrode surface. Alpha values observed in homogeneous reduction schemes are thought to be a composite of the transfer coefficients for dissociation of the encounter complex and for secondary electron transfer.[148] Similar distributions are obtained with heterogeneous electroreduction.[149]

The resulting electrogenerated anions can be obtained as lithium salts[150] or can be protonated or carboxylated,[151] depending on reaction conditions. Even in cases where persistent radicals are formed from homogeneous single electron donors, anions are produced ultimately with electrochemical methods, eqn 68.[152]

$$2NaNp^{\dot{-}} + RX \longrightarrow R^{\ominus} + X^{\ominus} \qquad (68)$$

Apparent n=1 values for the reduction of a variety of allylic halides, for example, were shown to be caused by subsequent S_N2 attack of the anion on the starting material, eqn 69.[153] Only

+2e⁻ ; S_N2 ; (n=1) (69)

with allylic halides with substantial steric blockage of the anion could the real n-2 value be demonstrated, eqn 70. Electrochemical

Br ; +2e⁻ ; (n=2) (70)

routes are especially useful in preparing alkyl anions with charge localized at unusual sites, e.g., at a bridgehead carbon.[154]

The primary electron transfer can occur to the sigma or pi system, producing an anion radical. In general, the lifetimes of aliphatic anion radicals are extremely short (e.g., $CH_3Br^{\dot{-}}$ half lives about 3 nanoseconds),[155] but those of stabilized aryl anion radicals can be very long. The half life of the 2-bromothiazine radical anion, for example, is 5 seconds,[156] sufficiently long to permit detailed electrochemical study. Radical anion lifetimes are often dependent on solvent, temperature, ion pairing association, and the availability of a reactive proton or other electrophile in the electrolyte. The course of an electrochemical reaction can also be influenced by the electrode material: trifluoroacetic acid, for example, is cleanly reduced to trifluoroethane on platinized platinum,[157] but is dehalogenated on mercury in DMF.[158] The electrode material can even be used as a sacrificial reagent, as in the electroreduction of benzal chloride in the presence of CO_2, eqn 71.[159]

$$PhCHCl_2 \xrightarrow[\text{Al}, \; CO_2]{+e^-} \underset{\underset{Cl}{|}}{PhCH^{\ominus}} + AlCl_3 \quad (71)$$

$$\longrightarrow \underset{\underset{Cl}{|}}{PhCHCO_2H} + PhCH(CO_2H)_2$$

Theoretical studies justify this high lability of the C-X bond in the alkyl halide radical anion. Salem and coworkers have found that the odd electron dissociates from the methyl chloride radical anion in the gas phase, and in solution the incipient C-X cleavage encounters such a low barrier when solvated that pre-dissociation lifetimes of 10^{-8} to 10^{-4} sec are to be anticipated.[160] A more rigorous ab initio treatment by Clark estimates the barrier to be 19 kcal/mol.[161]

In the formation of the anion radical, the cathodic reduction potential becomes more negative and the extent of the electrochemical reversibility diminishes as the s character of the exocyclic bonding orbital increases.[154] The observed reduction potentials of benzyl halides correlated well with Hammett sigma$^-$ values, implying that the potential determining step involves CX breakage or a substantially stretched CX bond in the radical anion.[162] Extensive tables of alkyl halide reduction potentials compiled in the last two decades allow one to predict electroreduction potential ranges with confidence.[163,164]

Formation of organometallic reagents is often observed upon alkyl halide reduction on reactive electrodes.[163] These species are themselves often electroactive, with the metal-halogen bond often being cleaved by one electron redox catalysis, as in the reduction of trialkylgermanium halides, eqn 72.[165] Organometallic

$$R_3GeX \xrightarrow[\text{DME}]{+e^-, \text{ pyrene}} R_3Ge\cdot \xrightarrow{RH} R_3GeH \quad (72)$$

reagents can sometimes be used as radical traps in either direct or catalytic reductions of alkyl halides.[166]

Electrochemical reduction potentials are sensitive to the identity of the halide and to its steric environment. The hundred-fold greater sensitivity of the alkyl chlorides to structure than the corresponding alkyl bromides indicates the requirement for greater bond reorganization in C-Cl than in C-Br cleavages. Steric and aromaticity effects can also be seen in the 1.4 V difference in reduction potential associated with C-Cl cleavage in perchlorocyclopentadiene, eqn 73, compared with a

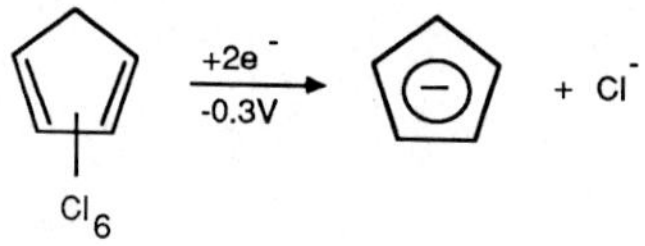

$$C_5Cl_6 \xrightarrow[-0.3V]{+2e^-} C_5Cl_5^- + Cl^- \qquad (73)$$

non-planar cyclopentenyl skeleton, eqn 74.[167] Steric effects have

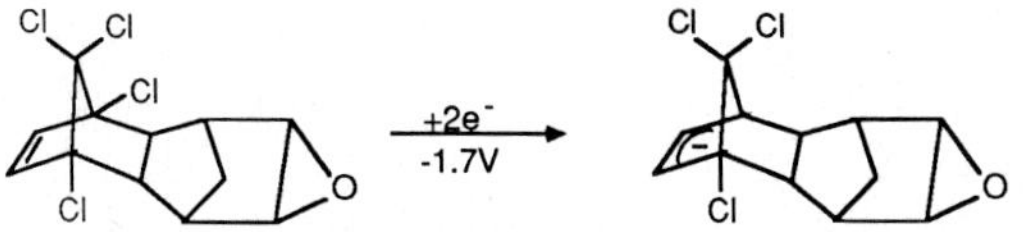

$$\xrightarrow[-1.7V]{+2e^-} \qquad (74)$$

also been suggested as important in determining the more facile reduction of axial than equatorial halides in cyclohexyl halides and in the dominant exo over endo halide reduction in bicyclic compounds.[163]

Electroreduction of vicinal dihalides provides a useful route to alkenes, eqn 75. As might be expected, good

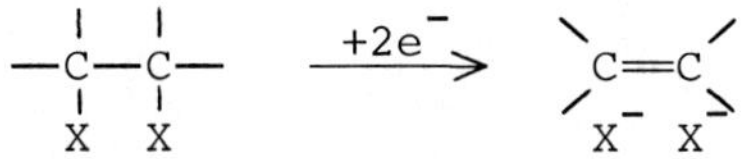

$$-\underset{X}{\overset{|}{C}}-\underset{X}{\overset{|}{C}}- \xrightarrow{+2e^-} >C=C< + X^- + X^- \qquad (75)$$

antiperiplanar pi alignment greatly assists in the elimination. When non-zero (or non -180°) dihedral angles are encountered, a significant cathodic shift in reduction potential is observed, Table 6. When a particularly strained double bond is formed (as at a bridgehead) unusually negative potentials are observed.[169] This route can be used either as a conformational probe, eqn 76,[170]

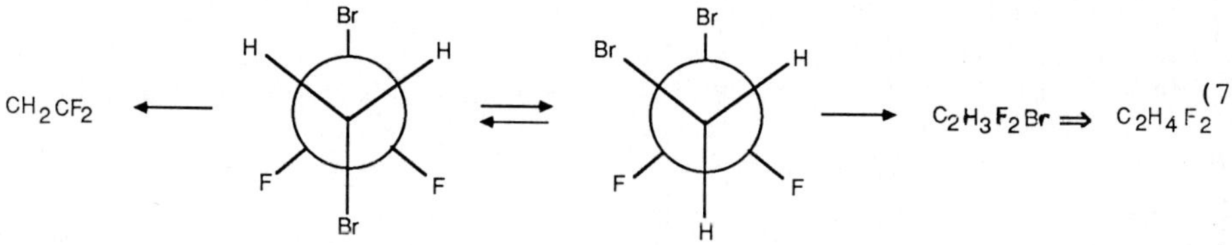

$$CH_2CF_2 \longleftarrow \quad \rightleftarrows \quad \longrightarrow C_2H_3F_2Br \Rightarrow C_2H_4F_2 \qquad (76)$$

or as an electrochemical deprotection method for dihalogen-protected alkenes, eqn 77.[171] The preferred orientation of adsorption

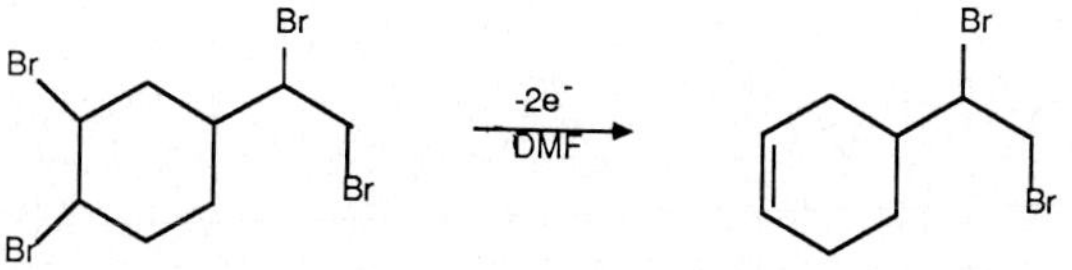

$$\xrightarrow[DMF]{-2e^-} \qquad (77)$$

TABLE 6

Reduction Potentials of 1,2-Dibromo Cyclohexanes of Varying Dihedral Angles

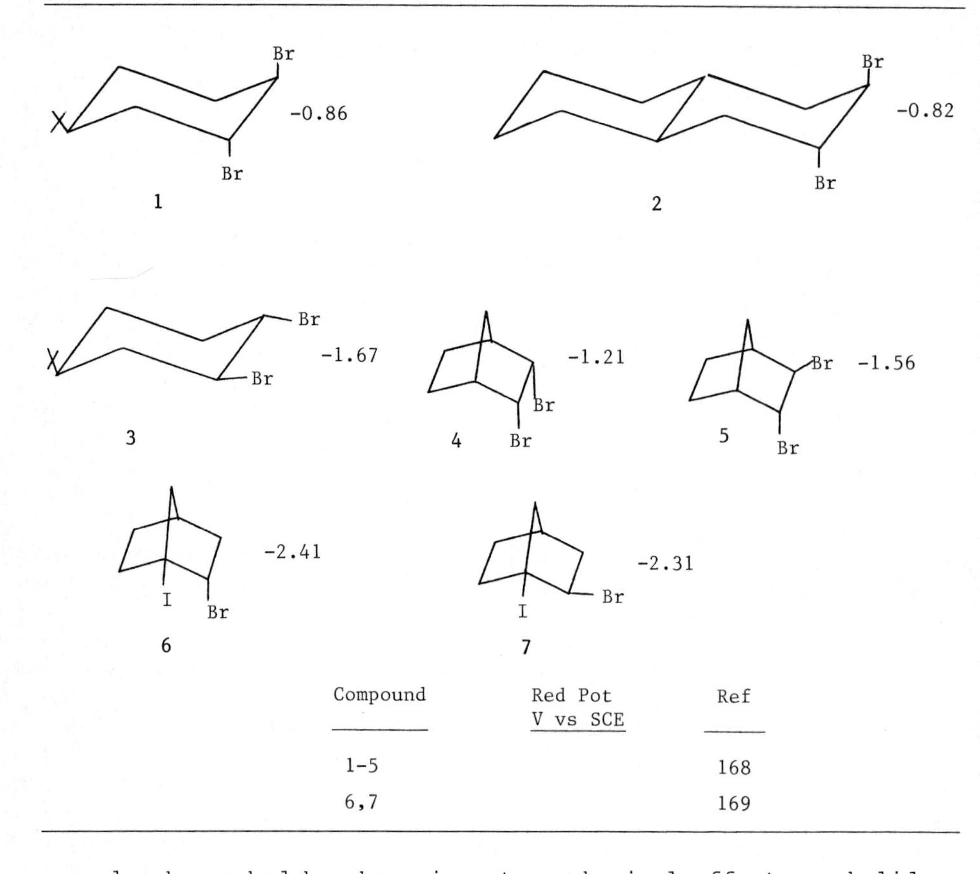

Compound	Red Pot V vs SCE	Ref
1-5		168
6,7		169

can also be probed by observing stereochemical effects on halide reductions as a function of substituent, Table 7.[172] Since the

TABLE 7

Stereochemistry of Electroreduction of Some Substituted Cyclopropyl Bromides

Ph, Ph, R, Br (cyclopropane) $\xrightarrow[Br^-, +H^+]{+2e}$ Ph, Ph, R, H (cyclopropane)

R	Stereochemistry
CO_2H	26-35% inversion
CO_2^-	31-38% retention
CO_2Me	30-50% inversion
CH_3	21% retention

two-electron reduction should produce a conformationally stable anion, the preferred face of protonation can be determined. Anions thus provide a sensitive stereochemical probe of the electrochemical double layer.[173]

Electroreduction of geminal dichlorides has also been reported. In the reduction of 9,9-dichlorofluorene, eqn 78, for

(78)

example, the products obtained ultimately depend on the current density (and hence on the relative rates of reduction), the rate of nucleophilic attack on the parent dihalide, and the availability of protons. No carbenes or carbene radical anions could be detected.[174]

Ring closure can be accomplished by intramolecular trapping of either the intermediate radical or the electrogenerated anion. For monobromides, this route thus provides an interesting contrast with hydride-induced cyclizations. The cyclization of bromoalkyl-denemalonates, eqn 79, is a net two-electron process, although the

$$(CH_2)_n\text{-ring-}C(E)_2 \xleftarrow[\text{cathode}]{Hg} BrCH_2(CH_2)_nC(H){=}C(E)_2 \xrightarrow{H^{\ominus}} (CH_2)_n\text{-ring-}C(E)_2 \qquad (79)$$

electrode potential is more positive than is required for direct reduction of the alkyl bromide.[175] Alpha, omega dihalides can also be electrochemically cyclized, sometimes with stereocontrol, although highest yields are obtained when three- or four-membered rings are formed.[176]

(B) Aryl Halide Reduction

Carbon-halogen bonds at vinyl or aryl positions can also be cleaved upon electroreduction, and the aryl radicals formed upon cleavage of the electrogenerated radical anions are often trapped by hydrogen donors or nucleophiles, allowing for an efficient substitution route, eqn 80.[177] The nucleophilic attack must

$$ArX \longrightarrow ArX^{\dot{-}} \longrightarrow Ar^{\cdot} + X^- \qquad (80)$$
$$Ar^{\cdot} \xrightarrow{+e^-} Ar^- \xrightarrow{RH} ArH$$
$$Ar^{\cdot} \xrightarrow{RH} ArH$$
$$Ar^{\cdot} \xrightarrow{Nuc^-} ArNuc^{\dot{-}} \xrightarrow{-e^-} ArNuc$$

dominate in a competition for further reduction of the aryl radical in order for high product yields to be observed. The rates of Ar-Cl cleavage correlate with standard reduction potentials, and extended Hückel calculations indicate that the efficiency of the electron transfer to the sigma* orbital from the initially populated pi* orbital depends on the C-X distance.[178] Since the sigma* level varies little with the identity of the arene, the standard potential of the electroreduction varies most dramatically with the pi* level, which is highly structure dependent. Rate constants for Ar-X cleavage greater than 10^4 are common, and good correlation of rates of homogeneous and heterogeneous electron exchange are obtained. Slow electron transfer to substituted halobenzenes and halopyridines are consistent with a required internal bond reorganization and solvent reorientation in approaching the transition state.[179] Parker and coworkers have attributed this shift in bond cleavage in the transition state to differential entropy of activation.[180] Thus, activation barriers for the cleavage of 9,10-dichloroanthracene, for 9,10-dibromoanthracene, and for p-nitrohalobenzene of about 15, 4, and 20 kcal/mole could be rationalized as reflecting varying degrees of bond cleavage in the transition state. Analogous cleavages are observed in halogenated benzophenones. The observation that p,p'-dibromobenzophenone is cleaved a thousand times faster than the corresponding dichloride, eqn 81, implies that bond cleavage is much further

$$\text{X-C}_6\text{H}_4\text{-CO-C}_6\text{H}_4\text{-X} \xrightarrow{+e^-} \text{X-C}_6\text{H}_4\text{-CO-C}_6\text{H}_4^{\bullet} + X^- \quad (81)$$

advanced in the brominated radical anion.[181]

A preparatively important consequence of aryl halide electroreduction is found in the electrocatalytic $S_{RN}1$ reaction, eqn 82.

$$ArX \longrightarrow ArX^{\dot{-}} \xrightarrow[-X^-]{} Ar\cdot \xrightarrow{Nuc^{\ominus}} ArNuc^{\dot{-}} \xrightarrow{ArX} ArX^{\dot{-}} \quad (82)$$

Here, the electrogenerated radical anion cleaves, producing an aryl radical which is rapidly attacked by a nucleophile, forming a substituted radical anion which back electron transfers to the starting halide and establishes a catalytic cycle.[182] Catalytic turnovers greater than 100 are common.[183] In the electrochemically induced substitution of 9-haloanthracene, eqn 83, the observed

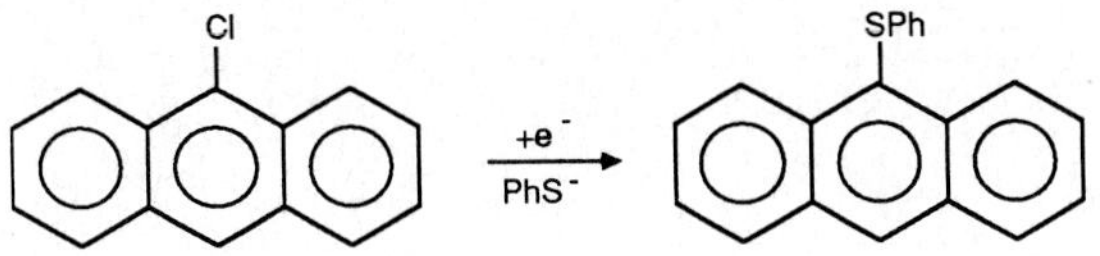

(83)

reactivity paralleled that of aryl radicals with nucleophiles, establishing that step as rate determining.[184] Parallel rate constants for this step could be attained by other electrochemical redox sequences, e.g., in the electroreduction of diazonium salts, eqn 84.[185] A detailed kinetic study indicates a large entropy

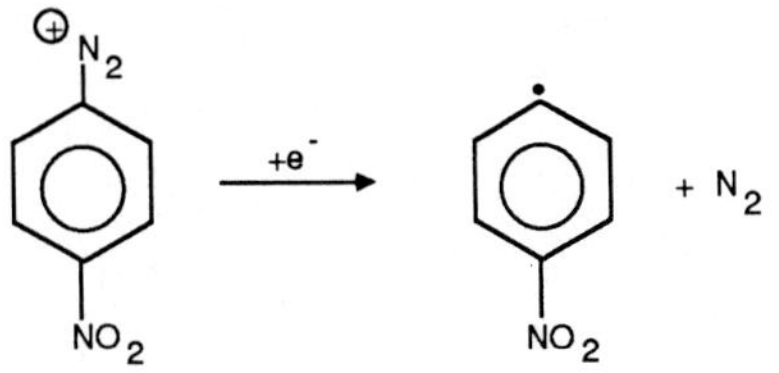

(84)

contribution in the key step of the $S_{RN}1$ reaction.[186]

Intramolecular attack can also occur, eqn 85,[187] although

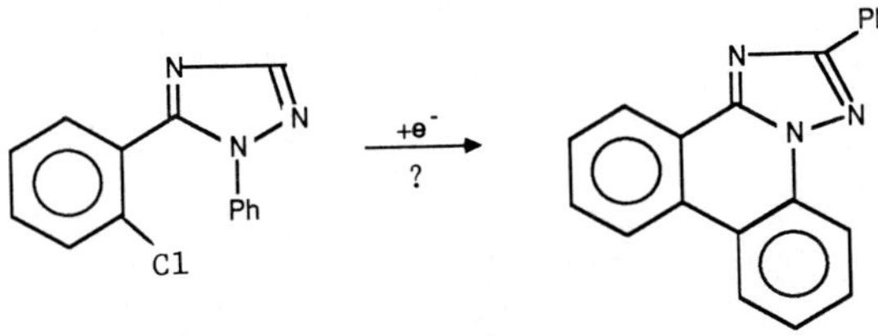

(85)

reductive dehalogenation can compete when the substrate adopts a conformation disfavoring cyclization. In fact, thermal cyclization may follow from the product of simple external nucleophilic capture, eqn 86.[188] So versatile is this procedure that even

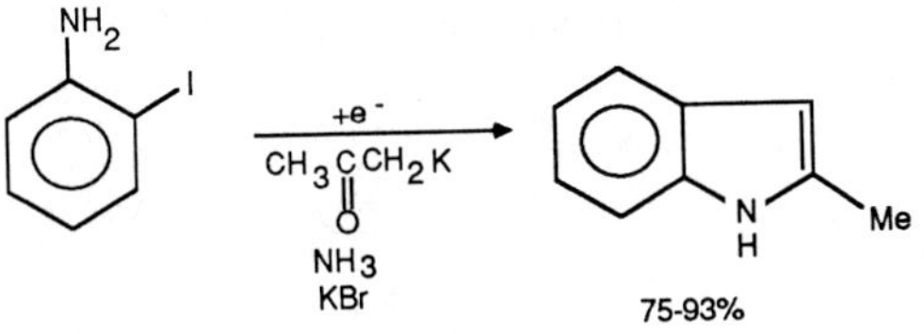

(86)

vinylation and arylation of iron porphyrins can be achieved in this method.[189] This clever variant involves a double electrochemical induction, forming _in situ_ the nucleophile and the aryl radical.

Since fluoride is a poor leaving group, reductive coupling products dominate the chemistry attained when aryl fluoride radical anions are electrogenerated. These species are themselves often thermally unstable and/or electroactive, and an interesting route to aryl dimer dianions is based on this approach, eqn 87.[190]

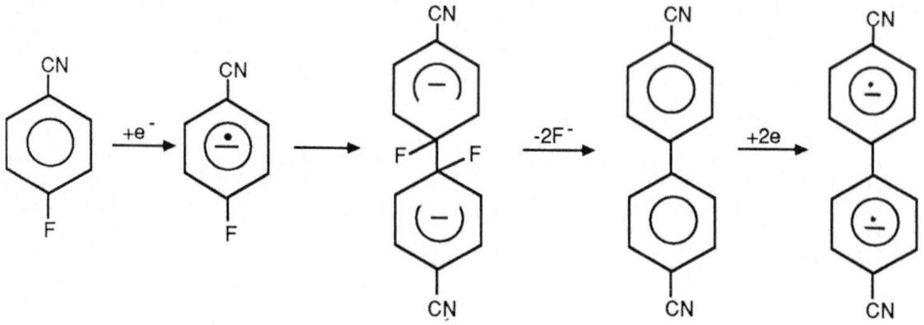

(87)

In the absence of an active nucleophile, reactivity parallel to that described earlier can be seen, and with controlled potential conditions, aryl halide cleavage can dominate over other possible reductive cleavages in molecules, eqn 88.[191]

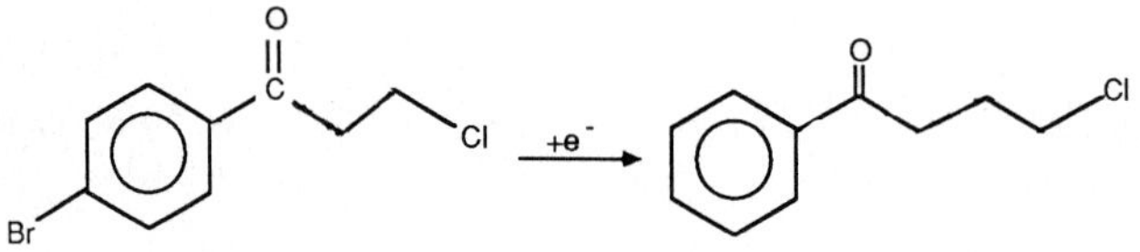

(88)

Although our attention here has been focussed on carbon-halogen cleavages, other reductive cleavages are also well-known upon electrochemical population of a sigma* orbital. We have seen one of these already (the electroreduction of aryldiazonium salt, eqn 84) and many other parallel conversions also occur. For example, benzyl ammonium salts, eqn 89,[192] and allylic ammonium

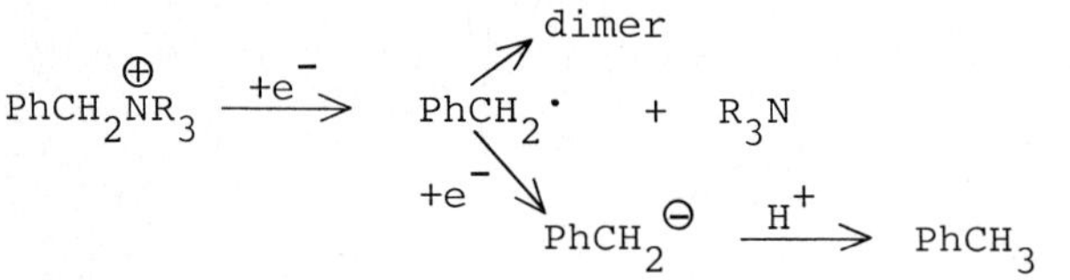

$$PhCH_2\overset{\oplus}{N}R_3 \xrightarrow{+e^-} PhCH_2^{\bullet} + R_3N \quad (\rightarrow \text{dimer}); \quad PhCH_2^{\bullet} \xrightarrow{+e^-} PhCH_2^{\ominus} \xrightarrow{H^+} PhCH_3 \qquad (89)$$

or phosphonium salts, eqn 90,[193] are cleaved upon electroreduction,

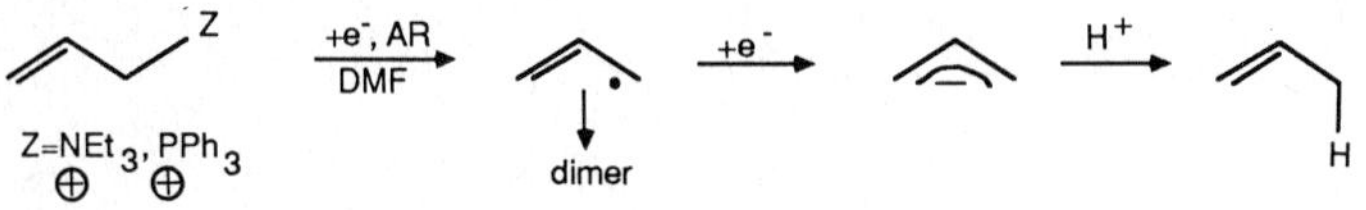

(90)

generating both the radical and the further reduced anion.

(C) Electrogenerated Bases

Clearly, the reductive cleavage reactions discussed above represent potential routes for the formation of anionic intermediates. Since these species are highly basic, electrochemical methods thus present an attractive method for introducing a desired quantity of base, in controlled fashion, into a reaction mixture. As with bases prepared in standard fashion, chemical catalysis can compete with reactions of bases as nucleophiles (i.e., in adduct formation).

Electrogeneration permits access to the conjugate bases of acids of high pK_a and to the reactions of these necessarily unstable anions. Electrogenerated bases and nucleophiles are thus one significant member of a larger class of electrogenerated reagents[194] whose role as indirect mediators of catalytic conversions is becoming of increasing importance.

The precursor to an electrogenerated base is called a probase. One of the first probases investigated resulted in the formation of the cyanomethylanion by reductive cleavage of a phosphonium salt, eqn 91.[195] The resulting anion could thus deprotonate the

$$Ph_3\overset{\oplus}{P}CH_2CN \xrightarrow{+2e^-} Ph_3P: + \overset{\ominus}{C}H_2CN \tag{91}$$

$$\overset{\ominus}{C}H_2CN \xrightarrow{Ph_3\overset{\oplus}{P}CH_2CN} Ph_3\overset{\oplus}{P}\overset{\ominus}{C}HCN$$

phosphonium salt, generating a ylid, which could be used in an "electro-Wittig" reaction.[196] Sulfonium[197] and ammonium[198] ylids have also been generated in parallel fashion by electrochemical methods.

Generation of carbanions by deprotonation with bases derived from azobenzene has been used frequently in organic electrosynthesis.[199] Azobenzene itself can be reduced to an intermediate base which, in turn, abstracts a proton from the desired acidic C-H parent. If azobenzene is reduced in the presence of fluorene containing an alkylating agent, eqn 92,[200] a mixture of mono- and

$$\text{fluorene} \xrightarrow[\text{HMPA, LiCl, MeI}]{\text{Ph-N=N-Ph},\ +e^-} \text{9-Me-fluorene} + \text{9,9-Me}_2\text{-fluorene} \tag{92}$$

dialkylation products can be isolated in 75% overall yield. The electrogenerated base abstracts a proton producing the fluorenyl anion which is alkylated in the electrochemical cell as it is formed. Depending upon proton availability, any of three anionic reduction products can be implicated as the active base: $PhN{=}NPh^{\cdot -}$, $PhNHN^-Ph$, or $PhN{=}NPh^{2-}$.[201] This same fluorenyl anion can be formed by reductive electrochemical cleavage of the methyl ether, eqn 93.[202]

$$\text{(fluorene, 9-R, 9-OMe)} \xrightarrow{+2e^-} \text{(9-R-fluorenyl anion)} + {}^{\ominus}OMe \quad (93)$$

Other bases of demonstrable synthetic utility can be generated electrochemically. For example, the succinimide anion can be formed either from N-halosuccinimide or from succinimide itself,[203] and the pyrrolidone anion can be produced from 2-pyrrolidone.[204] The latter ion catalyzes CCl_3^- addition to aldehydes and ketones and the Dieckmann or Thorpe reactions, eqns 94 and 95, respectively.[205] It can also be used as a source for

$$\text{E-N(CH}_2\text{CH(E)CH}_3\text{)CH}_2\text{CH}_2\text{CH}_2\text{CO}_2\text{Me} \xrightarrow[\text{2-pyrrolidone}]{+e^-} \text{(N-E-piperidinone, E, CH}_3\text{)} \quad (94)$$

$E = CO_2Me$

$$NCCH_2CH_2CH_2CH_2CN \xrightarrow[\text{2-pyrrolidone}]{+e^-} \text{2-amino-1-cyanocyclopentene } (NH_2, CN) \quad (95)$$

anions, which can be alkylated in high chemical yields, eqn 96.[206]

$$\text{1-(CO}_2\text{Me)naphthalene} \xrightarrow[\text{2-pyrrolidone, DMF, MeI}]{+e^-} \text{1-(CH(Me)CO}_2\text{Me)naphthalene} \quad (96)$$

This reaction was used recently in a key step in the synthesis of pyrethroids.

Diphenyldiazomethane is an interesting probase. Cyclic voltammetry indicated irreversible formation of the anion radical, and the presence on the reverse scan of the diphenylmethyl anion, eqn 97.[207] Whether this species was formed via an intermediate

$$Ph_2CN_2 \longrightarrow [Ph_2CN_2^{\overset{\cdot}{-}}] \xrightarrow[-N_2]{?} [Ph_2C^{\overset{\cdot}{-}}] \xrightarrow[+e^-]{H^+} Ph_2\overset{\ominus}{C}H \quad (97)$$

carbene anion radical is unclear.[208]

(D) Electrogenerated Nucleophiles

Carbanions generated by interaction with electroreduced pro-bases react in the normal nucleophilic sense. For example, the trichloromethyl anion formed by deprotonation of chloroform by electrogenerated base adds diastereoselectively to chiral branched aldehydes, eqn 98.[209] Enolate anions can be electrogenerated from

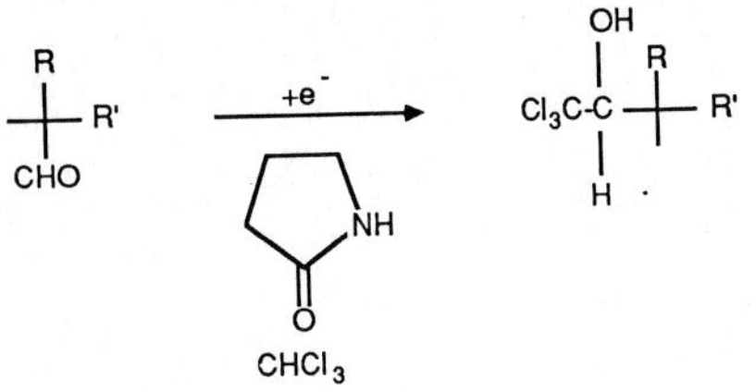

(98)

ketones bearing alpha hydrogens. For example, the anion of a beta-diketone was formed by the route shown in eqn 99.[210] The

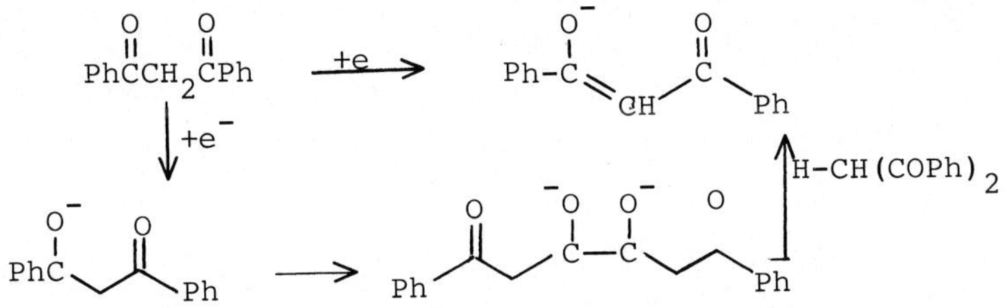

(99)

ketyl radical anion dimerizes forming the alkoxide of a beta-hydroxyketone, which deprotonates starting material to produce the dienolate. Such anions can be used in electrochemically induced Aldol condensations.[211] Ester enolates can be formed by reduction of the alpha-bromoester, eqn 100,[212,213] as can

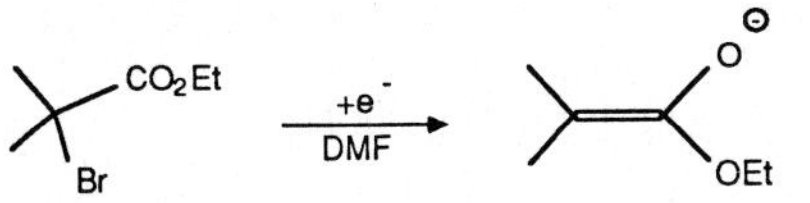

(100)

amidoenolates.[214] Allyl anions formed upon electroreduction are also known to effect nucleophilic addition, either directly, eqn 101,[215]

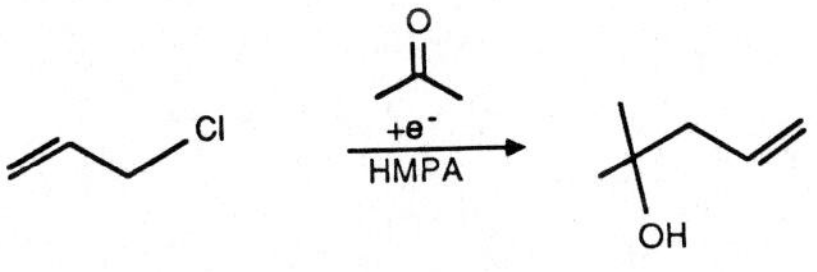

(101)

or in a Michael sense, eqn 102.[216] Since Michael reactions are

$$\text{(prenyl-X)} + \text{E-CH=CH-E} \xrightarrow{+e^-} \text{CH}_2\text{=CH-CH}_2\text{-CH(E)-CH}_2\text{E} \quad (102)$$

$E = CO_2Me$

catalytic in base, only a small amount of current need be passed to initiate the reaction cycle. Even the cyanomethyl anion has been electrogenerated by cathodic reduction in the presence of a suitable electrolyte, eqn 103.[217] Active olefins can fill the

$$PhCOCH_3 + CH_3CN \xrightarrow[CH_3CN,\ Et_4NBF_4]{+2e^-} Ph{-}C(OH)(CH_3){-}CH_2CN \quad (103)$$

role of the probase as well as that of a Michael acceptor. Ethyl acrylate can be reductively alkylated in the presence of diethylmalonate, eqn 104,[210] generating a nucleophilic anion $^-CHE_2$ which

$$CH_2E_2 + CH_2{=}CH{-}E \xrightarrow[DMF]{+2e^-} EtO_2CCH_2CH_2CH(CO_2Et)_2 \quad (104)$$

initiates a Michael addition. Homoenolates can be electrogenerated by reduction of the beta-bromocarbonyl compound, eqn 105.[218]

$$CH_3CHBrCH_2CO_2Et \xrightarrow[DMF]{+2e^-} CH_3\overset{\ominus}{C}HCH_2CO_2Et \quad (105)$$

Often these same reactions can be accomplished by conventional treatment with base, so that electrogenerated nucleophiles are most attractive in synthetic sequences which are themselves base-sensitive or in which only small steady state concentrations of base are desired. Under such circumstances, electrochemical methods may indeed be less harsh than the usual conventional synthetic routes.

V. REDUCTIONS VIA CARBANIONIC INTERMEDIATES

Electrogenerated radical anions, dianions, and monoanions are themselves interesting intermediates in net reduction sequences. In many cases, electrochemical routes to these anions may be preferable to conventional reagent-based reductions. In this section we provide a brief overview of the variety of functional group reductions and couplings which can be routinely accomplished by electrochemical routes.

(A) Hydrocarbon Reductions

Hydrocarbons lack the significant polarity of heteroatom-containing substrates and are therefore often reduced in homogeneous solution only under rigorous conditions, the negative charge of the radical anion having been forced onto carbon. Often hydrocarbon reduction potentials approach the cathodic limit attainable in electrochemical cells. Near this limit two extreme situations are observable: injection of solvated electrons into solution or reduction of the electrolyte cation to form the metal (or at a mercury electrode, an amalgam). Both the electrogenerated solvated electron[219] and the reducing amalgam[220] have found significant utility in preparatively important reductions.

The Birch reduction, eqn 106, which is conventionally

X X Na NH_3 EtOH (106)

X = alkyl, OR

accomplished with dissolving metals in liquid ammonia, can also be readily accomplished in divided electrochemical cells. Here the electrode, rather than the reducing metal, provides the necessary electron source.

A recent report verifies that even in undivided cells, 1,4-cyclohexadiene can be obtained from benzene, eqn 107.[221] This

$+e^-$ Hg cathode Steel anode Bu_4NOH 90% (107)

observation was somewhat surprising in view of the easy reoxidation/rearomatization expected of the product, but the high yields reported were achieved under conditions which favor anodic discharge of hydroxide rather than product oxidation. The stability of this product also stands in contrast to the further reduction observed in the presence of amines, eqn 108. Here the conjugated

$+2e^-$ LiCl $MeNH_2$ LiNHMe $+2e^-$ $MeNH_2$ (108)

two-electron reduction product is equilibrated to the 1,3-diene by lithium amide (an electrogenerated base). The conjugated diene can then be further reduced.[222] Other arenes can be similarly reduced, as in the reduction of naphthalene to the 1,4,5,8-tetra-hydro-derivative at a graphite electrode in liquid ammonia containing methanol as a proton donor, eqn 109,[223] or in the

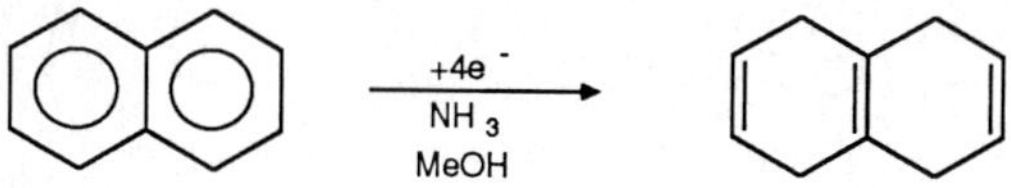

$$\xrightarrow[NH_3,\ MeOH]{+4e^-} \qquad (109)$$

reduction of phenol, anisole, or toluene to the corresponding cyclohexene derivatives.[224] Although electrochemical reduction of pyridinium salts proceeds in analogous fashion in buffered aqueous solutions, eqn 110,[225] that of aminopyrimidines is

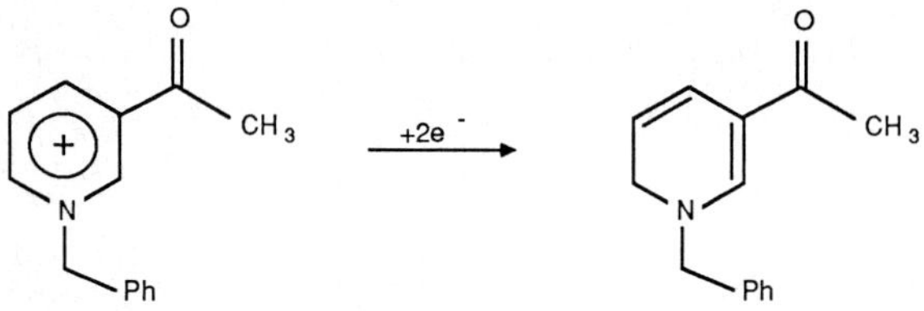

$$\xrightarrow{+2e^-} \qquad (110)$$

complicated by isomerization, ring-opening, and deamination, eqn 111.[226]

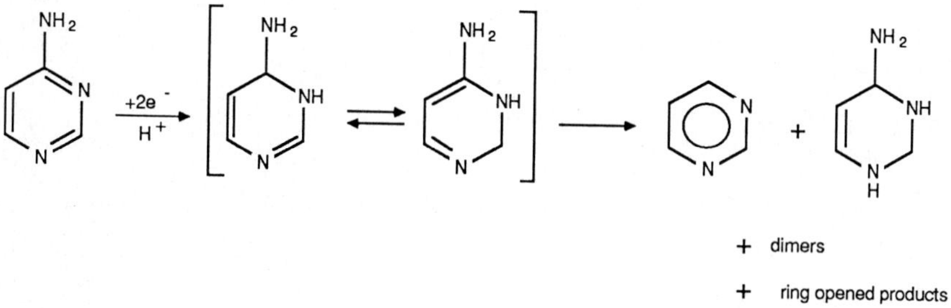

$$\xrightarrow[H^+]{+2e^-} \qquad (111)$$

+ dimers

+ ring opened products

With reducible conjugated aryl hydrocarbons, aromaticity is usually maintained. For example, in the electroreduction of a steroidal diene the nonconjugated monoene is obtained in good yield, eqn 112,[227] and in the reduction of phenylated ethylenes,

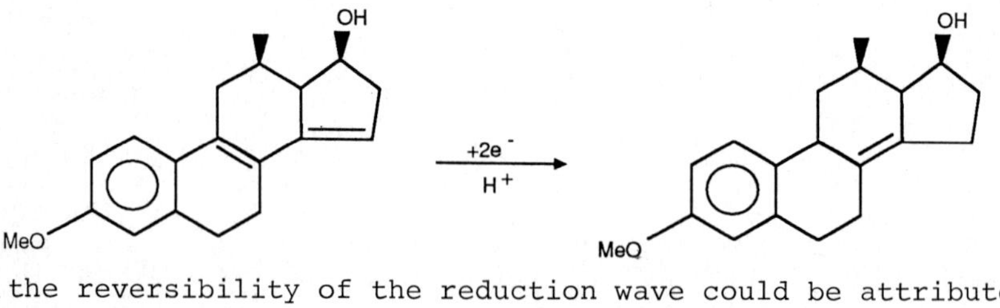

$$\xrightarrow[H^+]{+2e^-} \qquad (112)$$

the reversibility of the reduction wave could be attributed to the facility for protonation on the olefinic sites.[228] Peak potential dependence on pH was found to be related to the association of the hydrocarbon radical anion with the proton source.[229]

Even simple olefins can be reduced by electrochemical methods: these usually involve either the formation of solvated electrons or the electrochemical generation of active hydrogen. For example, cyclohexene and severely hindered alkenes like 2,3-dimethylbutene are readily reduced by solvated electrons in ethanolic HMPA.[230] Alternately, electrolysis of acidic solutions at cathodes made of the hydrogenation catalysts produces surface bound hydrogen, which effects hydrogenation in routes parallel to those observed in conventional catalytic hydrogenation. Thus, smooth platinum[231] or rhodium black[232] have been used to convert phenol to cyclohexanol electrochemically, and platinized platinum cathodes have been used to reduce allyl and crotyl alcohols to propene and butene, respectively.[233]

Surely, these electroreductions involve intermediate formation of radical anions, but mechanistic details after this first electrochemical step are quite complex. In the reduction of anthracene, for exammple, Saveant and coworkers have shown with linear sweep voltammetry and double potential step voltammetry that an ECE-disproportionation mechanism obtains,[234] although the protonation step is not a simple bimolecular sequence.[235] Furthermore, solvation was quite significant in controlling the relative importance of simple reduction vs. radical anion coupling. Chronopotentiometric measurements established that the anion radical itself did not dimerize,[236] but rather formed a one:one complex with water which then led to the product of reductive dimerization.[237] This specific solvation is critical to efficient dimerization,[238,239] and changes in both reaction order and mechanism are encountered when changes are made either in solvent or in other reaction conditions.[240]

The extent to which these electroreductions involve intermediate formation of the solvated electron has not been firmly established. The distinction is unclear even between reductions involving clearly indirect electrolytic transfer and electrochemical reductions. In indirect electroreductions, electrons are transferred from the electrode to a mediator, which then reduces the substrate of interest by homogeneous electron transfer, while conventional electroreduction involve charge transfer from the electrode to the substrate without any intermediate carriers. In fact, some chemists have suggested that solvated electrons may be the primary reductant in most cathodic reductions,[241] although this view has not met with wide acceptance.

Two models have been suggested to explain the stability of solvated electrons in polar liquids. In the first, an electron is trapped in a "cavity" in the bulk solvent in which solvent dipole orientation stabilizes the charged entity; in the second, the electron occupies an expanded hydrogen-like atomic orbital near the first solvent sphere of the electrolyte cation. In either model, the formation of solvated electrons would be accompanied by a considerable volume increase and by an enhanced electron mobility. Often lithium chloride is used as the electrolyte in solvated electron-mediated electroreductions (because of solubility and ion pairing considerations) and many electrode materials are suitable for this reaction. Since the rate of reaction of a solvated electron and a solvent molecule is generally low in the absence of a proton donor, reductions via solvated electrons would be most important in polar solvents of low protic activity where direct transfer to the substrate is blocked by energetic considerations.

(B) Carbonyl Reductions

The polarity inherent in a C=O double bond makes its electroreduction much more facile than that of its hydrocarbon analogues. Furthermore, the intermediate radical anion is sufficiently stable so that carbon-carbon coupling to afford 1,2-diols often accompanies simple reduction. The chemical challenge in conducting electroreductions of carbonyl compounds is thus to control the balance between direct reduction to alcohols and coupling of intermediate ketyl radicals or radical anions. The branching ratio between simple electroreduction and reductive hydrodimerization should be pH dependent, Scheme 1. Carbon-carbon bond

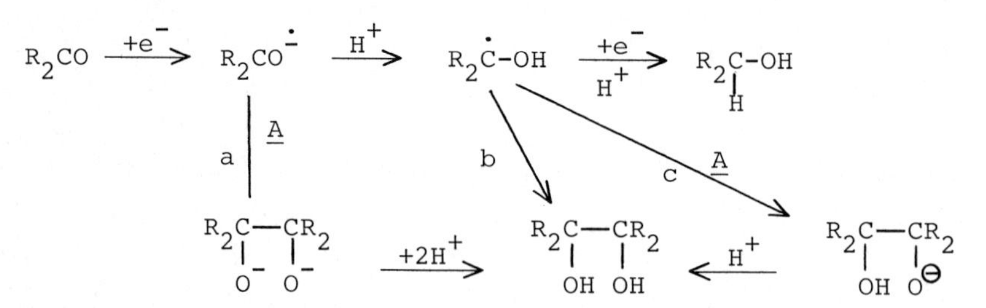

Scheme 1. Reduction vs Reductive Dimerization

formation via radical anion dimerization (path a) is disfavored by electrostatic considerations compared with dimerization of the hydroxymethyl radical obtained by radical anion protonation (path b) or by capture of the radical by the radical anion (path c). DIrect reduction to the alcohol is optimized ketyl (radical anion) is rapidly protonated, reduced, and reprotonated (path d).

We consider first reduction of carbonyls to alcohols. The successful reduction of carbonyls and alpha,beta-unsaturated carbonyls to the corresponding alcohols can be attained on many electrode materials under a wide variety of conditions. Thus, predicting the optimum conditions for a specific compound of interest tests the electrochemist's intuition and mastery of the art. Wide variations in yield and current efficiency are encountered. In the reduction of aldehydes to alcohols, e.g., eqn 113, hydrocarbon formation also sometimes accompanies formation

$$RCHO \xrightarrow[H^+]{+e^-} RCHOH + RCH_3 + \underset{\text{OH}}{RCH}-\underset{\text{OH}}{CHR} \qquad (113)$$

of alcohol and coupling product. Hydrocarbon formation is the main product of propionaldehyde reduction, in fact, at platinized platinum or at cadmium or as the temperature or acidity of the medium is increased.[242] In phosphate buffers, aldol condensation is observed: 2-methylpentane-1,2-diol is formed on lead and 2-methylpentenol is formed on zinc, eqn 114.[243] As the length of

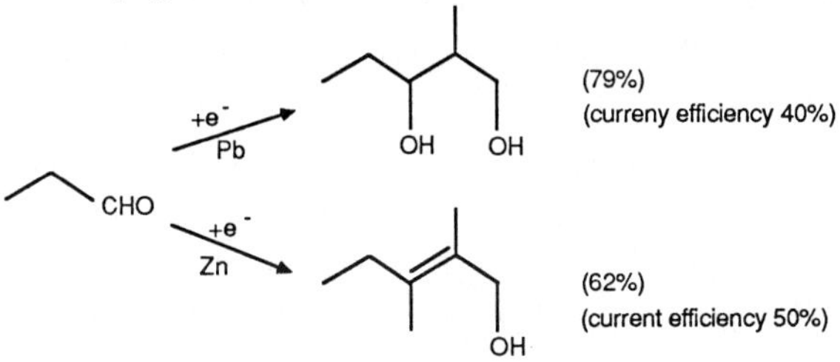

(114)

CHO; $+e^-$, Pb → (79%) (curreny efficiency 40%) OH OH; $+e^-$, Zn → (62%) (current efficiency 50%) OH

the alkyl chain of the aldehyde is increased, the fraction of coupling also increases. Reduction of the carbon-carbon double bond of unsaturated enones also can be accomplished in high yield, eqn 115.[244] The extent of reduction can also be controlled

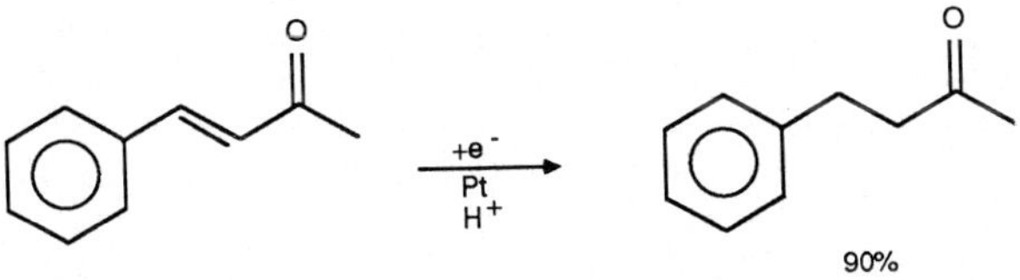

(115)

$+e^-$, Pt, H^+ → 90%

in subtle ways: two electron reduction of diacetylbenzene is observed on a dropping mercury electrode, whereas four electron reduction is observed on a mercury pool, eqn 116,[245] probably

$$COCH_3\text{-}C_6H_4\text{-}COCH_3 \xrightarrow[\text{Hg drop}]{+2e^-} COCH_3\text{-}C_6H_4\text{-}CHOHCH_3$$

$$COCH_3\text{-}C_6H_4\text{-}COCH_3 \xrightarrow[\text{Hg pool}]{+4e^-} CHOHCH\text{-}C_6H_4\text{-}CHOHCH \qquad (116)$$

because subsequent reduction is too slow compared with the lifetime of the mercury drop. By-products can sometimes be minimized by current reversal techniques, as in the high current efficiency obtained in the electroreduction of acetophenone to ethylbenzene.[246]

Since several excellent summaries of reaction conditions leading to simple electroreduction are available,[247-249] a detailed consideration of these features will not be given here. Two recent approaches to controlled reduction are, however, worth mention. First, the inclusion of a compound of interest within the cavity of beta-cyclodextrin can influence the chemical reactivity of electrogenerated intermediates in a profound way. Electroreduction of complexed ethyl cinnamate, for example, yields the dihydrocompound with only 23% of the dimers, a result which stands in contrast to the high yield of dimer formation observed upon electroreduction of the free substrate.[250] Inclusion complexes also can affect the regiochemistry and stereochemistry of dimer formation and can offer, with covalent attachment of an appropriate electrophore, an unusual type of electrocatalyst.[241]

Second, modest levels of asymmetric induction in carbonyl reductions can sometimes be observed on chiral electrode surfaces or with chiral ketones. For example, the asymmetric reduction of (-)-menthylphenylglyoxylate has been reported on a mercury cathode in the presence of strongly adsorbed alkaloids.[252] Preferential adsorption of one prochiral face of the carbonyl was thought to be responsible for the observed effect. Similarly, diastereomeric secondary alcohols were produced in unequal amounts when ketones bearing chiral centers at the alpha position were electroreduced,[253] the ratio of isomers having been thought to be controlled by the conformation of the adsorbed ketone prior to reduction.

It is thus obvious that electrochemical methods can be used as synthetic routes for simple carbonyl reduction, but since other reagent-based methods are also available for these conversions, it is reductive coupling of carbonyls in which electrochemical methods offer unique synthetic advantage. Scheme 1 suggests several routes to cathodic coupling involving initial formation of the ketyl radical anion. Parker has shown that in acetonitrile a one:one complex (through which dimerization can occur) exists between water and acetophenone ketyl.[254] The bond forming step is nearly thermoneutral and is driven by entropy, with reaction rates forthe dimerization being quite solvent dependent.[255] Unlike the hydrocarbon dimerizations discussed earlier, these carbonyl hydrodimerizations often appear to involve direct dimerization of the electrogenerated radical anion.[256] The relative stability of these intermediates provides a rationalization for the observation that highest chemical yields of coupling products are found in alkaline media and that low to moderate yields of pinacol coupling are attained with aliphatic carbonyl compounds, with better yields being observed with aromatic carbonyl compounds.[2] Intramolecular coupling in 1,3-diones can also be achieved, providing, for example, a convenient route to 1,2-cyclopropanediols, eqn 117.[257]

OAc OAc

$+2e^-$ (THF, Ac_2O) (117)

Optically active pinacols can be attained under appropriate conditions, although optical purity is typically quite poor. For example, asymmetric induction in pinacol formation can be observed if the coupling takes place in a chiral medium.[258] Diastereoselection in the reductive coupling of acetophenone in acetonitrile is destroyed by the presence of water, which breaks up tight ion pairs between the cation and ketyl radical anion.[259] That adsorption effects on the electrode are important is established by the observation that negligible optical induction was observed when acetophenone, 4-acetylpyridine, or ethyl phenylglyoxylate were reduced at a modified carbon electrode in which optically active groups were attached to the graphite by a conformationally flexible covalent tether.[260]

With alpha,beta-unsaturated carbonyls, a stereoisomeric mixture of dimeric products coupled at the beta position are usually obtained, as in the intramolecular hydrodimerization shown in eqn 118.[261] If the beta position is blocked sterically,

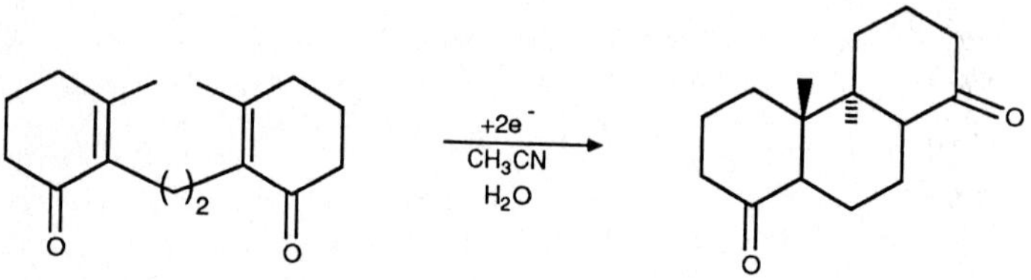

$$\xrightarrow[\substack{CH_3CN \\ H_2O}]{+2e^-} \qquad (118)$$

coupling can occur at the carbonyl carbon, eqn 119.[262] With more

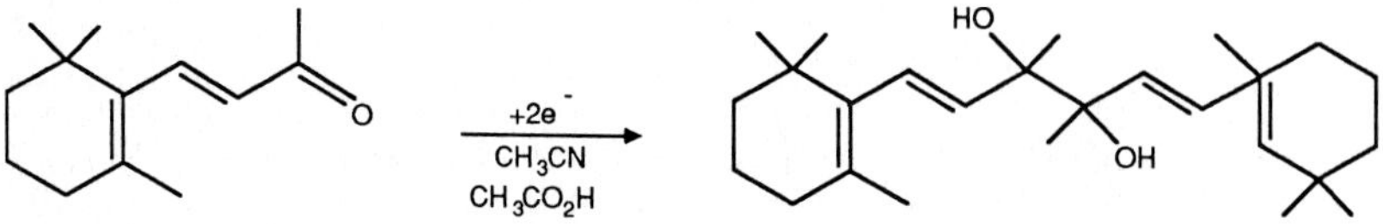

$$\xrightarrow[\substack{CH_3CN \\ CH_3CO_2H}]{+2e^-} \qquad (119)$$

complex unsaturated carbonyls, reaction conditions can significantly alter product regiochemistry. In the electrolytic coupling of pyrones, for example, cathodic cyclization occurs, giving a complex mixtures of products, eqn 120.[263] In completely aprotic media,

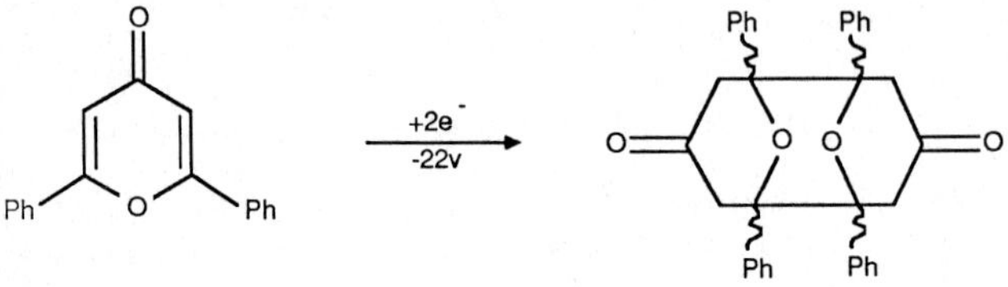

$$\xrightarrow[-22v]{+2e^-} \qquad (120)$$

ring opening also occurs.[264] With sterically encumbered thiopyrones, the opposite regiochemistry and desulfurization is observed, eqn 121,[265] and in the hydrodimerization of coumarin,

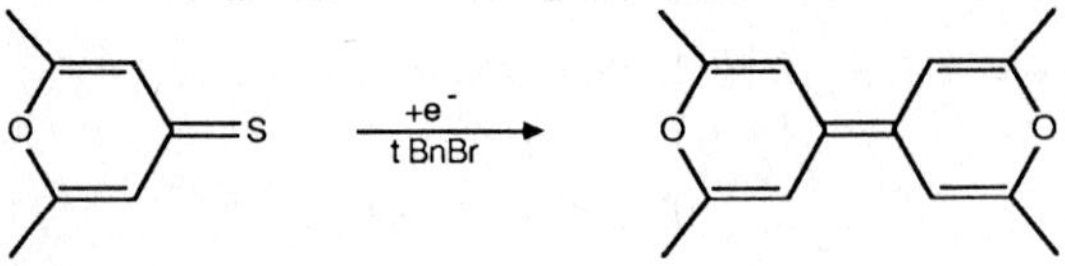

$$\xrightarrow[t\,BnBr]{+e^-} \qquad (121)$$

coupling via the radical, the radical anion, and the dianion (formed by disproportionation of the radical anion) have been implicated.[266] Furthermore, the presence of metal ions can profoundly influence the regiochemistry of quinone reductive couplings.[267]

As with hydrodimerizations of hydrocarbons and simple carbonyl compounds, radical anions are involved in coupling, but often with additional mechanistic complexity. Dimerization can result either by combination of radical anions[268] or by attack of the radical anion on the unsaturated starting material.[269] The branching between simple reduction and carbon-carbon bond formation can sometimes be controlled by judicious choice of the electrolyte.[270] Stereoselection is also controlled by the availability of a proton source.[271] Cyclization of the dianionic intermediates formed by radical anion dimerization can sometimes be observed,[272-274] as can geometrical isomerization.[275]

Radical anions have also been implicated in mixed reductive couplings or in trapping by electrophiles. Single electron electroreduction of aldehydes, eqn 122,[276]

$$\mathrm{ArCHO} \xrightarrow[\mathrm{CCl_4/CHCl_3}\ \text{in DMF}]{+e^-} \mathrm{ArC(H)(OH)CCl_3} \qquad (122)$$

or enones, eqn 123,[277] give rise to ketyls which participate in

$$\text{indenone} \xrightarrow[\mathrm{Rx}]{+e^-} \text{2-R-2-Me-3-methylene-indan-1-one} \qquad (123)$$

chain electron transfer and subsequent radical-ketyl coupling. These same ketyls can be formed chemically by electron transfer from closed shell anions, e.g., eqn 124.[278,279]

$$\mathrm{Ph_3CLi} + \mathrm{{}^tBuCH{=}CHC(O){}^tBu} \longrightarrow [\text{ketyl}] \longrightarrow \mathrm{Ph_2CH{-}C_6H_4{-}CH({}^tBu){-}CH_2C(O){}^tBu} \qquad (124)$$

As with direct reductions of carbonyl compounds, the practical application of electrochemical methods to hydrodimerization is an art. Many of these subtleties are discussed in extensive review articles,[249,251] and the interested reader is urged to consult these sources for guidance in choosing optimal experimental conditions for the reductions.

(C) Reduction of Nitro Compounds

Cathodic reduction of nitro compounds was one of the first electrochemical reactions to be studied, and it was with these substrates that Haber first proposed that organic electrochemistry could be defined by controlling an applied potential. Several important summaries of the electroreduction of nitro compounds are available,[289,281] and we discuss here only those aspects of nitro electroreduction relevant to anion production. Like azo compounds,[282] imines,[283] and other multiply bound nitrogen-containing compounds[284] nitro compounds exhibit a complex reduction chemistry. Aryl nitro compounds have been most extensively studied, producing nitroso compounds, azoxybenzene, and ultimately aniline, eqn 125.[285]

$$ArNO_2 \xrightarrow[+2H^+,\ -H_2O]{+2e^-} ArNO \xrightarrow{+e^-} ArNO^{\dot{-}} \xrightarrow{H^+} \tfrac{1}{2}Ar\text{-}\overset{\oplus}{N}(O^{\ominus})\text{=}NAr \qquad (125)$$

Controlled potential electrolysis has revealed the sequence of anionic intermediates shown as Scheme 2.[286] Here, electro-

$$PhNO_2 \xrightarrow{-0.45V} PhNO_2^{\dot{-}} \xrightarrow{-1.25V} PhNO_2^{=} \xrightarrow{ROH} Ph\text{-}N(OH)O^{\ominus} \xrightarrow{-OH^-} Ph\text{-}\ddot{N}\text{=}O$$

$$PhNO \xrightarrow{-0.2V} PhNO^{\dot{-}} \xrightarrow{-1.20V} PhNO^{=} \longrightarrow Ph\text{-}\ddot{N}^{\ominus}\text{-}OH \xrightarrow{-H^-} PhNO$$

Scheme 2. Intermediates in the Electroreduction of Nitrobenzene

reduction forms the radical anion, which is reduced at more negative potentials to the dianion. Upon protonation, the resulting monoanion is formed, from which hydroxide is lost to form the neutral nitroso compound. This species is in turn reduced to the mono- and dianion and protonated to a species, which can be reconverted to the nitroso compound, can act as a nucleophile toward electrophilic reagents, or can be further reduced to aniline. The radical anion can be observed in the presence of radical spin traps[287] or it can participate in electron transfer with alkyl halides as a source of alkyl radicals or anions, Scheme 3.[288] The counter ion and the solution ionic strength can strongly influence ion pairing of the radical anion in aprotic solvents, effects which can be easily noted by shifts of the half-wave potential and in the esr spectrum of the radical anion.[289]

$$RNO_2 \longrightarrow RNO_2^{\dot{-}} \xrightarrow{RX} RNO_2 + RX^{\dot{-}} \longrightarrow R\cdot$$

$$RNO_2^{\dot{-}} \xrightarrow{RX} R{-}N(OR)O^{\cdot} + X^- \xrightarrow{+e^-} R{-}N(OR)O^-$$

$$RNO_2^{\dot{-}} \xrightarrow{R\cdot} RNO_2 + R^-$$

Scheme 3. Trapping of Organic Nitro Radical Anions with Alkyl Halide

If the radical anion of the aryl nitro compound bears a halogen atom, the primary electron transfer can initiate C-X cleavage. The fraction of $-NO_2$ vs. ring reduction then reflects the strength of the C-X bond in the radical anion, with 75% aniline formation being observed with fluoro- and chloronitrobenzene, but much lower yields with bromo- or iodonitrobenzenes, eqn 126.[290] Analogous observations have also been made with

$$o\text{-}XC_6H_4NO_2 \xrightarrow[\text{Hg, 50:50 PrOH:}H_2O]{+2e^-} o\text{-}XC_6H_4NH_2 \;(75\%\text{ for X = F, Cl}) + C_6H_5NO_2 \;(\text{major for X = Br, I}) \tag{126}$$

nitrohalothiophenes.[291]

In polar protic media, the intermediate radical anion or dianion of non-halogenated nitrocompounds is protonated to a hydroxylamine. Intramolecular trapping of this intermediate provides a useful route to heterocycle synthesis. For example, nitroketones have shown to provide pyrroline-N-oxides, pyrrolines, or pyrrolidines, eqn 127.[292] Similarly, benzothiazinedioxides are

$$O_2N(CH_2)_3COR \xrightarrow[H_2O/NH_3]{4e^-,\ 4H^+} HON H(CH_2)_3COR \xrightarrow{-H_2O} \text{pyrroline-N-oxide} \xrightarrow[+2H^+]{+2e^-} \text{pyrroline} \xrightarrow[H^+]{2e^-} \text{pyrrolidine} \tag{127}$$

obtained by electroreduction of suitable nitrophenylsulphones, eqn 128.[293]

$$o\text{-}(RCOCH_2SO_2)C_6H_4NO_2 \xrightarrow[+4H^+]{+4e^-} \text{4-hydroxy-3-R-4H-1,4-benzothiazine 1,1-dioxide} \tag{128}$$

Nitroolefins can be similarly reduced, via radical anionic intermediates, to the corresponding oxime, eqn 129.[294] Again,

$$PhCH{=}CHNO_2 \xrightarrow[\text{Pt, } H^+/MeOH]{+e^-} PhCH_2CH{=}NOH \qquad (129)$$

intramolecular trapping can produce unusual heterocycles, e.g., eqn 130.[295]

$$R\text{-}C(=O)\text{-}C(Ph){=}C(Ph)NO_2 \xrightarrow[\text{H}_2\text{O, EtOH, pH 1.4}]{+e^-} \text{(Ph, Ph, R; N-OH, O=C-R)} \longrightarrow \text{3,4-diphenyl-5-R-isoxazole} \qquad (130)$$

The synthetic utility of radical anions of nitro compounds is thus clear, and as before, these intermediates can also be produced by electron transfer from closed shell anions. Buncel and Menon have shown, for example, that upon treatment of p-nitrotoluene with one of several anions, the corresponding benzyl anion, radical anion, or coupled radical anion products can be observed, eqn 131.[296] The high stability of the radical anions

$$p\text{-}CH_3C_6H_4NO_2 + \begin{matrix} H^- \\ \phi_3C^- \\ \text{diensyl} \\ \text{fluotenyl} \end{matrix} \longrightarrow p\text{-}{}^{-}CH_2C_6H_4NO_2 + [p\text{-}CH_3C_6H_4NO_2]^{\cdot -} + [O_2NC_6H_4CH{=}CHC_6H_4NO_2]^{\cdot -} \qquad (131)$$

and dianions derived from nitro compounds thus makes them attractive species in routes for generating or reacting carbanions.

(D) Reduction of Acid Derivatives

Stable ketyl radical anions are encountered not only in the electroreduction of aldehydes and ketones, but also with organic substrates at the next higher formal oxidation level. Carboxylic acids and their derivatives (esters, amides, acid halides, hydrazides, nitriles, and cyclic analogues (anhydrides, imides, lactones, lactams) have a long and well-documented[297-299] history in organic electrochemistry, but mechanistic details are still often lacking. Since these compounds bear an electron-releasing heteroatom adjacent to the C=O double bond, their reduction potentials are often more negative than those of the expected initial reduction product (aldehyde). If acid derivatives are to be selectively reduced to aldehydes, the acid derivative must be activated by a strong electron withdrawing group near the carboxyl group and there must be no group present in the molecule which is more easily reducible. Furthermore, the aldehyde must be protected as it is formed, usually as a hydrate or acetal. If these criteria are met, the radical anionic intermediates formed in the electroreduction of carboxylic acid derivatives can be used as interesting chemical intermediates for controlled electroreduction.

For simple acids, the requirement for activation implies that oxalic or lactic acids will be more easily reduced than simple carboxylic acids. Benzoic acid, for example, is reduced only in strong acid, implying direct formation of the ketyl radical and avoiding intermediate formation of the radical anion, eqn 132.[300]

$$ArCO_2H \rightleftharpoons Ar{-}C(={}^{\oplus}OH)OH \xrightarrow{+e^-} Ar\dot{C}(OH)_2 \xrightarrow{+e^-} Ar\overset{\ominus}{C}(OH)_2 \rightarrow ArCH(OH)_2 \rightleftharpoons ArCHO \quad (132)$$

Substituents on the ring induce linear shifts in the reduction potentials according to a normal Hammett relationship.[300]

Perhaps surprisingly on electrostatic grounds, anionic borate complexes of carboxylates can be effectively electroreduced, eqn 133.[301] With halogenated unsaturated carboxylic acids,

$$\text{bis(salicylato)borate}^{\ominus} \xrightarrow[2H^+]{+4e} \text{HO–}\overset{\ominus}{C}\text{(aryl)–O–BS}_2 \rightleftharpoons {}^{\ominus}O\text{–CH(aryl)–O–BS}_2 \rightarrow O{=}\overset{\ominus}{B}< \; + \; o\text{-}{}^{\ominus}OC_6H_4CHO \quad (133)$$

dehalogenation occurs, unless the electroreduction is conducted at highly negative potentials where a kinetic preference for reduction of the carboxylate ultimately produces alcohol, eqn 134.[302]

$$C_6F_5CO_2H \xrightarrow{-1.2V} p\text{-}H\text{-}C_6F_4CO_2H \qquad C_6F_5CO_2H \xrightarrow{>-1.3V} C_6F_5CH_2OH \tag{134}$$

These same criteria hold in understanding the reductive electrochemistry of acid derivatives. Both cyclic voltammetric[303] and esr[304] evidence support the formation of stable radical anions upon reduction of esters and other derivatives. Acid chlorides often undergo fragmentation, a cleavage which has found synthetic utility, eqn 135.[305] Unusual rearrangements of the intermediate

$$2ArCOCl \xrightarrow{+2e^-} 2Ar\dot{C}O \longrightarrow ArC(=O)\text{-}C(=O)\text{-}Ar \xrightarrow{+2e^-} Ar\text{-}C(O^-)=C(O^-)\text{-}Ar \xrightarrow{ArCOCl} ArC(=O)O(Ar)C=C(Ar)OC(=O)Ar \;(cis) + ArC(=O)O(Ar)C=C(Ar)OC(=O)Ar \;(trans) \tag{135}$$

radical anions are occasionally observed, especially when stabilized by the presence of heteroatoms as in the dithioester shown in eqn 136.[306]

$$o\text{-}C_6H_4(C(=O)SAr)_2 \xrightarrow[DMF]{+e^-} \text{[cyclic } O^- \text{, SAr / SAr intermediate]} \longrightarrow \left[\text{3,3-bis(SAr)phthalide}\right]^{\cdot -} \tag{136}$$

(E) Reductive Eliminations

The electrochemical formation of radical anions and dianions activates molecules for elimination. One significant consequence of such a sequence is in the design of electrochemically sensitive protecting groups. Depending on the electron demand of the potential leaving group, these routes also provide another method for the electrochemical generation of anions discussed in Section IV.

The critical feature involved in the reductive cleavage of a bond A-B is the ability of either A or B to accommodate negative charge as an unpaired electron is localized on the other fragment. This, of course, provides the driving force for the C-X$^-$ cleavage of alkyl and aryl halides discussed earlier, as well as for the primary step in many of the sequences considered for electrochemical generation of bases and nucleophiles.

It is important to realize that such cleavages are not limited to halides, however. Cleavage of aryl sulfides, for example, is often as chemically efficient as that of aryl halides, eqn 137.[307] In fact, catalytic enhancement of the rate can often

$$Ph_3C\text{-}S\text{-}C_6H_4\text{-}NO_2 \xrightarrow{+e^-} Ph_3C\text{-}S\text{-}[C_6H_4]^{\bullet -}\text{-}NO_2 \longrightarrow Ph_3C\bullet + {}^{-}S\text{-}C_6H_4\text{-}NO_2 \quad (137)$$

be observed as the reaction proceeds.[308]

The N-alkyl bond of pyridinium salts can often be cleaved upon electroreduction, dependening on the stability of the radical formed. Thus, 2,4,6-trialkylpyridinium salts cleave when the N-substituent is allyl or benzyl, but form stable radical when a simple alkyl group is attached to nitrogen, eqn 138.[309] Thus,

$$\text{2,4,6-}Ph_3C_5H_2N^+\text{-}R \xrightarrow{+e^-} \text{2,4,6-}Ph_3C_5H_2N^{\bullet}\text{-}R \longrightarrow \text{2,4,6-}Ph_3C_5H_2N^{-} + R\bullet \quad (138)$$

stable for R = isopropyl; R = allyl or benzyl

stable coupled pyridinium salts can provide access either to the stable radical cation or diradical upon electroreduction, eqn 139.[310]

$$C_5H_5N^+\text{-}(CH_2)_n\text{-}N^+C_5H_5 \xrightarrow{+e^-} C_5H_5N^{\bullet}\text{-}(CH_2)_n\text{-}N^+C_5H_5 \xrightarrow{+e^-} C_5H_5N^{\bullet}\text{-}(CH_2)_n\text{-}N^{\bullet}C_5H_5 \quad (139)$$

These in turn can be used as electron donors to acceptors of interest.

Phenoxides are comparably cleaved, perhaps by a two-electron process, if the accompanying radical is not sufficiently stable, eqn 140.[311] The ease of cleavage parallels the pKa of the phenol.

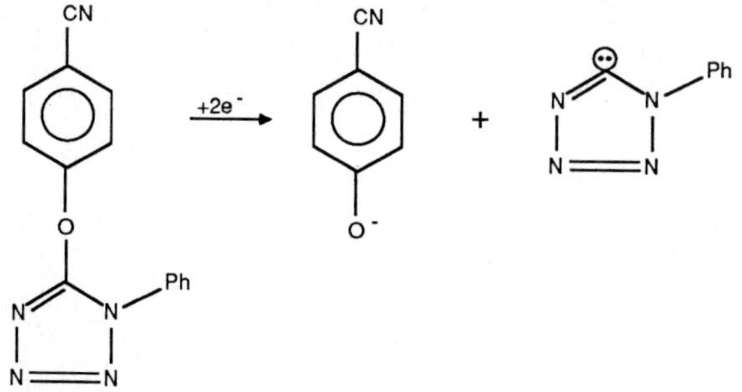

(140)

In other cases, dimerization of the radical anion occurs before loss of the leaving group, as in the reductive coupling of diaryl ethers, eqn 141.[312]

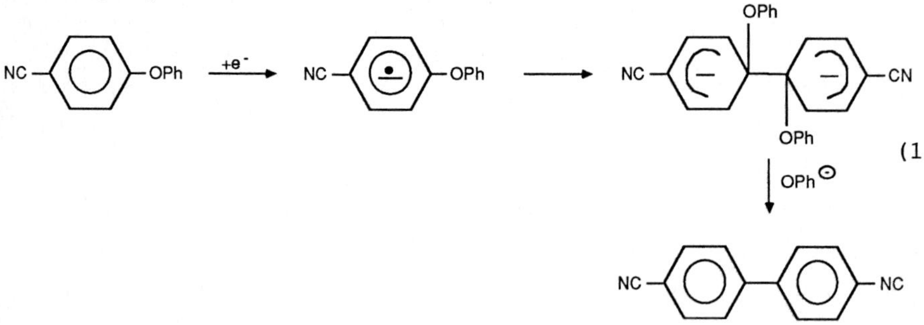

(141)

Tosylate can also be lost upon electroreduction, eqn 142,[313]

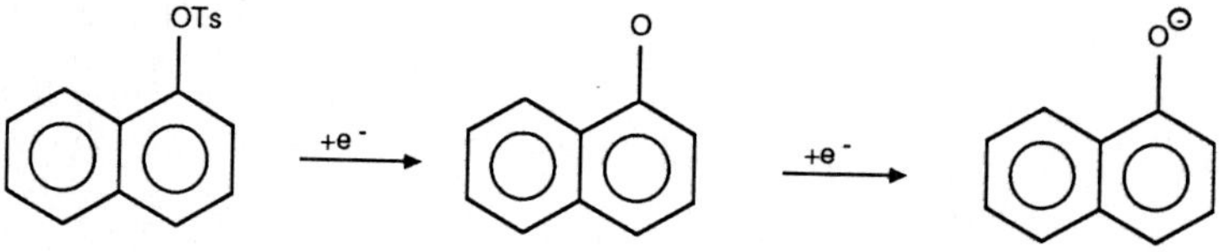

(142)

and coupling can occur between sites bearing tosylate substituents in ditosylates, eqn 143.[314] Such reactions also occur with

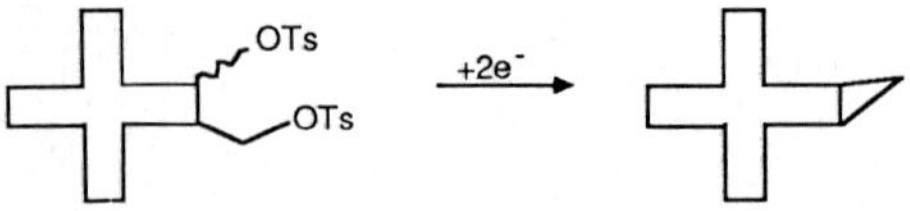

(143)

sulfones, as shown in eqns 144[315] and 145.[316] Parallel reductive cleavages of benzohydrazides[317] and diazo compounds,[318] for example, often provide access to unusual reactive intermediates. Radical coupling of unsaturated olefins bearing vinyl leaving groups can produce substituted butadienes by a coupling-elimination sequence.[319]

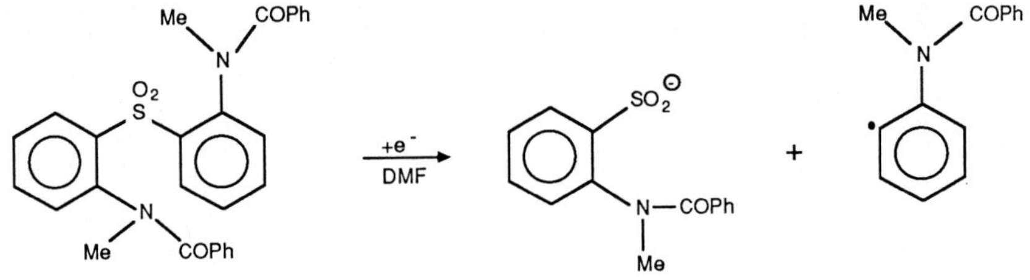

(144)

(145)

Reductive eliminations to unmask double bonds are a common use of these electrochemical conversions. Several arbitrarily chosen examples are shown in eqns 146 - 149. Respectively, these involve cleavage of a dibromide,[320] a cyclic carbonate,[321] a dioxalate,[322] and a vicinal acetate thioether.[323] Occasionally,

(146)

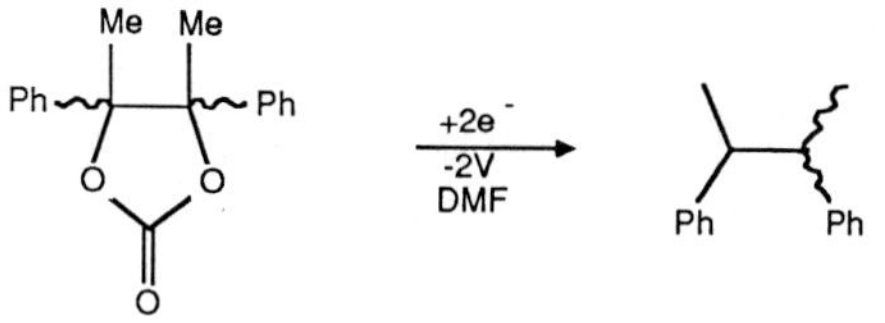

(147)

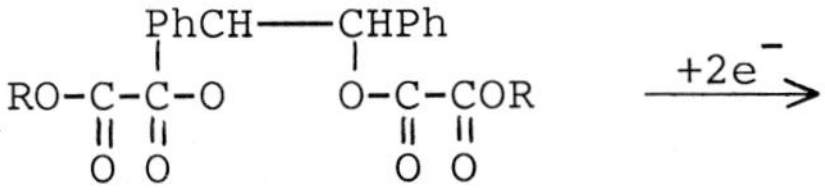

$$\text{RO-C(=O)-C(=O)-O-CH(Ph)-CH(Ph)-O-C(=O)-COR} \xrightarrow{+2e^-} \text{PhCH=CHPh} \quad (148)$$

$$\text{PhCH(SPh)-CH(OAc)Ph} \xrightarrow[-1.7V]{+2e^-} \text{PhCH=CHPh} \quad (149)$$

further reduction will occur, as in the formation of the alkane from the diacetate, eqn 150.[324]

$$\text{PhCH(OAc)-CH(OAc)-Ph} \xrightarrow[-2.2V,\ H^+]{+2e^-} \text{PhCH}_2\text{CH}_2\text{Ph} \quad (150)$$

(F) Reduction of Organometallic Compounds

The investigation of the electroreduction chemistry of organometallic compounds is a relatively new area. Reactions are still being discovered and few mechanistic details regarding the preferred modes of reaction are known with sufficient confidence to predict general behavior. Nonetheless, the same principles useful for understanding the chemistry of organic compounds are also applicable for compounds containing metal atoms.

We have seen in Section I that the normal polarization of a carbon-metal bond permits us to think of alkali metal salts as anions, whose oxidation chemistry is understandable as a route to reactive free radicals. In compounds formed when heavy metals bind with the anionic site, removal of electrons cannot be thought of as arising from ionic intermediates. Instead, the carbon-metal bond must be thought of as a highly polar covalent linkage. Like other polar linkages, reduction populates a sigma antibonding orbital and causes bond fragmentation. Thus, triarylgermanium halides, for example, suffer reductive cleavage to form a triarylgermyl radical, eqn 151.[325] As was seen with hydrocarbons,

$$Ar_3GeX \xrightarrow{+e^-} Ar_3Ge\cdot \quad + \quad X^- \qquad (151)$$

X = halide

catalytic enhancement of the cleavage occurs as the reaction proceeds. Similarly, organomercurials suffer two electron reduction to form anions, eqn 152,[326] a reaction parallel to those

$$R_2Hg \xrightarrow[\text{Pt rotating ring}]{+2e^-} 2R^{\ominus} \xrightarrow[CH_3CN]{} 2RH \qquad (152)$$

discussed earlier as a method for forming electrogenerated nucleophiles. No correlation was observed between the ease of reductive cleavage and the pKa of the anionic fragment. Anion formation could be monitored by observing anodic currents as a rotating ring counter-electrode or by isolation of the protonated (neutral) hydrocarbon. Even ferrocene, a species which can be reversibly reduced to the radical anion at potentials less negative than -30 mV, undergoes two electron reduction at more negative potentials.[327] In DMF, reductive cleavage occurs, giving the cyclopentadienide anion and an iron-based radical.[328]

Metal-metal dimers containing iron-iron bonds are also fragmented by electroreduction.[329] Electroreduction of metallocarbenes causes electron transfer toward the ligand.[330] Metal acyl complexes are electroreduced via a two electron step to form the metal anion and the aldehyde, eqn 153. No mechanistic details are yet known.[331]

$$CH_3-\overset{O}{\overset{\|}{C}}-Re(CO)_5 \xrightarrow[H^+]{+2e^-} CH_3-\overset{O}{\overset{\|}{C}}-H + {}^{\ominus}Re(CO)_5 \qquad (153)$$

Reductive electrochemistry has also proved useful for the synthesis of organometallics. For example, the reductive electrolysis of a mixture of tin tetrachloride in the presence of alkyl halide gives good yield of the tetraalkyltin, eqn 154.[332]

$$SnCl_4 + MeCl \xrightarrow{+8e^-} Me_4Sn + 4Cl_2 \qquad (154)$$

Parallel routes have been used to prepare carbon-metal bonds with many other elements, including, for example, boron, antimony, lead, and mercury. Cathodic reduction of electron-deficient olefins at metal electrodes can also produce novel carbon-metal bonds with current efficiencies typically greater than 50%, eqn 155.[333]

$$CH_2{=}CH{-}CN \xrightarrow[Sn,\ H^+]{+4e^-} (NC{-}CH_2CH_2)_4Sn \qquad (155)$$

VI. INDIRECT ELECTROREDUCTIONS

Direct electrogeneration of anionic intermediates from organic compounds with high negative reduction potentials is extremely difficult. Because of the potential importance of such conversions, however, indirect processes, often involving inorganic reagents, have been used.[8] Sometimes the electroreduction generates an active reducing metal as in the indirect reduction of nitrobenzene by electrolysis of metal halides, eqn 156.[334]

$$MX_n \xrightarrow{+ne^-} M \quad (PhNO_2 \xrightarrow{H^+} PhNH_2) \qquad (156)$$

M = Cu, Fe, Sn, Zn

In other cases, electroreduction merely alters the oxidation level of an inorganic complex or enzymatic cofactor as in Weinkamp and Steckhan's indirect reduction of NAD^+ to NADH by electrolysis of $Rh(bpy)_3^{3+}$.[335]

Inorganic reagents can thus react chemically to promote radical anion formation. An example of such assistance can be seen in the indirect electroreduction of allylic alcohols in the presence of iodide at mercury electrodes, eqn 157.[336]

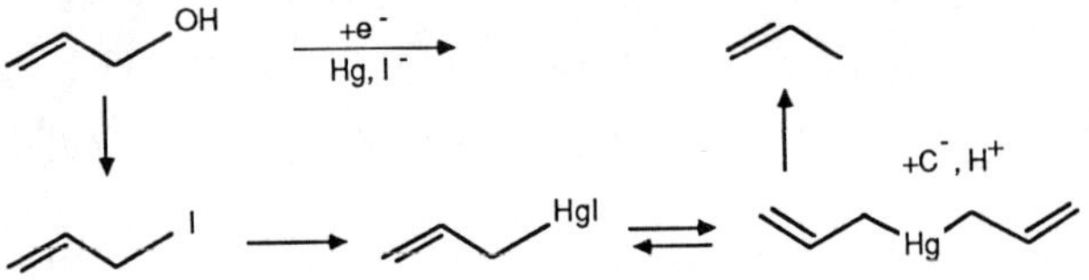

(157)

Chromium salts have proven to be especially useful as an indirect route to radical anions. Facile reductive coupling of benzyl chloride can be observed, for example, in the presence of Cr(II) salts, arising exclusively via the radical dimer.[337] Similarly, polyene carbonyl compounds in DMF give good yields of pinacol in the presence of $CrCl_3$, a chromium complex of the ketone having undergone the critical electroreduction step, eqn 158.[338]

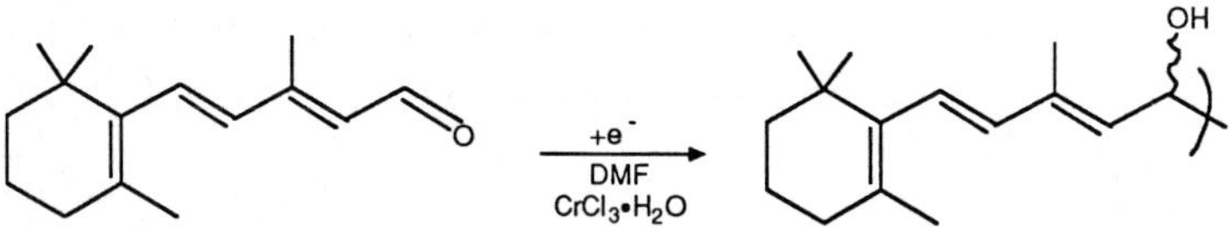

(158)

Easily reducible organic substrates can also act as electron transfer reagents to other species for which electron exchange across the electrode surface is kinetically retarded. Alkyl halides[339,340] and epoxides[341] have thus been reduced in the presence of quinones and easily reduced arenes. Even if the electron transfer between the reduced electrocatalyst and the ultimate substrate is not energetically favored, reaction can still be observed if the substrate radical anion decomposes rapidly. This route has been exploited, for example, in Saveant and coworkers' catalysis of aryl halide reduction by arenes, eqn 159.[342]

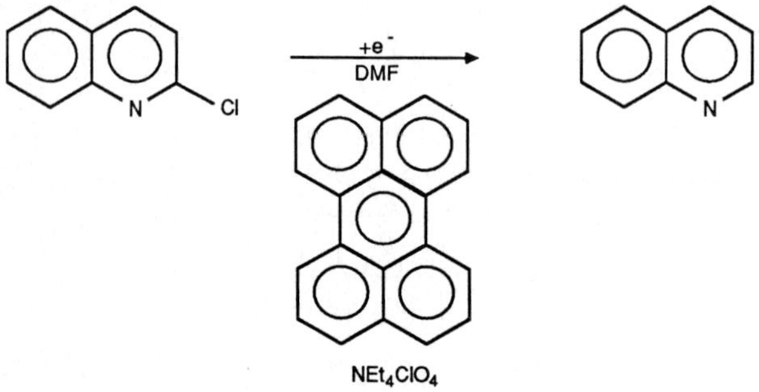

(159)

Potentials as much as 2 V positive of the direct electroreduction potential have been achieved for alkyl halide reductions in the presence of persistent radicals as electrocatalysts.[343] Redox catalysis of bianthrone reduction by quinones,[344] of the cathodic coupling of sulfonium ions in the presence of reducible spin traps,[345] and of the reduction of disulphides by arene radical anions[346] have also been reported.

Indirect catalysis has been especially assisted by the recent availability of highly oriented pyrolytic graphite electrodes, on which special cathodic behavior is observed.[347] These electrodes can be thought of as extended arene electrocatalysts. Such materials have been used as lamellar reducing agents for functionalized alkyl halides[348] and, when complexed with chiral alkaloids, as electrodes for the asymmetric reduction of chiral carbonyls.[349]

Thus, organic radical anions can be formed by indirect routes even if their neutral precursors have appreciable overpotentials for direct electron reduction. If rates of electron exchange across conventional inert electrodes are retarded by thermodynamic or kinetic considerations, the use of electroactive mediators can provide an attractive alternative mode for forming anions.

VII. SUMMARY

We have shown in this review how electrochemical techniques can be used by carbanion chemists to several purposes. Characterization of pi distributions in carbanions and direct determination of hydrocarbon pKas can be made by electrochemical study, as can identification of conformational equilibria in either ground state neutral substrates or their reduced anionic derivatives. We have seen synthetic applications of carbanion oxidation to achieve reduction, coupling, and cyclization by novel routes. Here formation of the organometallic precedes reductive elimination to form the reduced product. Electroreduction as a route to the formation of anions opens new methods for in situ generation of bases and nucleophiles useful for synthetic or mechanistic studies. Finally, either direct or indirect electroreduction of organic substrates can provide efficient routes to radical anions, dianions, or anions. Since electrochemical methods can be a powerful tool, they deserve careful use more routinely in physical organic chemistry.

VIII. ACKNOWLEDGEMENT

Our research program on carbanion electrochemistry has been supported by the National Science Foundation and by the Robert A. Welch Foundation.

REFERENCES

1. F. Fichter, "Organische Elektrochemie" in K. F. Bonhoeffer, "Die Chemische Reaktion," Vol. 4, T. Steinkopff, Leipzig, 1942.
2. H. J. Schaefer, Angew. Chem. 20 (1981) 911.
3. M. M. Baizer and H. Lund, "Organic Electrochemistry," Marcel Dekker, New York, 1983.
4. C. K. Mann and K. K. Barnes, "Electrochemical Reactions in Nonaqueous Systems," Marcel Dekker, New York, 1971.
5. A. J. Bard and H. Lund, "Encyclopedia of Electrochemistry of the Elements," Vol. XI-XV, Marcel Dekker, New York, 1984.
6. A. J. Bard and L. Faulkner, "Electrochemical Methods" Wiley, New York, 1980.
7. C. L. Perrin, Prog. Phys. Org. Chem 3 (1965) 165.
8. L. Eberson and K. Nyberg, Adv. Phys. Org. Chem. 12 (1976) 2 and references cited therein.
9. L. Eberson and H. Schaefer, Top. Curr. Chem. 21 (1971) 5.
10. G. J. Hoijtink, Adv. Electrochem. and Electrochem. Eng. 7, (1970) 1.
11. B. S. Jensen and V. D. Parker, J. Am. Chem. Soc. 97 (1975) 5211.
12. V. D. Parker, Pure Appl. Chem. 51 (1979) 1021.
13. V. Svanholm and V. D. Parker, J. Chem. Soc., Perkin II (1973) 1594.
14. (a) G. J. Hoijtink, E. de Boer, P. H. van der Meij and J. P. Wejland, Rec. Trav. 75 (1956) 487; (b) J. Jagur-Grodzinski, M. Feld, S. L. Yang and M. Szarc, J. Phys. Chem. 69 (1965) 628.
15. K. Izutsu, S. Sakura and T. Fujinaga, Bull. Chem. Soc. Japan 46 (1973) 2148.
16. L. A. Avaca and A. Bewick, J. Electroanal. Chem. 41 (1973) 405.
17. B. Goldberg and A. J. Bard, J. Phys. Chem. 75 (1971) 3281.
18. R. Dietz, in ref. 3, p. 237.
19. V. D. Parker, Acta Chem. Scand. B 35 (1981) 583.
20. R. Dietz and M. E. Peover, Trans. Far. Soc. 62 (1966) 3535.
21. M. Svaan and V. D. Parker, Acta Chem. Scand. B 35 (1981) 559.
22. M. Svaan and V. D. Parker, Acta Chem. Scand. B 36 (1982) 351.
23. M. Svaan and V. D. Parker, Acta Chem. Scand. B 36 (1982) 357.
24. M. Svaan and V. D. Parker, Acta Chem. Scand. B 36 (1982) 65.
25. I. Bergman, "Polarography", Macmillan, London (1964), p. 925.
26. R. D. Allendoerfer and P. H. Rieger, J. Am. Chem. Soc. 87 (1965) 2336.
27. S. Hayano and M. Fujihara, Bull. Chem. Soc. Japan 44 (1971) 2046.
28. A. Misono, T. Osa and T. Yamagishi, Bull. Chem. Soc. Japan 44 (1971) 1496.
29. R. Breslow, Pure Appl. Chem. 40 (1974) 493.
30. R. Breslow and R. Goodin, J. Am. Chem. Soc. 98 (1976) 6076.
31. (a) A. K. Hoffman, W. G. Hodgson, D. L. Maricle and W. H. Jura, J. Am. Chem. Soc. 86 (1964) 631; (b) A. J. Fry and R. L. Krieger, J. Org. Chem. 41 (1976) 54.
32. R. Breslow and J. Grant, J. Am. Chem. Soc. 99 (1977) 7745.

33. B. Jaun, J. Schwarz and R. Breslow, J. Am. Chem. Soc. 102 (1980) 5741.
34. A. J. Bard and A. Merz, J. Am. Chem. Soc. 101 (1979) 2959 and references cited therein.
35. D. M. LaPerriere, W. F. Carroll, B. C. Willett, E. C. Torp and D. G. Peters, J. Am. Chem. Soc. 101 (1979) 7561.
36. C. P. Andrieux, L. Nadjo and J. M. Saveant, J. Electroanal. Chem. 26 (1970) 147.
37. T. Psarras and R. E. Dessy, J. Am. Chem. Soc. 88 (1966) 5132.
38. D. J. Schiffrin, Dis. Far. Soc. 56 (1973) 1975.
39. J. M. Kern and P. Federlin, Tetrahedron 34 (1978) 661.
40. J. M. Kern and P. Federlin, J. Electroanal. Chem. 96 (1979) 209.
41. J. M. Kern and P. Federlin, Tetrahedron Lett. (1977) 837.
42. P. Lockert and P. Federlin, Tetrahedron Lett. (1973) 1109.
43. H. W. van der Born and D. H. Evans, J. Am. Chem. Soc. 96 (1974) 4296.
44. M. A. Fox and R. C. Owen, J. Am. Chem. Soc. 102 (1980) 6559.
45. M. A. Fox and N. J. Singletary, J. Org. Chem. 47 (1982) 3412.
46. S. Banks, C. L. Ehrlich and J. A. Zubieta, J. Org. Chem. 44 (1979) 1454.
47. S. Banks, A. Schepartz, P. Grammatteo and J. Zubieta, J. Org. Chem. 48 (1983) 3458.
48. S. Banks, C. L. Ehrlich and J. A. Zubieta, J. Org. Chem. 46 (1981) 1243.
49. F. G. Bordwell and A. H. Clemens, J. Org. Chem. 46 (1981) 1035.
50. F. G. Bordwell and M. Bausch, private communication.
51. S. Banks, R. P. Marcantonio and C. H. Bushweller, J. Org. Chem. 49 (1984) 5091.
52. E. Ahlberg and V. D. Parker, Acta Chem. Scand. B 37 (1983) 723.
53. E. Aharon-Shalom, J. Y. Becker and I. Agranat, Nouv. J. Chim. 3 (1979) 643.
54. N. Acton, D. Hon, J. Schwarz and T. J. Katz, J. Org. Chem. 47 (1982) 1011.
55. D. Wilhelm, T. Clark and P. Schleyer, Chem. Commun. (1983) 211.
56. M. A. Fox and C. C. Chen, Chem. Commun. (1985) 23.
57. H. Dietrich, W. Mahdi, D. Wilhelm, T. Clark and P.v.R. Schleyer, Angew. Chem. Inter. Ed. 23 (1984) 621.
58. R. Breslow, R. Grubbs and S. I. Murahashi, J. Am. Chem. Soc. 92 (1970) 4139.
59. R. Breslow, D. Murayama, S. I. Murahashi and R. Grubbs, J. Am. Chem. Soc. 95 (1973) 6688.
60. G. L. Bitman, E. C. Perevolova, I. I. Skorokhodov and I. F. Myaskovskii, Zh. Fiz. Khim 43 (1969) 1290.
61. M. R. Wasielewski and R. Breslow, J. Am. Chem. Soc. 98 (1976) 4222.
62. M. A. Fox, Kabir-ud-Din, D. Bixler and W. S. Allen, J. Org. Chem. 44 (1979) 3208.
63. R. Breslow and K. Balasubramanian, J. Am. Chem. Soc. 91 (1969) 5182.
64. R. Breslow and S. Mazur, J. Am. Chem. Soc. 95 (1973) 584.
65. G. D. Luer and D. E. Bartak, J. Org. Chem. 47 (1982) 1238.
66. M. R. Feldman and W. C. Flythe, J. Org. Chem. 43 (1978) 2597.
67. J. M. Leal, T. Teherani and A. J. Bard, J. Electroanal. Chem. 91 (1978) 275.
68. C. G. Screttas, J. Chem. Soc., Perkin II (1975) 165.

69. A. Merz, R. Tomahogh and P. Iversen, Angew. Chem. 18 (1979) 938.
70. Y. Miura, A. Yamamoto and M. Kinoshita, Electrochim. Acta 12 (1984) 1731.
71. A. M. de P. Nicholas and D. R. Arnold, Can. J. Chem. 60 (1983) 2165.
72. B. A. Olsen and D. H. Evans, J. Am. Chem. Soc. 103 (1981) 839.
73. B. A. Olsen, D. H. Evans and I. Agranat, J. Electroanal. Chem., 136 (1982) 139.
74. T. Matsue, D. H. Evans and I. Agranat, J. Electroanal. Chem. 163 (1984) 137.
75. O. Hammerich and V. D. Parker, Acta Chem. Scand., B35 (1981) 395.
76. D. H. Evans and A. Fitch, J. Am. Chem. Soc. 106 (1984) 3039.
77. D. H. Evans and N. Xie, J. Electroanal. Chem. 133 (1982) 367.
78. D. H. Evans and N. Xie, J. Am. Chem. Soc. 105 (1983) 315.
79. T. Matsue and D. H. Evans, J. Electroanal. Chem. 168 (1984) 287.
80. D. H. Evans and R. W. Busch, J. Am. Chem. Soc. 104 (1982) 5057.
81. P. Neta and D. H. Evans, J. Am. Chem. Soc. 103 (1981) 7041.
82. A. J. Klein and D. H. Evans, J. Am. Chem. Soc. 101 (1979) 757.
83. J. Eriksen, K. A. Joergensen, J. Linderberg and H. Lund, J. Am. Chem. Soc. 106 (1984) 5083.
84. B. J. Huebert and D. E. Smith, J. Electroanal. Chem. 31 (1971) 333.
85. B. S. Jensen, A. Ronlan and V. D. Parker, Acta Chem. Scand. B 29 (1975) 394.
86. A. J. Fry, C. S. Hutchins and C. L. Ching, J. Am. Chem. Soc. 97 (1975) 591.
87. W. E. Britton, J. P. Ferraris and R. L. Soulen, J. Am. Chem. Soc. 104 (1982), 5322.
88. L. B. Anderson, J. F. Hansen, T. Kakihana and L. A. Paquette, J. Am. Chem. Soc. 93 (1971) 161.
89. B. S. Jensen, R. Lines, P. Pagsberg and V. D. Parker, Acta Chem. Scand. B 31 (1977) 707.
90. H. C. Wang, G. Levin and M. Szwarc, J. Am. Chem. Soc. 99 (1977) 2642.
91. E. Laviron and Y. Mugnier, J. Electroanal. Chem. 93 (1978) 69.
92. G. Mabon, G. LeGuillanton and J. Simonet, J. Electroanal. Chem. 130 (1981) 387.
93. S. P. Mulliken, Ann. Chem. J. 15 (1892) 523.
94. F. Adickes, W. Brunnert and O. Lucher, J. Prakt. Chem. 130 (1931) 163.
95. H. G. Thomas, M. Streukens and R. Peek, Tetrahedron Lett. (1978) 45.
96. R. Brettle and J. G. Parkin, Chem. Commun. (1967) 1352.
97. D. A. White, J. Electrochem. Soc. 124 (1977) 1177.
98. C. T. Bahner, Ind. Eng. Chem. 44 (1952) 317.
99. R. Bauer and H. Wendt, J. Electroanal. Chem. 80 (1977) 395.
100. J. L. Morgat and R. Pallaud, Compt. Rend. Acad. Sci. 260 (1965) 5579.
101. (a) H. Schaefer, Angew Chemie Inter. Ed. 20 (1981) 911; (b) H. Schaefer and A. A. Azrak, Chem. Ber. 105 (1972) 2398.
102. T. Chiba, M. Okimoto, H. Nagai and Y. Takata, J. Org. Chem. 44 (1979) 3519.
103. H. Schaefer, Chem. Ing. Tech. 42 (1970) 164.

104. S. Torii, K. Uneyama, T. Onishi, Y. Fugita, M. Ishiguro and T. Nishida, Chem. Lett. (1980) 1603.
105. T. Shono, "Electroorganic Chemistry as a New Tool in Organic Synthesis", Springer-Verlag, Berlin, 1984.
106. For reviews, see (a) A. K. Vijh and B. E. Conway, Chem. Rev. 67 (1967) 623; (b) J. H. P. Utley, in "Technique of Electroorganic Synthesis", N.L. Weinberg, ed., Wiley, New York, 1974, p. 793; (c) L. Eberson in "Chemistry of Carboxylic Acids and Esters", S. Patai, ed., Wiley, New York, 1969, p. 53.
107. B. C. L. Weedon, Quart. Rev. 6 (1952) 380.
108. L. Eberson and J. H. P. Utley, in ref. 3 p. 449.
109. K. Knolle and H. J. Schaefer, Angew. Chem., Inter. Ed. 14 (1975) 758.
110. E. J. Corey, N. L. Bauld, R. T. LaLonde, J. Casanova and E. T. Kaiser, J. Am. Chem. Soc. 82 (1960) 2645.
111. J. G. Traynham and J. S. Dehn, J. Am. Chem. Soc. 89 (1967) 2139.
112. T. L. Staples, J. Jagur-Grodzinski and M. Swarc, J. Am. Chem. Soc. 91 (1969) 3721.
113. D. Lelandais, C. Bacquet and J. Einhorn, Tetrahedron 37 (1981) 3131.
114. T. Imagawa, S. Sugita, T. Akiyama and M. Kawanisi, Tetrahedron Lett. 22 (1981) 2569.
115. H. H. Westberg and H. J. Dauben, Tetrahedron Lett. (1968) 5123.
116. T. Shono, J. Hayashi, H. Omoto and Y. Matsumura, Tetrahedron Lett. (1977) 2667.
117. J. H. P. Utley and G. B. Yates, J. Chem. Soc., Perkin II (1978) 395.
118. A. F. Vellturo and G. W. Griffin, J. Org. Chem. 31 (1966) 2241.
119. E. J. Corey and J. Casanova, J. Am. Chem. Soc. 85 (1963) 165.
120. J. E. Barry, M. Finkelstein, E. A. Mayeda and S. D. Ross, J. Am. Chem. Soc. 98 (1976), 8098.
121. R. N. Renaud, P. J. Champagne and M. Sauvard, Can. J. Chem. 57 (1979) 2617.
122. E. S. Wallis and F. H. Adams, J. Am. Chem. Soc. 55 (1933) 3838.
123. G. E. Hawkes, J. H. P. Utley and G. B. Yates, J. Chem. Soc., Perkin II (1976) 1709.
124. J. P. Coleman, R. Lines, J. H. P. Utley and B. C. L. Weedon, J. Chem. Soc., Perkin II (1974) 1064.
125. B. Krauetler and A. J. Bard, J. Am. Chem. Soc. 100 (1978) 239.
126. B. Krauetler, C. D. Jaeger and A. J. Bard, J. Am. Chem. Soc. 100 (1978) 4903.
127. M. A. Fox, Accts. Chem. Res. 16 (1983) 314.
128. H. Harada, T. Sakata and T. Ueda, J. Am. Chem. Soc. 107 (1985) 1773.
129. J. F. Ambrose and R. F. Nelson, J. Electrochem. Soc. 115 (1968) 1159.
130. A. J. Frank, K. Honda and A. Diaz, J. Photochem. 29 (1985) 195.
131. G. Cauquis and M. Gewies, Bull. Soc. Chim. Fr. (1967) 3220.
132. N. L. Weinberg and E. A. Brown, J. Org. Chem. 31 (1966) 4054.
133. N. Clauson-Kaas, F. Limborg and K. Glens, Acta Chem. Scand. 6 (1952) 531.
134. P. G. Gassman and B. L. Fox, J. Org. Chem. 32 (1967) 480.
135. H. Lund, Adv. Het. Chem. 12 (1970) 213.

136. H. Lund in ref. 3, p. 533.
137. J. M. Bobbitt, C .L. Kulkarni and J. P. Willis, Heterocycles 15 (1981) 495.
138. M. A. Fox, Chem. Rev. 79 (1979) 253.
139. M. A. Fox and R. C. Owen, Adv. Chem. Ser. 146 (1981) 337.
140. M. A. Fox and Kabir-ud-din, J. Phys. Chem. 83 (1979) 1800.
141. J. R. Hohman and M. A. Fox, J. Am. Chem. Soc. 104 (1982) 401.
142. P. V. Kamat, M. A. Fox and A. J. Fatiadi, J. Am. Chem. Soc. 106 (1984) 1191.
143. P. V. Kamat and M. A. Fox, Chem. Phys. Lett. 102 (1983) 379.
144. P. V. Kamat and M. A. Fox, J. Electrochem. Soc. 131 (1984) 1032.
145. H. Lund and H. S. Carlsson, Acta Chem. Scand. B32 (1978) 505.
146. H. S. Carlsson and H. Lund, Acta Chem. Scand. B34 (1980) 409.
147. M. von Stackelburg and W. Stracke, Z. Elektrochim. 53 (1949) 118.
148. S. Bank and D. A. Juckett, J. Am. Chem. Soc. 97 (1975) 567.
149. S. Banks and W. K. S. Cleveland, Heterocycles 18 (1982) 145.
150. P. K. Freeman and L. L. Hutchinson, J. Org. Chem. 45 (1980) 1924.
151. P. Fuchs, V. Hess, H. H. Hess and H. Lund, Acta Chem. Scand. B35 (1981) 185.
152. C. P. Andrieux, A. Merz, J. M. Saveant and R. Tomahogh, J. Am. Chem. Soc. 106 (1984) 1957.
153. A. J. Bard and A. Merz, J. Am. Chem. Soc. 101 (1979) 2959.
154. R. S. Abeywickrema and E. W. Della, J. Org. Chem. 46 (1981) 2353.
155. P. P. Infelta and R. H. Schuler, J. Phys. Chem. 76 (1972) 987.
156. K. Alwair and J. Grimshaw, J. Chem. Soc., Faraday Trans. 2 (1973) 1811.
157. R. Woods, Electrochim. Acta 15 (1970) 815.
158. A. Inesi, L. Rampazzo and A. Zeppa, J. Electroanal. Chem. 69 (1976) 203.
159. G. Silvestri, S. Gambino, G. Filardo, C. Greco and A. Gulotta, Tetrahedron Lett. 25 (1984) 4307.
160. E. Canadell, P. Karafiloglon, and L. Salem, J. Am. Chem. Soc. 102 (1980) 855.
161. T. Clark, Chem. Commun. (1984) 93.
162. D. D. Tanner, J. A. Plambeck, D. W. Reed and T. W. Mojelsky, J. Org. Chem. 45 (1980) 5177.
163. L. G. Feoktistov, in ref. 3, p. 259.
164. M. D. Hawley, in ref. 5, Vol. XIV.
165. G. Dabosi, M. Martineau and J. Simonet, J. Electroanal. Chem. 139 (1982) 211.
166. J. J. Barber and G. M. Whitesides, J. Am. Chem. Soc. 102 (1980) 239.
167. L. G. Feoktistov and A. S. Solonar, Zh. Obschch. Khim. 37 (1967) 986.
168. J. Zavada, J. Krupicka and J. Sicher, Coll. Czech. Chem., Commun. 28 (1963) 1664.
169. E. Stamm, L. Walder and R. Keese, Helv. Chim. Acta 61 (1978) 1545.
170. L. G. Feoktistov, M. M. Gol'din, V. R. Polishchuk and L. S. German, Elektrokhim. 9 (1973) 67.
171. U. Husstedt and H. J. Schaefer, Synthesis (1979) 964.
172. C. K. Mann, J. L. Webb and H. M. Walborsky, Tetrahedron Lett. (196) 2249.
173. J. T. Andersson and J. H. Stocker, in ref. 3, p. 905.

174. F. M. Triebe, K. J. Borhani and M. D. Hawley, J. Electroanal. Chem. 107 (1980) 375.
175. S. T. Nugent, M. M. Baizer and R. D. Little, Tetrahedron Lett. 23 (1982) 1339.
176. R. Gerdil, Helv. Chim. Acta 53 (1970) 210.
177. C. Amatore, J. Badoz-Lambling, C. Bonnel-Huyghes, J. Pinson, J. M. Saveant and A. Thiebault, J. Am. Chem. Soc. 104 (1982) 1979.
178. C. P. Andrieux, J. M. Saveant and D. Zarn, Nouv. J. Chim. 8 (1984) 107.
179. C. P. Andrieux, C. Blocman, J. M. Saveant and J. M. Dumas-Bouchiat, J. Am. Chem. Soc. 101 (1979) 3431.
180. V. D. Parker, Acta Chem. Scand. B35 (1981) 595; (1981) 655.
181. B. Aalstad and V. D. Parker, Acta Chem. Scand. B 36 (1982) 47.
182. C. Amatore, J. Pinson, J. M. Saveant and A. Thiebault, J. Am. Chem. Soc. 103 (1981) 6930.
183. C. Amatore, M. A. Otman, J. Pinson, J. M. Saveant and A. Thiebault, J. Am. Chem. Soc. 106 (1984) 6318.
184. a) C. Amatore, J. M. Saveant and A. Thiebault, J. Am. Chem. Soc. 104 (1982) 817; b) O. R. Brown and J. A. Harrison, J. Electroanal. Chem. 21 (1969) 387.
185. V. D. Parker, Acta Chem. Scand. B35 (1981) 533.
186. M. Tilset and V. D. Parker, Acta Chem. Scand. B 36 (1982) 311.
187. J. Grimshaw and S. A. Hewitt, Proc. Roy. Ir. Acad. B83 (1983) 93.
188. K. Boujlel, J. Simonet, G. Roussi and R. Beugelmanns, Tetrahedron Lett. 23 (1982) 173.
189. D. Lexa and J. M. Saveant, J. Am. Chem. Soc. 104 (1982) 3503.
190. J. Heinze, Angew. Chem., Internat. Ed. 23 (1984) 831.
191. A. J. Fry, M. A. Mitnick and R. G. Reed, J. Org. Chem. 35 (1970) 1232.
192. R. N. Gedye, Y. N. Sadana and R. Eng, J. Org. Chem. 45 (1980) 3721.
193. J. S. Mayell and A. J. Bard, J. Am. Chem. Soc. 85 (1963) 421.
194. J. Simonet, in ref. 3, p. 843.
195. J. H. Wagenknecht and M. M. Baizer, J. Org. Chem. 31 (1966) 3885; J. Electrochem. Soc. 113 (1967) 1095.
196. V. L. Pardini, L. Roullier and J. H. P. Utley, J. Chem. Soc., Perkin II (1981) 1520.
197. a) T. Shono, T. Akazawa and M. Mitani, Tetrahedron 29 (1973) 817; b) P. Beak and T. A. Sullivan, J. Am. Chem. Soc. 104 (1982) 4450.
198. P. E. Iversen, Tetrahedron Lett. (1971) 55.
199. M. M. Baizer and J. P. Anderson, J. Org. Chem. 30 (1965) 1351.
200. T. Troll and M. M. Baizer, Electrochim. Acta 20 (1975) 33.
201. S. Cheng and M. D. Hawley, J. Electroanal. Chem. 133 (1982) 57.
202. C. Nuntnarumit, F. M. Triebe and M. D. Hawley, J. Electroanal. Chem. 126 (1981) 145.
203. J. E. Barry, M. Finkelstein, W. M. Moore, S. D. Ross, L. Eberson and L. Jonsson, J. Org. Chem. 47 (1982) 1292.
204. T. Shono, S. Kashimura, K. Ishizaki and O. Ishige, Chem. Lett. (1983) 1311.
205. T. Shono, ref. 105, p. 158.
206. T. Shono, S. Kashimura and H. Nogusa, J. Org. Chem. 49 (1984) 2043.

207. F. M. Triebe, J. H. Barnes, M. D. Hawley and R. N. McDonald, Tetrahedron Lett. 22 (1981) 5145.
208. D. Bethell and V. D. Parker, J. Chem. Soc., Perkin II (1982) 841.
209. T. Shono, N. Kise and T. Suzumoto, J. Am. Chem. Soc. 106 (1984) 259.
210. M. M. Baizer, J. L. Chruma and D. White, Tetrahedron Lett. (1973) 5209.
211. T. Shono, S. Kashimura and K. Ishizaki, Electrochim. Acta 29 (1984) 603.
212. A. Inesi, A. Zeppa and E. Zeuli, J. Electroanal. Chem. 137 (1982) 103.
213. A. Inesi, L. Rampazzo and A. Zeppa, J. Electroanal. Chem. 122 (1981) 233.
214. I. Carelli, A. Inesi, M. A. Casedi, B. DiRienzo and F. M. Moracci, J. Chem. Soc., Perkin II (1985) 179.
215. S. Satoh, H. Suginome and M. Tokuda, Bull. Chem. Soc. Japan 56 (1983) 1791.
216. S. Satoh, H. Suginome and M. Tokuda, Bull. Chem. Soc. Japan 54 (1981) 3456.
217. B. F. Becker and H. P. Fritz, Justus Liebigs Ann. (1976) 1015.
218. A. Inesi, A. Zeppa and E. Zeuli, J. Electroanal. Chem. 126 (1981) 175.
219. H. Lund in ref. 3, Chapter 27, p. 873.
220. H. Lund in ref. 3, Chapter 28, p. 829.
221. J. P. Coleman and J. H. Wagenknecht, J. Electrochem. Soc. 128 (1981) 322.
222. R. A. Benkeser, E. M. Kaiser and R. L. Lambert, J. Am. Chem. Soc. 86 (1964) 5272.
223. D. Ginsburg and W. J. W. Mayer, US Patent 4,251,332 (CA 94:164824) 1981.
224. R. A. Misra, A. K. Yadav, Bull. Chem. Soc. Japan 55 (1982) 347.
225. F. M. Moracci, S. Tortorella, B. DiRienzo and I. Carelli Syn. Commun. 11 (1981) 329.
226. B. Czochralska and P. J. Elving, Electrochim. Acta 26 (1981) 1755.
227. H. Kasch and K. Ponsold, Chem. Abs. 95:81 (1981) 317.
228. T. Troll and M. M. Baizer, Electrochim. Acta 19 (1979) 951.
229. O. Lerflaten and V. C. Parker, Acta Chem. B 36 (1982) 193.
230. H. W. Sternberg, R. E. Markby, I. Wender and D. M. Mohilner, J. Am. Chem. Soc. 91 (1969) 4191.
231. R. A. Misra and B. L. Sharma, Electrochim. Acta 24 (1979) 727.
232. L. L. Miller and L. Christensen, J. Org. Chem. 43 (1978) 2059.
233. G. Horanyi, G. Inzelt and K. Torkas, J. Electroanal. Chem. 101 (1979) 101.
234. C. Amatore, M. Gareil and J. M. Saveant, J. Electroanal. Chem. 147 (1983) 1.
235. C. Amatore, M. Gareil and J. M. Saveant, J. Electroanal. Chem. 176 (1984) 377.
236. C. Amatore, J. Pinson and J. M. Saveant, J. Electroanal. Chem. 137 (1982) 143.
237. C. Amatore, J. Pinson and J. M. Saveant, J. Electroanal. Chem. 139 (1982) 193.
238. J. M. Saveant, Acta Chem. Scand. B 37 (1983) 365.
239. O. Hammerick and V. D. Parker, Acta Chem. Scand. B 37 (1983) 379.

240. V. D. Parker, Acta Chem. Scand. B 37 (1983) 163.
241. D. C. Walker, Anal. Chem. 39 (1967) 896.
242. G. Horanyi and M. Novak, Acta Chim. Sci. Hung. 75 (1973) 369.
243. V. G. Khomyakov, A. P. Tomilov and B. G. Soldatov Elektrokhim. 5 (1969) 850.
244. S. Ishiwata and T. Nozaki, J. Pharm. Soc. Japan 71 (1951) 1257.
245. J. H. Kargin, O. Manousek and P. Zuman, J. Electroanal. Chem. 12 (1966) 443.
246. D. Pletcher and R. Mohammad, Electrochim. Acta 26 (1981) 819.
247. D. H. Evans, in ref. 5, vol. XII, p. 3.
248. L. G. Feoktistov and H. Lund in ref. 3, Chapter 9.
249. M. M. Baizer and L. G. Feoktistov in ref. 3, Chapter 10.
250. C. Z. Smith and J. H. P. Utley, Chem. Commun. (1981) 492.
251. C. Z. Smith and J. H. P. Utley, Chem. Commun. (1981) 792.
252. M. Jubalt, E. Raoult and D. Peltier, Electrochim. Acta 26 (1981) 287.
253. T. Nonaka, Y. Kusayanagi and T. Fuchigami, Electrochim. Acta 25 (1980) 1679.
254. V. D. Parker, Acta Chem. Scand. B 37 (1983) 169.
255. V. D. Parker, Acta Chem. Scand. B 38 (1984) 189.
256. V. D. Parker and O. Lerflaten, Acta Chem. Scand. B 37 (1983) 403.
257. T. J. Curphey, C. W. Amelotti, T. P. Layloff, R. L. McCartney and J. H. Williams, J. Am. Chem. Soc. 91 (1969) 2817.
258. D. Seebach, H. A. Oei and H. Daum, Chem. Ber. 110 (1977) 2316.
259. W. J. M. Tilborg and C. J. Smit, Recl. Trav. Chim. 97 (1978) 89.
260. L. Horner and W. Brich, Liebigs Ann. Chem. (1977) 1354.
261. L. Mandrell, R. F. Daley and R. A. Day, J. Org. Chem. 41 (1976) 4087.
262. R. E. Sioda, B. Terem, J. H. P. Utley and B. C. L. Weedon, J. Chem. Soc., Perkin I (1976) 561.
263. G. Mabon, G. le Guillanton and J. Simonet, Chem. Comm. (1982) 571.
264. G. Mabon, G. le Guillanton and J. Simonet, Nouv. J. Chim. 7 (1983) 305.
265. G. Mabon and J. Simonet, Tetrahedron Lett. 21 (1984) 193.
266. J. Sarrazim, J. Simonet and A. Tallec, Electrochim. Acta 27 (1982) 1763.
267. D. H. Evans and D. A. Griffith, J. Electroanal. Chem. 136 (1982) 149.
268. P. Margaretha and V. D. Parker, Chem. Lett. (1984) 681.
269. O. Lerflaten, V. D. Parker and P. Margaretha, Monatsh. Chem. 115 (1984) 697.
270. C. Amatore, R. Guidelli, M. R. Moncelli and J. M. Saveant, J. Electroanal. Chem. 148 (1983) 25.
271. M. A. Orliac-LeMoing, J. Delaunay and J. Simonet, Nouv. J. Chim. 8 (1984) 217.
272. O. Lerflaten and V. D. Parker, Acta Chem. Scand. B 36 (1982) 225.
273. J. Delaunay, A. Lebond and G. le Guillanton, Electrochim. Acta 27 (1982) 287.
274. B. Aalstad and V. D. Parker, Acta Chem. Scand. B 36 (1982) 187.
275. V. D. Parker, Acta Chem. Scand. B 37 (1983) 393.
276. T. Shono, N. Kise, A. Yamazaki and H. Ohmizu, Tetrahedron Lett. 23 (1982) 1609.

277. J. DeLaunay, M. A. Orliac-LeMoing and J. Simonet, Chem. Comm. (1983) 820.
278. W. C. Still and A. Mitra, Tetrahedron Lett (1978) 2659.
279. H. O. House and P. D. Weeks, J. Am. Chem. Soc. 97 (1975) 2785.
280. W. Kemula, T. M. Krygowski, ref. 5, p. 78, 132.
281. H. Lund in ref. 3, Chapter 8.
282. G. A. Elenien, N. Ismail, J. Reiser and K. Wallenfels, Liebigs Ann. Chem. (1981) 1598.
283. L. Horner and D. H. Skaletz, Justus Liebigs Ann. Chem. (1975) 1210.
284. H. Lund, Adv. Heter. Chem. 12 (1970) 213; H. Lund and I. Tabakovic, Adv. Heter. Chem. 36 (1984) 235.
285. L. J. T. Janssen, E. Barendrecht, Electrochim. Acta 26 (1981) 1831.
286. W. H. Smith and A. J. Bard, J. Am. Chem. Soc. 97 (1975) 5203.
287. G. Gronchi, P. Courbis, P. Tordo, G. Mousset and J. Simonet, J. Phys. Chem. 87 (1983) 1343.
288. P. Martigny, J. Simonet and G. Mousset, J. Electroanal. 148 (1983) 51.
289. (a) W. R. Fawcett and A. Lasia, J. Phys. Chem. 82 (1978) 1114; (b) B. G. Chankan, W. R. Fawcett and A. Lasia, J. Phys. Chem. 81 (1977) 1476.
290. J. Marquez and D. Pletcher, Electrochim. Acta 26 (1981) 1751.
291. I. M. Sosonkin, G. N. Strogov, T. K. Ponomareva, A. N. Domarev, A. A. Glushkova and G. N. Freedlin, Khim. Get. Soedin (1981) 195.
292. M. Carrion, R. Hazard, M. Jubault and A. Tallec, Tetrahedron Lett. 22 (1981) 3961.
293. C. P. Maschmeier, H. Tanneburg and M. Matschiner, Z. Chem. 21 (1981) 219.
294. T. Shono, H. Hamaguchi, H. Mikami, H. Nogusa and S. Kashimura, J. Org. Chem. 48 (1983) 2103.
295. C. Bellec, P. Maitte, J. Armand and C. Viele, Can. J. Chem. 59 (1981) 527.
296. E. Buncel and B. C. Menon, J. Am. Chem. Soc. 102 (1980) 3499.
297. L. Eberson and J. H. P. Utley in ref. 3, Chapter 11, p. 375.
298. F. D. Popp and H. P. Schultz, Chem. Rev. 62 (1962) 19.
299. L. Eberson and K. Nyberg in ref. 8, Vol. XII, p. 261.
300. M. D. Birkett and A. T. Kuhn, Electrochim. Acta 19 (1974) 49.
301. J. C. Hoffmann, P. M. Robertson and N. Ibl, Tetrahedron Lett. (1972) 3433.
302. P. Carrahar and F. G. Drakesmith, Chem. Comm. (1968) 1562.
303. J. P. Coleman in "The Chemistry of Acid Derivatives, Supplement B", S. Patai, ed., Wiley, New York, 1979, p. 782.
304. (a) M. Hirayama, Bull. Chem. Soc. Japan 40 (1967) 1822; (b) R. E. Sioda and W. S. Koski, J. Am. Chem. Soc. 89 (1967) 475.
305. A. Guirado, F. Barba, C. Manzanera and M. D. Velasco, J. Org. Chem. 47 (1982) 142.
306. K. Praefcke, C. Weichsel, M. Falsig and H. Lund, Acta Chem. Scand. B 34 (1980) 403.
307. G. Capobianco, G. Farnia, M. G. Severin and E. Vianello, J. Electroanal. Chem. 165 (1984) 251; 136 (1982) 197.
308. L. Griggio, J. Electroanal. Chem. 140 (1982) 155.
309. J. Grimshaw, S. Moore, N. Thompson and J. Trocha-Grimshaw, Chem. Comm. (1983) 783.
310. J. Grimshaw, S. Moore and J. Trocha-Grimshaw, Acta Chem. Scand. B 37 (1983) 485.
311. U. Akbulut, L. Toppare and J. H. P. Utley, J. Chem. Soc., Perkin II (1982) 391.

312. M. D. Koppang, N. F. Woolsey and D. E. Bartak, J. Am. Chem. Soc. 107 (1985) 4692.
313. K. Itaya, M. Kawai and S. Toshima, J. Am. Chem. Soc 100 (1978) 5996.
314. T. Shono, Y. Matsumura, K. Tsubata and Y. Sugihara, J. Org. Chem. 47 (1982) 3090.
315. H. Lund and C. DeGrand, Chem. Rev. Heb. Seances C 287 (1978) 535.
316. K. Ankner, B. Lamm and J. Simonet, Acta Chem. Scand B 31 (1977) 742.
317. H. Lund, Electrochim. Acta 28 (1983) 395.
318. D. Bethell, L. J. McDowall and V. D. Parker, Chem. Comm. (1984) 308.
319. G. Mabon, C. Moinet and J. Simonet, Chem. Comm. (1981) 1040.
320. J. B. Kerr, L. L. Miller and M. R. Van De Mark, J. Am. Chem. Soc. 102 (1980) 3383.
321. J. H. Riley, D. W. Sopher, J. H. P. Utley and D. J. Walton, J. Chem. Res. Syn. 12 (1982) 326.
322. D. W. Sopher and J. H. P. Utley, Chem. Comm. (1981) 134.
323. P. Martiguy and J. Simonet, J. Electroanal. Chem. 81 (1977) 409.
324. P. Martiguy, M. A. Michel and J. Simonet, J. Electroanal. Chem. 73 (1976) 373.
325. G. Dabosi, M. Martineau and J. Simonet, J. Electroanal. Chem. 139 (1982) 211.
326. K. P. Butin, M. T. Ismail and O. A. Reutov, J. Organomet. Chem. 174 (1979) 157.
327. Y. Mugnier, C. Moise, J. Tirouflet and E. Laviron, J. Organomet. Chem. 189 (1980) C49.
328. N. El Murr, A. Chaloyard and E. Laviron, Nouv. J. Chim. 2 (1978) 15.
329. S. G. Davies, S. J. Simpson and V. D. Parker, Chem. Comm. (1984) 352.
330. J. P. Battion, D. Lexa, D. Mansuy and J. M. Saveant, J. Am. Chem. Soc 105 (1983) 207.
331. L. I. Denisovich, A. A. Ioganson, S. P. Gubin, N. E. Kolobova and K. N. Anisinov, Akad. Nauk. SSSR (1968) 258.
332. W. Sundermeyer and W. Verbekk, Angew Chem. Inter. Ed. 5 (1966) 1.
333. A. P. Tomilov and L. V. Kaabak, J. Appl. Chem. USSR 32 (1959) 2677.
334. N. E. Gunawardena and D. Pletcher, Angew Chem. 94 (1982) 786.
335. R. Wienkamp and E. Steckhan, Angew. Chem. 94 (1982) 786.
336. T. Lund and H. Lund, Acta Chem. Scand. B 38 (1984) 387.
337. J. Wellmann and E. Steckhan, Synthesis (1978) 901.
338. D. W. Sopher and J. H. P. Utley, Chem. Comm. (1979) 1087.
339. K. Boujlel, P. Martiguy and J. Simonet, J. Electroanal. Chem. 144 (1983) 437.
340. W. E. Britton and A. J. Fry, Anal. Chem. 47 (1975) 95.
341. H. Lund and E. Holboth, Acta Chem. Scand. B30 (1976) 895.
342. C. P. Andrieux, C. Blocman, J. M. Dumas-Bouchiat, F. M. Halla and J. M. Saveant, J. Am. Chem. Soc. 102 (1980) 3806.
343. C. P. Andrieux, A. Merz, J. M. Saveant and R. Tomahogh, J. Am. Chem. Soc. 106 (1984) 1957.
344. D. H. Evans and N. Xie, J. Electroanal. Chem. 133 (1982) 367.
345. P. Martiguy, J. Simonet, G. Mousset and J. Vigneron, Nouv. J. Chim. 7 (1983) 299.
346. J. Simonet, M. Carrion and H. Lund, Liebigs Ann. Chem. (1981) 1665.
347. J. Berthelot and J. Simonet, Electochim. Acta 29 (1984) 1181.

348. J. Berthelot, M. Jubauld and J. Simonet, Chem. Comm. (1982) 759.
349. J. Berthelot, M. Jubault and J. Simonet, Electrochim. Acta 28 (1983) 1719.

CHAPTER 3

STRUCTURES OF ORGANODIALKALI METAL COMPOUNDS ("DIANIONS")

ERLING GROVENSTEIN, JR.
School of Chemistry, Georgia Institute of Technology, Atlanta, Georgia

CONTENTS

I. INTRODUCTION 175
II. STRUCTURES FROM X-RAY DIFFRACTION 177
A. CRYSTAL STRUCTURES OF ORGANOLITHIUM COMPOUNDS 177
B. CRYSTAL STRUCTURES OF ORGANOSODIUM AND HEAVIER ALKALI METAL COMPOUNDS 189
III. THEORETICAL CONSIDERATIONS 192
A. SIMPLE ELECTROSTATIC MODELS 192
B. QUANTUM MECHANICAL CALCULATIONS 198
1. CALCULATIONS UPON COMPOUNDS OF KNOWN CRYSTAL STRUCTURE 199
2. CALCULATIONS UPON COMPOUNDS OF UNKNOWN STRUCTURE 203
IV. CONCLUSIONS 214
REFERENCES 215

I. INTRODUCTION

From a historical perspective, the subject of organodialkali metal compounds or "dianions" has sometimes been in bad repute. On the one hand the experimental difficulties in this area of pyrophoric compounds, unstable toward air, moisture, and, frequently, solvent (but sometimes unstable without solvent) are such that the compounds have in many cases not been purified but were characterized only by derivatization or by spectral properties. This state of affairs can lead to a misunderstanding of composition and structure.

For example, carbonation of amylsodium preparations with gaseous carbon dioxide can give as much as 50% yield of *n*-butylmalonic acid in addition to the expected *n*-caproic acid.[1] This result suggested that reaction of amyl chloride with sodium gave much amylidenedisodium ($C_5H_{10}Na_2$) in addition to amylsodium ($C_5H_{11}Na$). However subsequent work in which carbonation was effected by forcing the organosodium reagent onto crushed solid carbon dioxide gave _n_-caproic acid contaminated with only traces of _n_-butylmalonic acid.[2] The interpretation is that the dicarboxylic acid is a product of a secondary reaction according to equation (1).

$$RCH_2Na \xrightarrow{CO_2} RCH_2CO_2Na \xrightarrow{RCH_2Na} RCH(Na)CO_2Na \xrightarrow{CO_2} RCH(CO_2Na)_2 \quad (1)$$

In another example, the black solid from the reaction of benzene with cesium at 28°C was found to contain cesium in an amount near that expected for phenylcesium and was tentatively suggested to be phenylcesium.[3] A subsequent investigator reported that the black compound had the empirical composition of $C_6H_6Cs_6$ and was evidently a loose addition compound of benzene with cesium.[4] But was the "loosely bound" cesium distinguished in fact from unreacted cesium? More recent investigators, who measured the amount of cesium extracted from excess liquid Cs-K-Na alloy by benzene in tetrahydrofuran (THF), reported the initial formation of cesium benzenide (C_6H_6Cs) which dimerized and gave finally dicesium biphenylide ($C_{12}H_{10}Cs_2$).[5]

Modern spectrometric investigations can sometimes lead to questionable interpretations. The reaction of pyrene with sodium or potassium in THF gives, after several days at room temperature, a species whose ^1H and ^{13}C NMR spectra were interpreted to be those of a pair of unsymmetrical tetraanions (made unsymmetrical by the location of the four alkali metal cations).[6] A reinvestigation suggests that the proposed pyrene tetraanion is really 1-hydropyrenyl monoanion, _i.e._, the monoprotonated derivative of pyrene dianion.[7] The proton may have come from solvent impurities or likely the solvent itself since dianions of many aromatic hydrocarbons have been found to react with ethereal solvents at room temperature.[8]

Dianions have sometimes been thought to be theoretically unlikely species because of strong electron-electron repulsion brought about by having two units of negative charge on one molecule. This viewpoint fails to consider that the terms "carbanion" and "dianion" are but convenient abbreviations for the organoalkali metal compound. In the absence of a cation even the mononegative ions of molecules such as formaldehyde, 1,3-butadiene, benzene, and naphthalene are unstable with respect to autodetachment of the additional electron; these temporary anions, which can be detected in the gas phase by electron transmission spectroscopy, possess lifetimes of only some 10^{-12} to 10^{-15} seconds.[9]

It is the purpose of this review to show that many organodialkali metal compounds are now well established species from the viewpoint of both experiment and theory.

II. STRUCTURES FROM X-RAY DIFFRACTION

The structures of organoalkali metal compounds have frequently been deduced by derivatization and spectral properties. As noted above, these methods are not always infallible and, limited as they frequently are to the organic moiety, tell little about the role of the alkali metal in stabilization of the structure. X-ray diffraction provides the most comprehensive experimental method for determining structure of crystalline compounds.

(A) Crystal Structures of Organolithium Compounds

In understanding the structure of organolithium compounds it is necessary to recognize that they are ordinarily associated as $(RLi)_n$ with n commonly 2, 4, or 6 but rarely 1. Note, however, that the usual chemical nomenclature ignores the degree of association. As an example the crystal structure[10] of methyllithium reveals tetramer units, $(CH_3Li)_4$, with lithium atoms at the corners of a tetrahedron (**1**) at a Li-Li distance of 2.68 ±

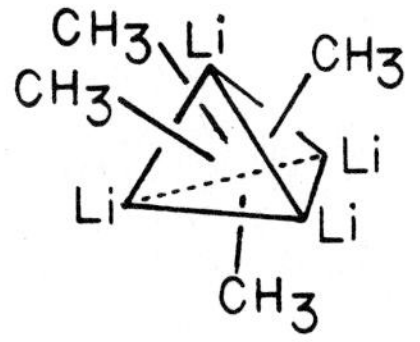

CH3 Li CH3
CH3 Li
Li Li
CH3

(**1**)

0.05 Å and methyl groups centered over the facial planes of the tetrahedron at a Li-C distance of 2.31 ± 0.05 Å. The methyl groups of each tetramer unit are quite close (Li-C, 2.36 ± 0.05 Å) to lithium atoms of adjacent tetramer units such that crystalline methyllithium is a non-volatile polymer.

Methyllithium and N,N,N',N'-tetramethylethylenediamine (tmeda) react with formation of $(CH_3Li)_4(tmeda)_2$. A single-crystal x-ray diffraction study[11] reveals that methyllithium tetramer units persist (Li-Li, 2.56-2.57 Å; Li-C, 2.23-2.27 Å) with these units now linked through Li-tmeda-Li bridges by coordination of lithium to nitrogen in a monodentate fashion. It is possible that the basic tetrahedral structure of methyllithium persists in solution in diethyl ether or tetrahydrofuran with coordination now of lithium to oxygen of the solvent; molecular weight studies[12] confirm a tetrameric state in these solvents. Schleyer[13], however, suggests that in sufficiently good solvents the tetrahedral structure may transform into a planar, eight-membered ring structure (**2**) which he calculates is only 12.4 kcal/mole

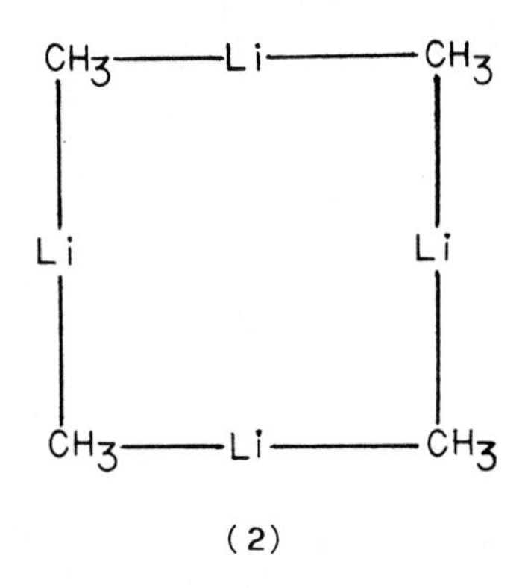

(**2**)

higher in energy than the tetrahedral structure for the isolated molecule in the vapor phase. Note that the cyclic structure has two open coordination sites per lithium while the tetrahedral structure has only one; hence solvation should tend to favor the cyclic array.

In another example, the crystal structure of phenyllithium etherate[14], $(PhLi{\cdot}Et_2O)_4$, reveals a tetrahedral arrangement approximately like that of (**1**) but with phenyl replacing methyl and an ether coordinated at the corners of the tetrahedron onto each lithium (average C-Li, 2.33 Å). With tmeda as solvate, however, a dimeric structure for $(PhLi{\cdot}tmeda)_2$ as in (**3**) is observed[15] with Li-C bonds of 2.21 and 2.28 Å. In (**3**) the bidentate ligand tmeda completes the tetrahedral coordination about each lithium. In cyclohexane-diethyl ether at 26°C phenyllithium appears to be tetrameric according to NMR studies[16] but in diethyl ether at ~ 60°C or in THF at room temperature it is largely dimeric.[17] With the tridentate ligand pentamethyldiethylenetriamine (pmdta), phenyllithium forms a compound of

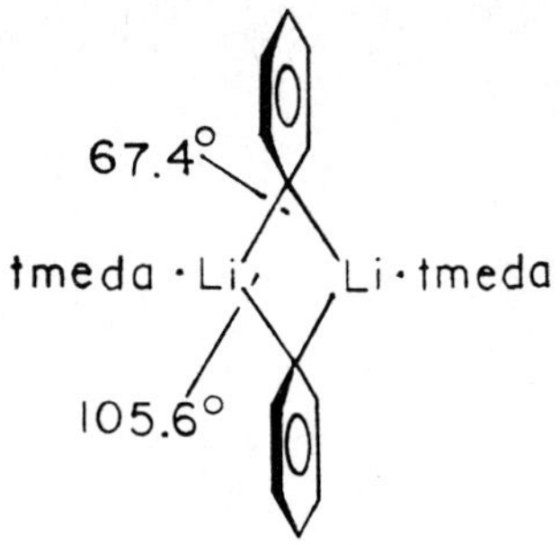

67.4°

tmeda · Li   Li · tmeda

105.6°

(3)

composition PhLi·(pmdta) whose crystal structure[18] reveals that phenyllithium is now monomeric with a 2.14 Å Li-C bond and with the three nitrogens of pmdta completing the coordination sphere of lithium. Note that with increasing numbers of oxygen or nitrogen atoms as ligands for lithium, the carbon-lithium bond becomes shorter. This fact suggests that the last carbon-lithium bond will be hardest to break. In fact no crystalline organolithium compounds are known in which a lithium cation is completely separated by ligands from a carbanion with localized negative charge; such a compound would be expected to self-destruct in view of the expected high reactivity of such a "naked" carbanion.

The crystal structure[19] of 2,2'-di(lithium-tmeda)biphenyl (**4**)

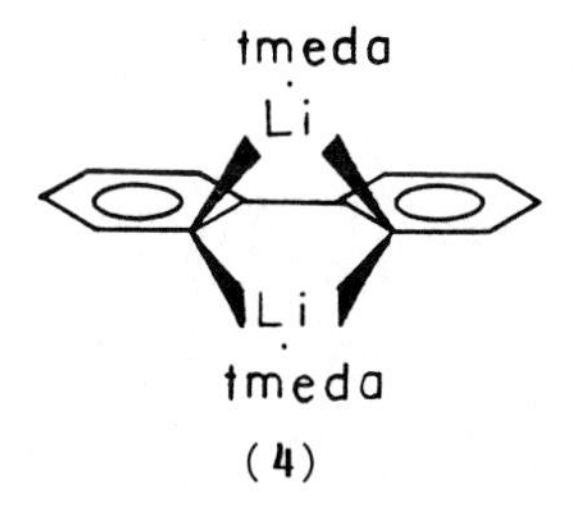

(**4**)

is remarkably similar to that of $(PhLi \cdot tmeda)_2$ (**3**); the Li-C bonds are 2.12 and 2.15 Å, somewhat shorter than in (**3**), and the Li-Li separation is 2.55 Å compared to 2.49 Å in (**3**). Thanks to the lithium cations, the negative charges of the two anionic centers of (**4**) are as effectively shielded from one another as in the dimer of phenyllithium.

The aromatic 10π electron dilithium pentalenide-dimethoxyethane complex, $C_8H_6Li_2 \cdot (DME)_2$, has the structure (**5**) as shown by X-ray analysis.[20] Structure (**5**) is somewhat similar to the

dilithium adduct of naphthalene whose crystalline tetramethylethylenediamine complex[21] has the structure (**6**). Some notable differences exist between (**5**) and (**6**). For the aromatic 10π electron system (**5**) the ring atoms are very nearly planar with approximately constant C-C bond lengths of 1.42 to 1.46 Å. For the anti-aromatic 12π electron system (**6**) carbons 1, 4, 5, and 8 are about 0.15 Å on either side of the average plane of the other ring atoms, being on the same side of the plane as the nearest lithium ion; the bond lengths vary from 1.34 Å for C(2)-C(3) to 1.45 for C(4a)-C(8a). The distortion of the naphthalene ring in the dianion from the planar arrangement of the neutral molecule is believed to reduce anti-aromaticity and is in the direction to enhance favorable electrostatic interaction between the lithium cations and the dianion. In dilithium pentalenide the lithium ions are almost centered over the five-membered rings but are slightly closer to C(1) and C(3) (average 2.22 Å) than to C(3a) and C(6a) (average 2.31 Å). In dilithium naphthalenide the lithium ions are not centered over the six-membered rings but are shifted toward the C(2)-C(3) and C(6)-C(7) bonds. Thus one lithium ion is about 2.26 Å from C(2) and C(3), 2.32 Å from C(1) and C(4), and 2.66 Å from C(4a) and C(8a). Since HMO theory predicts that the negative changes for the pentalenide are at C(1), C(3), C(4), and C(6) and for the naphthalenide at C(1), C(4), C(5), and C(8) and lesser at C(2), C(3), C(6) and C(7), the lithium ions are observed to be located near the center of negative charge on each 5- or 6-membered ring with the lithium ions on opposite sides of the dianions to reduce cation-cation repulsion while maximizing cation-anion attraction.

Bis(tetramethylethylenediaminelithium) anthracenide[22] has the structure (**7**) similar to the corresponding derivative of

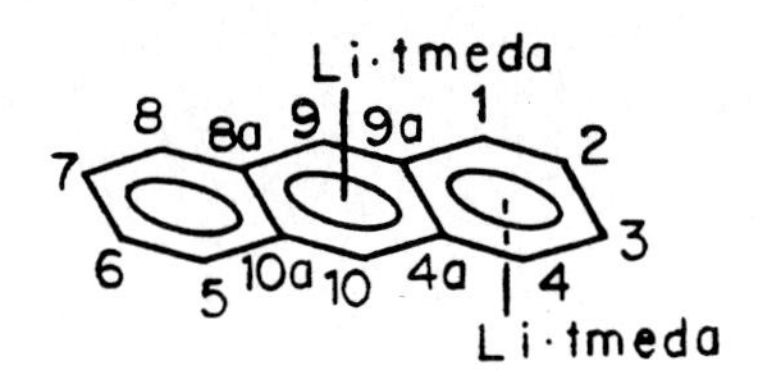

(**7**)

naphthalene (**6**). According to HMO calculations for the isolated dianion, the highest charge density should be at C(9) and C(10), the next highest at C(1) and C(4), followed by C(2) and C(3), with least at C(4a) and C(9a). Hence to maximize cation-anion interactions one lithium ion lies above the central ring (ring of highest charge density) at a distance of essentially 2.34 Å from C(9) and C(10) but slightly closer to C(8a) and C(10a) (2.45 Å) than to C(4a) and C(9a) (2.58 Å). The second lithium ion is below an adjacent ring some 2.39 Å from C(1) and C(4), 2.20 Å from C(2) and C(3), and 2.69 Å from C(4a) and C(9a). The lithium ions (likely aided by the anti-aromaticity of the 16π electron system) distort the aromatic rings from planarity. For the central ring, C(9) and C(10) move up 0.17 Å from the average plane of the other central ring carbon atoms toward the Li^+. For the outer ring complexed with Li^+, C(1) and C(4) move downward 0.08 Å from the plane of the other carbon atoms of the ring toward the Li^+; whereas for the outer ring not complexed with Li^+, the downward displacement of C(5) and C(8) is only some 0.02 Å. Likely the non-centrosymmetric location of Li^+ over the central ring results from repulsion with the second Li^+ which in turn is displaced toward C(2) and C(3); the latter displacement is aided by the higher charges at C(2) and C(3) than at C(4a) and C(9a).

Bis(tetramethylethylenediaminelithium) acenaphthylenide has the structure[23] (**8**). The five-membered ring has a larger negative charge density than the six-membered rings[20] and evidently binds both lithium cations tightly. The lithium cations are not quite centrosymmetric with respect to the five-membered ring but are closer to C(1), C(2), and C(2a) than to the remaining ring atoms. The acenaphthylene group is nearly planar.

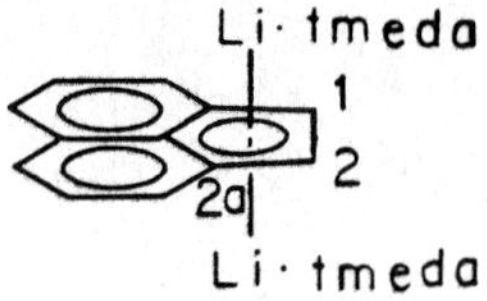

Li·tmeda
1
2
2a
Li·tmeda

(**8**)

The stilbene dilithium adduct has the general structure[24] (**9**)

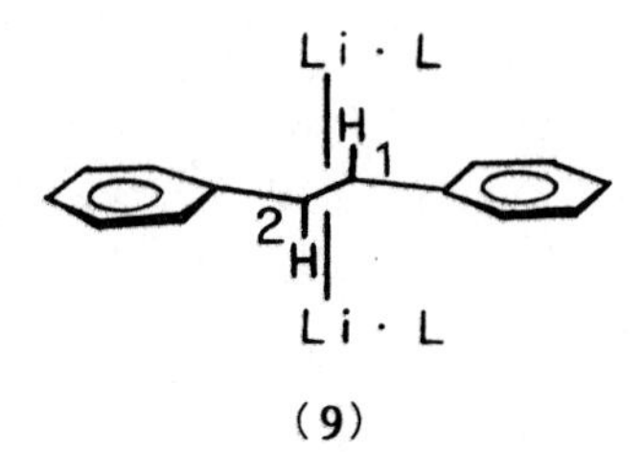

(**9**)

both when the ligand L is the bidentate tetramethylethylenediamine (tmeda) or the tridentate pentamethyldiethylenetriamine (pmdta). In both crystalline complexes the stilbene dianion moiety is planar. The average Li-C(1) or C(2) distance for the tmeda complex is 2.16 Å, for the pmdta complex 2.34 Å. For the tmeda complex the two lithium cations lie essentially on a line through the midpoint of the C(1)-C(2) bond and perpendicular to the stilbene plane, i.e., above and below the center of negative charge, but in the pmdta complex the lithium ions are somewhat displaced from this position (likely to give a more compact unit cell). The dipole of the lithium-pmdta fragment is directed toward the center of negative charge (between the two carbons of highest negative charge) as expected to maximize electrostatic attraction. The C(1)-C(2) bond has lengthened about 0.1 Å in the dianion compared to the same bond in trans-stilbene.

$\Delta^{9,9'}$-Bifluorene (**10**) forms a bis(lithium tetramethylethylenediamine) complex[25] which is rather similar to (**9**) except that the fluorenyl groups are twisted to each other by an angle of 48°, as as compared to 42° in the neutral hydrocarbon (**10**). The Li-C(9) and Li-C(9') distances are 2.29 Å, while the Li-C(10) and Li-C(13') distances are 2.55 and 2.57 Å. Again the C(9)-C(9') bond is increased 0.1 Å in the dianion compared to the neutral

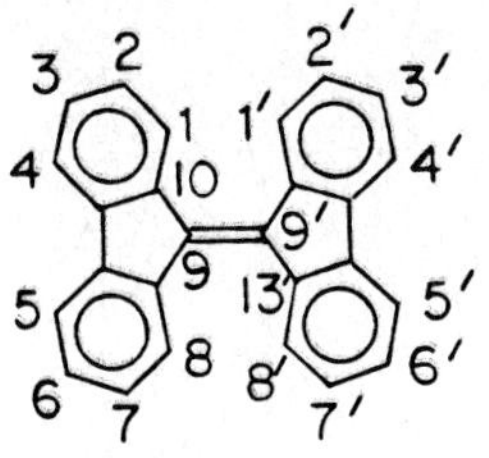

(10)

Bis(tetramethylethylenediaminelithium) 1,2-diphenylbenzocyclobutadienide exists in two crystalline modifications[26]. In one, (11), the lithium ions are approximately over and under the middle

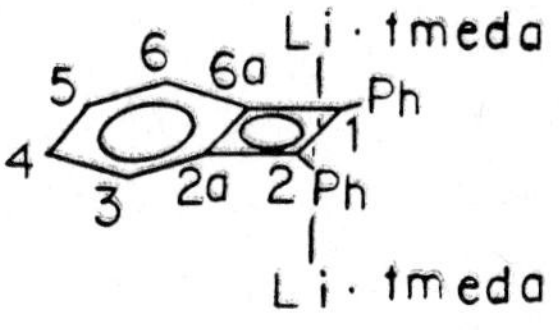

(11)

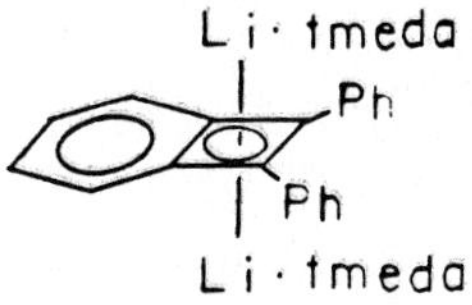

(12)

of the C(1)-C(2) bond, with Li-C(1) and Li-C(2) distances of 2.14-2.22 Å. In the other modification, (12), the lithium ions are approximately centrosymmetric with respect to the four-numbered ring, with Li-C distances of 2.23-2.37 Å. Evidently the potential energy surface for interaction of the lithium cation with the delocalized dianion is rather flat and several minima occur. This example serves as a warning that crystal packing effects can alter the location of the cation relative to the anion in crystalline organoalkali metal compounds.

Bis(tetramethylethylenediaminelithium) hexatrienide has structure (13) which is similar to that of dilithium pentalenide

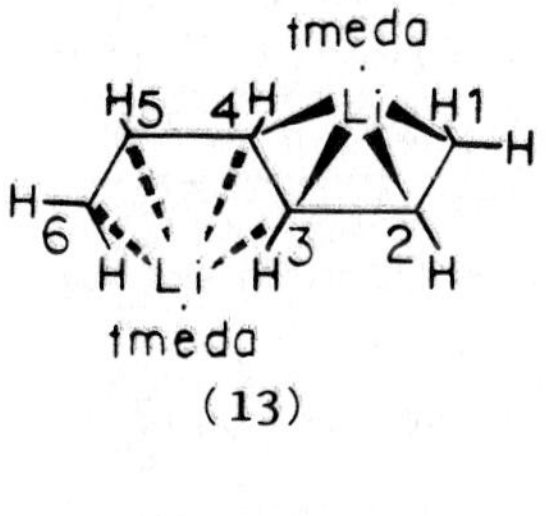

(13)

(**5**) and dilithium naphthalenide (**6**). Compound (**13**) has a center of symmetry at the midpoint of the C(3)-C(4) bond. All of the carbon atoms and four of the hydrogen atoms of the hexatriene moiety are within 0.02 Å of coplanarity. The internal hydrogen on C(1) and the hydrogen on C(3) are 0.87 and 0.17 Å respectively out of the plane, both bending away from the nearest lithium ion. By calculations the negative charge is largest on C(1), C(3), C(4), and C(6) and hence (**13**) is roughly approximated as two allyl anions joined by a single bond [C(3)-C(4) is 1.46 Å compared to 1.36 and 1.38 Å respectively for C(1)-C(2) and C(2)-C(3)]. In agreement the C-Li bonds are 2.27, 2.21, and 2.26 Å respectively for C(1), C(2), and C(3) but 2.40 Å for C(4).

Bis(tetramethylethylenediaminelithium) <u>trans</u>, <u>trans</u>-1,4-diphenylbutadienide has the structure[28] (**14**) in which lithiums "doubly bridge" the <u>cis</u> conformer of the dianion. The lithiums, however, have off-center locations somewhat as in (**13**). Thus the bottom lithium of (**14**) is closest to and interacts chiefly with

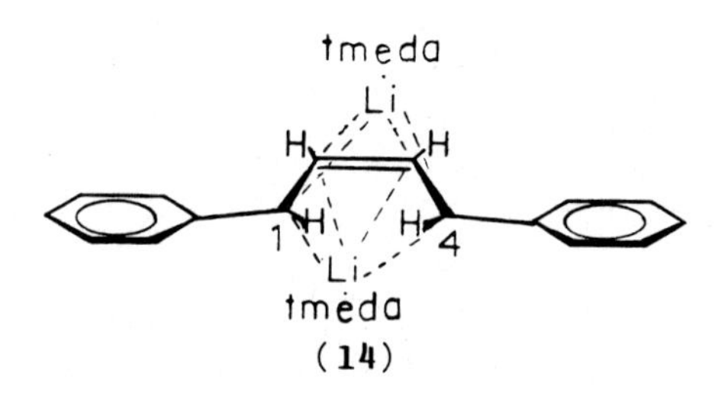

(**14**)

C(1), C(2), and C(3) and the top lithium with C(2), C(3), and C(4). The related [<u>o</u>-$C_6H_4(CHSiMe_3)_2$][Li(tmeda)]$_2$ has the nearly symmetrically bridged structure[29] (**15**) in which Li-C(α) distances

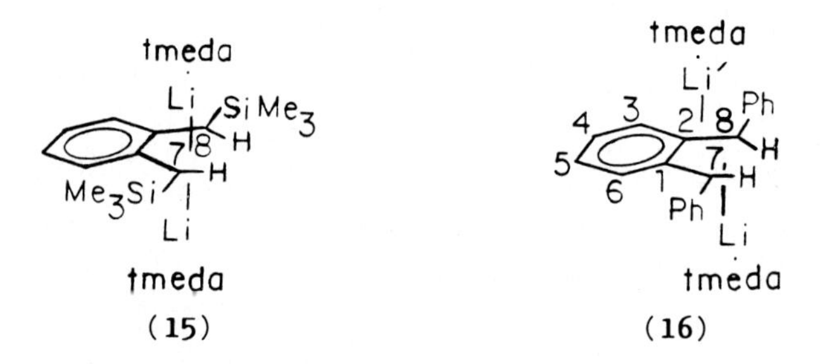

(**15**) (**16**)

are 2.34-2.41 Å and the Li-C(aromatic) are 2.32-2.42 Å. In contrast to (**14**) and (**15**), the dilithium compound from metalation of <u>o</u>-dibenzylbenzene has structure[30] (**16**) in which one lithium is coordinated to four carbon atoms [Li-C(7), 2.23 Å; Li-C(1), 2.36 Å; Li-C(2), 2.43 Å; Li-C(8), 2.40 Å] as in (**14**) and the other essentially to only three carbons [Li'-C(3), 2.59 Å; Li'-C(2), 2.22 Å; Li'-C(8), 2.19 Å] as in benzyllithium[31]. It has been suggested that the reason[32] for the different locations of lithium in (**14**), (**15**), and (**16**) is the difference in charge density at the terminal carbons C(1) and C(4) for (**14**) or C(7) and C(8) for (**15**) and (**16**). The compound (**15**) with largest charge density at the terminal carbons is symmetrically bridged by lithium. Compound (**14**) has an additional benzenoid ring for charge delocalization from the terminal carbons and is unsymmetrically bridged. Finally in (**16**) with yet an additional benzenoid ring for charge delocalization, one lithium is unsymmetrically bridged between the terminal carbons and the second lithium bridges a terminal benzylic carbon to the ortho position on a benzene ring; moreover the two benzylic groups in (**16**) tilt so as to orient the benzylic carbon orbitals toward the lithium cations.

The compound 3,6-di(lithio-THF)-2,2,7,7-tetramethyl-3,4,5-octatriene from lithium reduction of 1,4-di-t-butyldiacetylene exists as a dimer of structure[33] (**17**). The lithium atoms which

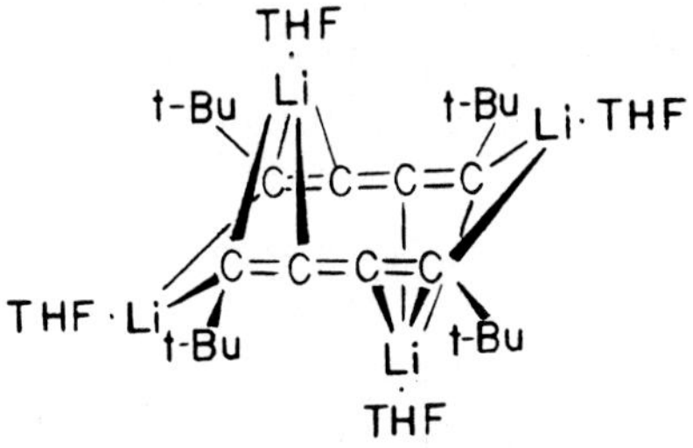

(**17**)

bridge the ends of the triene moieties have Li-C bond lengths of of 2.17-2.18 Å while the lithium atoms which bridge the "π bonds" have Li-C bonds of 2.19-2.24 Å. The lithium atoms at the ends of the triene are unusual in coordinating only three ligands. The coordination number of lithium, while frequently four, is governed primarily by steric factors;[34] the t-butyl groups of (**17**) likely prevent coordination of additional THF.

The dilithiation of dibenzylacetylene in presence of tetramethylethylenediamine gives[34] dilithiodibenzylacetylene-bis(tmeda) (**18**). This compound can be thought of as

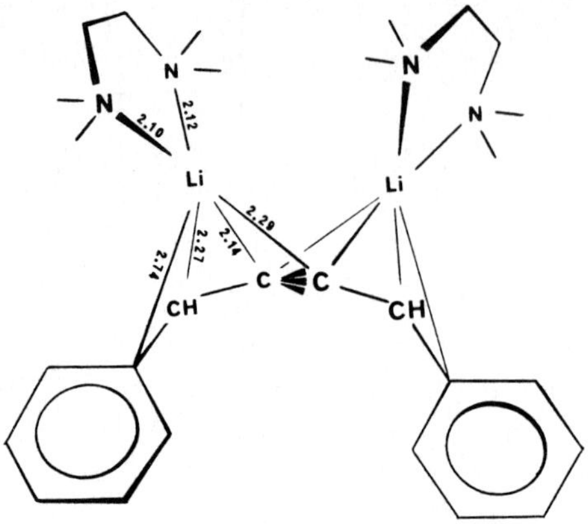

(18)

(From W. N. Setzer and P. v. R. Schleyer: Adv. Organomet. Chem. 24(1925) 390. Used by permission of the authors and the publisher Academic Press, Inc.)

consisting of two orthogonal allenyl anions bridged by lithium ions.

The dilithiation of 2,2,8,8-tetramethyl-3,6-nonadiyne (**19**)

$$\underline{t}\text{-Bu-C}\equiv\text{CCH}_2\text{C}\equiv\text{C-}\underline{t}\text{-Bu}$$

(**19**)

in diethyl ether gives a tetrameric complex which is reported[34] to have structure (**20**). This compound appears to be held together by lithium-acetylenic π interactions as well as by lithium-lithium interactions.

Bis(tetramethylethylenediaminelithium) tribenzylidenemethanide has the structure[35] (**21**) in which one lithium bridges C(3) and C(4) [Li-C(3), 2.23 Å; Li-C(4), 2.30 Å; Li-C(2), 2.29 Å]. On the opposite side the other lithium bridges the C(1)-C(2) bond [Li-C(1), 2.30 Å; Li-C(2), 2.34 Å] and, more weakly, interacts with the nearest <u>ortho</u>-carbon of the phenyl ring attached to C(4) [Li-(<u>o</u>-C), 2.67 Å] and with C(4) itself [Li-C(4), 2.63 Å]. The three carbon-carbon bonds of the trimethylenemethane moiety are no longer of equal length in (**21**) [C(1)-C(2), 1.39 Å; C(2)-C(3), 1.46 Å; C(2)-C(4), 1.44 Å]. The x-ray structure[36] of the related

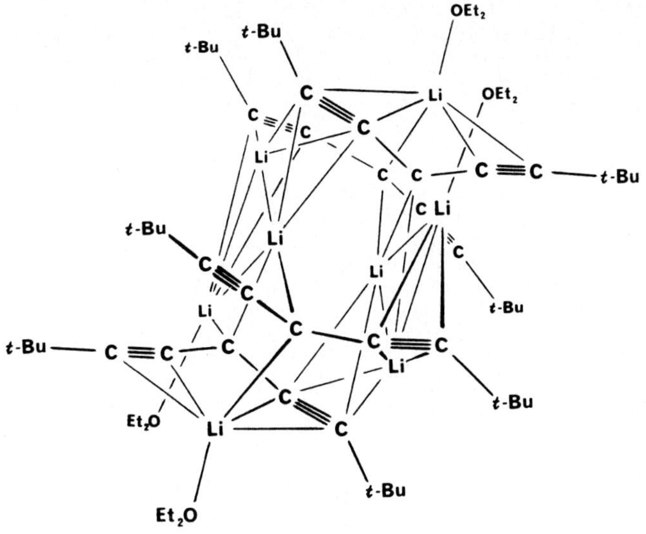

(20)

(From W. N. Setzer and P. v. R. Schleyer: Adv. Organomet. Chem. 24 (1985) 388. Used by permission of the authors and the publisher Academic Press, Inc.)

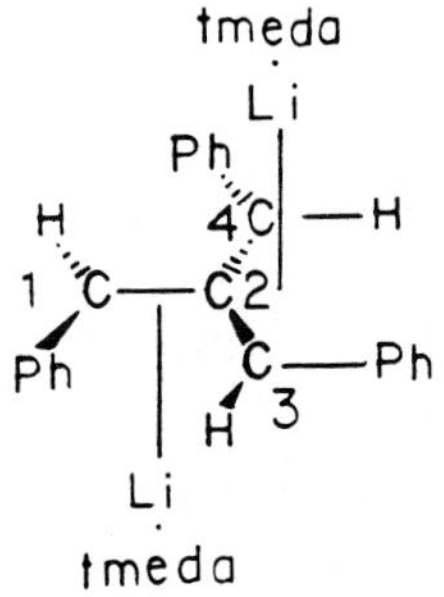

(21)

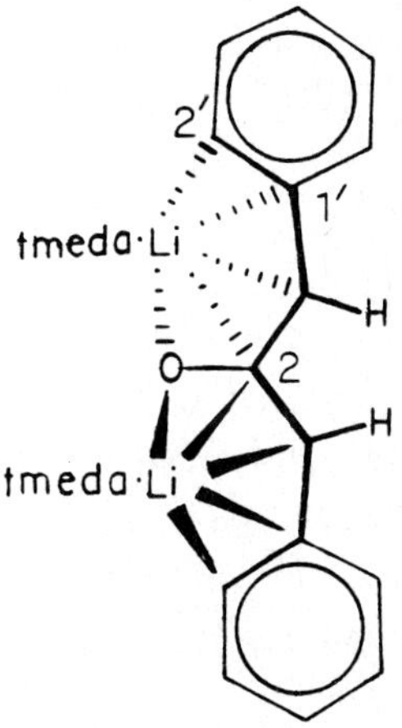

(22)

"Y-conjugated dianion", 1,3-dilithiodibenzylketone$\cdot$(tmeda)$_2$, is summarized in (22). The molecule has approximate C_2 symmetry with both phenyl groups twisted ca. 20° in opposite directions out of the plane of the central atoms. Each lithium is coordinated strongly to oxygen (Li-O, 1.86 Å) and is in close contact with four carbon atoms (Li-C, 2.42 to 2.73 Å).

The structure of the benzophenonedilithium complex[37] with tetrahydrofuran and tetramethylethylenediamine is shown schematically in (**23**). According to x-ray analysis this

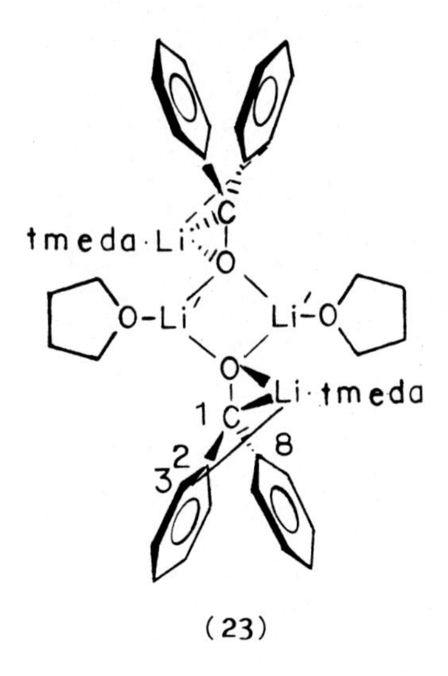

(**23**)

compound exists as a dimer with an inversion center through a planar four-membered ring consisting of the oxygens of two benzophenone moieties linked by lithium (Li'-O, 1.87 Å). The Li'-Li' distance, 2.45 Å, is short enough to suggest Li-Li bonding but, as in other cases[38], may be more apparent than real. The dihedral angle between the planar four-membered ring and the plane of the benzophenone moiety [OC(1)C(2)C(8)] is 110.4°. The O-Li(tmeda) bond length is 1.89 Å, C(1)-Li(tmeda) 2.24 Å, C(2)-Li(tmeda) 2.56 Å, and C(3)-Li(tmeda) 2.58 Å.

Dilithioferrocene-pentamethyldiethylenetriamine[38] is also dimeric as shown in (**24**). Again there is a crystallographic inversion center in the middle of a four atom group [C(1), Li,

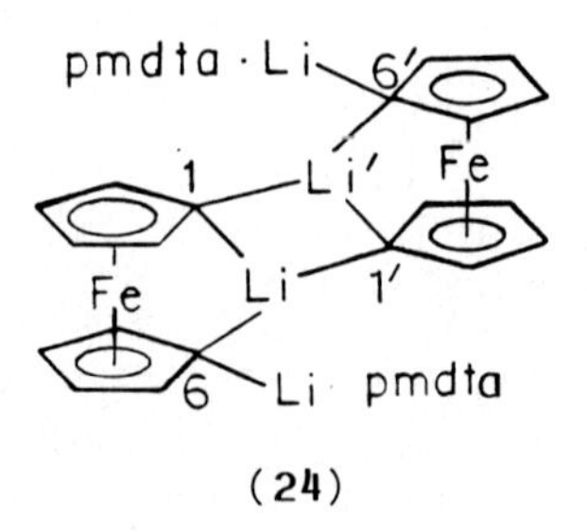

(**24**)

C(1'), Li'] which bridges the monomeric units. The carbon atoms C(1), C(6) and C(1') are 2.15, 2.12, and 2.19 Å from the bridging lithium Li in a roughly trigonal array. The Li-Li' distance is 2.37 Å, comparable to that (2.40 Å) in cyclohexyllithium.[39] It is thought that there is little Li-Li' bonding. Instead Li interacts with the nearest iron; the Li-Fe separation is 2.67 Å or about that expected for the sum of the covalent radii of the two metals.[38] The geometry of the lithium attached to the tridentate ligand pmdta is roughly tetrahedral with the C(6)-Li(pmdta) bond 2.11 Å.

(B) Crystal Structures of Organosodium and Heavier Alkali Metal Compounds

In contrast to organolithium compounds only a few of the heavier organoalkali metal compounds have had their crystal structures determined. Methylsodium[40] occurs in $(CH_3Na)_4$ units with Na at the corners of a tetrahedron and methyl on the face in a structure analogous to that of methyllithium (**1**) with a Na-Na distance of 3.15 ± 0.06 Å and Na-C of 2.61 ± 0.07 Å. Again the methyl groups of each tetramer unit are quite close (Na-C, 2.74 ± 0.04 Å) to sodium of adjacent tetramer units.

In contrast to methyllithium and methylsodium, methylpotassium has a NiAs-type crystal structure[41] in which potassium ions are surrounded in the lattice by an octahedral array of six methyl anions while the methyl anions are surrounded by six nearest potassiums in a trigonal-prismatic arrangement (**25**). The K-K distances are now 4.14 and 4.28 Å while the K-CH_3 (to center

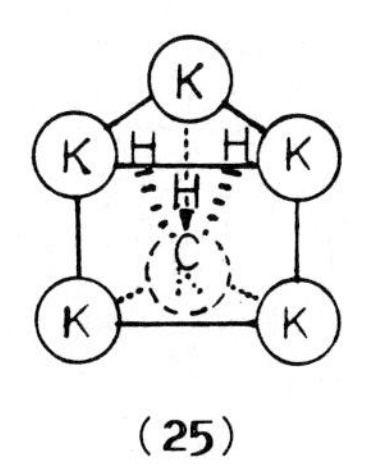

(**25**)

of gravity of the vibrating CH_3 group) is 3.22 Å (or 3.17 Å for the estimated[40] K-C distance). Methylrubidium and methylcesium have crystal structures[42] like that of methylpotassium. The Rb-Rb distances are 4.28 and 4.49 Å while the Rb-CH_3 is 3.36 Å; the Cs-Cs distances are 4.50 and 4.70 Å, the Cs-CH_3 3.53 Å.

A neutron powder-diffraction study[43] of disodium acetylide gave the $^{-}C{\equiv}C^{-}$ bond distance as 1.200 ± 0.06 Å or similar to that in CaC_2 (1.19 Å) and acetylene (1.20 Å). Each Na^+ ion is encompassed by six contacting C^- ions, two each at 2.62, 2.65, and 2.77 Å. Also each C^- has six Na^+ ions as closest neighbors. In monosodium acetylide[42] the $^{-}C{\equiv}C$ distance is 1.27 Å and the $Na^{+}\cdots{}^{-}C$ distances of closest approach are 2.62 and 2.71 Å.

Bis(potassium-diglyme) 1,3,5,7-tetramethylcyclooctatetraenide, $[K(CH_3OCH_2CH_2)_2O]_2[C_8H_4(CH_3)_4]$, has the structure[45] (**26**) with a

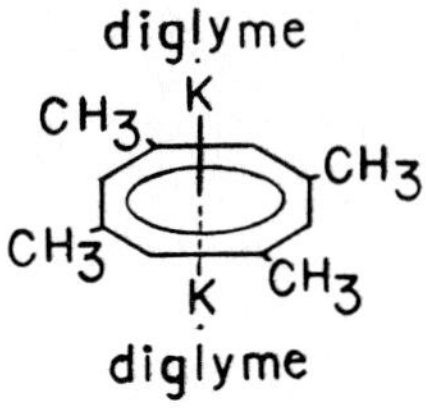

(**26**)

center of inversion through the middle of the eight-carbon ring which is essentially planar with all ring C-C bond lengths of 1.41 ± 0.02 Å. In contrast cyclooctatetraene itself is non-planar with alternating double and single bonds. Thus reaction with potassium has converted the non-aromatic $4n$ π electron hydrocarbon into an aromatic $2 + 4n$ dianion. In (**26**) the potassium nuclei are both 2.38 Å from the ring center or are 3.00 ± 0.03 Å from all the ring carbon atoms.

Dipotassium cyclooctatetraenide·diglyme, $K_2(C_8H_8)\cdot(CH_3OCH_2CH_2)_2O$, has the crystal structure[46] sketched in (**27**). The potassium ions lie on a crystallographic mirror plane with the cyclooctatetraenide ions and the diglyme ligands perpendicular to this plane. This structure is rather similar to (**26**) in that two potassium ions are centered one on either side of the cyclooctatetraenide ion; however only the potassium ions [K(2) of (**27**)] on the outside of the double column of cyclooctatetraenide ions are coordinated to diglyme. The potassium ions [K(1) of (**27**)] inside the double column are coordinated only to cyclooctatetraenide ions. In a $K_2(C_8H_8)$ unit the average K-C distance is 2.98 ± 0.02 Å for K(1) and 3.05 ± 0.02 Å for K(2). The shortest K-C distances of potassium ion to two other $K_2(C_8H_8)$ units are 3.26 ± 0.02 Å and 3.31 ± 0.02 Å for the distances

K(1')-C(1) and K(1')-C(4) respectively in (**27**). The C-C bond lengths in the cyclooctatetraenide ion are nearly constant at 1.40 ± 0.04 Å.

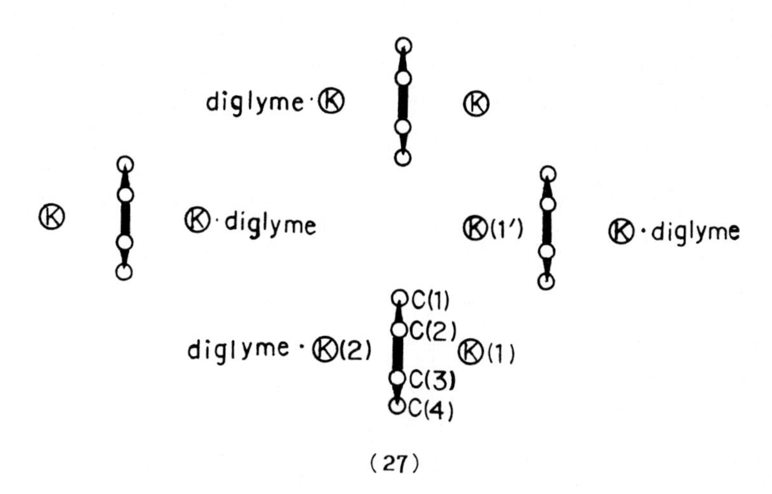

(**27**)

Dirubidium cyclooctatetraenide·diglyme,[47] $Rb_2(C_8H_8)\cdot(CH_3OCH_2CH_2)_2O$, is isomorphous with the potassium compound (**27**). In a $Rb_2(C_8H_8)$ unit the average Rb-C distance is 3.10 ± 0.02 Å for Rb(1) [numbering as in (**27**)] and 3.15 ± 0.02 Å for Rb(2). The shortest Rb-C distances of rubidium ion to two other $Rb_2(C_8H_8)$ units are 3.37 ± 0.01 Å for Rb(1')-C(1) and 3.43 ± 0.01 Å for Rb(1')-C(4). The cyclooctatetraenide ion is essentially a regular octagon with C-C bond lengths of 1.40 ± 0.03 Å.

Bis[tri(tetrahydrofuran)sodium] p-terphenylide, $[Na\cdot(THF)_3]_2(p\text{-}C_6H_5C_6H_4C_6H_5)$, has the structure[48] (**28**). The sodium ions are

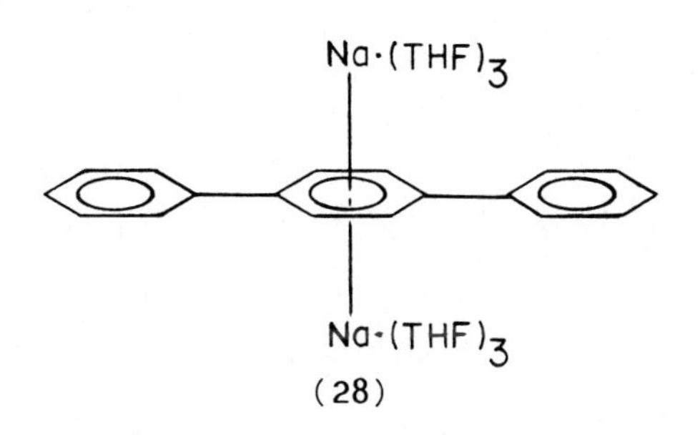

(**28**)

situated above and below the center of the middle ring of the dinegative terphenylide ion at a distance of 2.43 Å. Each sodium ion is further coordinated to three tetrahydrofuran molecules whose oxygen atoms lie at the corners of an equilateral triangle whose plane is approximately parallel to the plane of the central ring. The average Na-O distance is 2.30 Å. Compound (**28**) is regarded as a tight ion pair of two externally solvated sodium ions in contact with a dinegative terphenylide ion. It is of interest that the two sodium ions lie above and below the center of negative charge of the *p*-terphenylide anion and moreover, according to HMO calculations, the central ring bears a larger negative charge than the terminal phenyl rings of the *p*-terphenylide anion.

III. THEORETICAL CONSIDERATIONS

(A) Simple Electrostatic Models

Pauling[49] gives the electronegativity of Li, Na, K, Rb, and Cs as 1.0, 0.9, 0.8, 0.8, and 0.7 respectively and the electronegativity of carbon as 2.5. On the basis of Pauling's curve relating electronegativity difference and amount of ionic character, the percentage of ionic character increases from 43% to 50% on going from Li-C to Cs-C bonds. Pauling notes that estimates such as this can have only approximate significance. Royer[50] suggests that if the electronegativity difference between two elements is greater than about 1.5 units, the bond between them should "probably be considered essentially ionic." In this spirit the author has described the X-ray diffraction structures of organoalkali metal compounds in the previous section as containing alkali metal cations and organic anions. Moreover, as has been noted in the previous section, the packing of cations and anions in the crystal lattice of organoalkali metal compounds can frequently be understood in terms of simple electrostatic interactions between ions but the predictions of simple electrostatic calculations are, of course, dependent upon the type of calculation performed and are not always accurate, especially for organolithium compounds.[23-25]

Pauling's percent ionic character was part of his qualitative valence-bond description of molecules and was taken as a measure of the contribution of ionic structure (**30**) versus the covalent structure (**29**). Contributions from structures such as (**30**) give

$$\underset{(\mathbf{29})}{\text{R-M}} \longleftrightarrow \underset{(\mathbf{30})}{\text{R:}^{-}\ \text{M}^{+}}$$

added stability to the bonding of the metal M to carbon. Moreover purely ionic bonding may be as strong as covalent bonding.

Streitwieser[51] has proposed an electrostatic model of methyllithium tetramer consisting of a collection of four positive and four negative point changes arranged as two interpenetrating tetrahedra subject only to Coulombic forces [see the arrangement of Li and CH_3 in (**1**)]. He was able to demonstrate that the electrostatic energy of this model has a minimum value when the ratio of the two tetrahedral edges was 0.783. In the X-ray structure of methyllithium tetramer reported by Weiss and Henchen[10] the edges of the interpenetrating tetrahedra are 2.68 ± 0.05 Å for Li-Li and 3.68 ± 0.05 Å for CH_3-CH_3 or a ratio of 0.73 ± 0.02. Hence Streitwieser's model fits the experimental results reasonably well especially if the unit negative charge of the pyramidal methyl anion is not centered at the carbon nucleus but 0.26 Å from carbon. This is obviously not a complete model;[52] only the ratio of the edges was calculated not the length of the edges themselves. In this model there is nothing to prevent the collection of point charges from collapsing to a point!

In another calculation highly pertinent to the topic of organodialkali metal compounds, Streitwieser and Swanson[53] note that in an ion triplet consisting of a dianion and two alkali metal cations the geometry of the triplet is likely to be as in (**31**). The electrostatic energy (E) in hartrees of this assembly

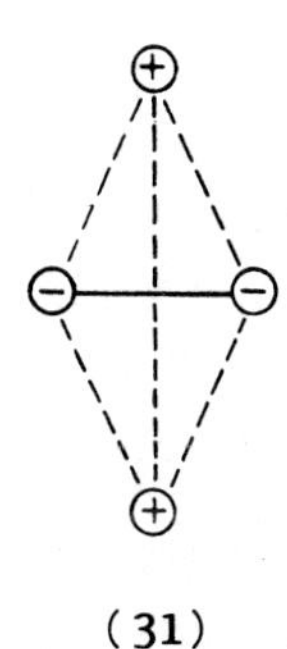

(**31**)

of point charges is given by equation (2):

$$E = 1/R_{--} + 1/R_{++} - 4/R_{+-} \quad (2)$$

where R_{--}, R_{++}, and R_{+-} are the distances between negative charges, positive charges, and positive to negative charges respectively (in bohr units). This simple Coulomb treatment shows that the proximity of charges in an ion triplet of this type provides stabilization provided that the ratio of R_{--}/R_{++} ranges from 0.145 to 6.91. From eq. (2) it may be demonstrated the electrostatic stabilization is at a maximum for the ratio of R_{--}/R_{++} equal to 0.577. The range of R_{--}/R_{++} ratios for which net electrostatic stabilization is calculated covers a wide range of chemically significant structures modeled after (**31**), *e.g.* (**3**), (**8**), (**9**), (**10**), (**11**), (**12**), (**15**), (**26**), (**27**), and (**28**) in the previous section. Thus the belief that organodialkali metal compounds are unlikely species because of strong electron-electron repulsion in a "dianion" neglects the strong electrostatic stabilization[54] provided by the cation in an ion triplet such as (**31**).

We have noted that ion triplets sometimes occur in a less symmetrical form than (**31**). The structure may approximate (**32**)

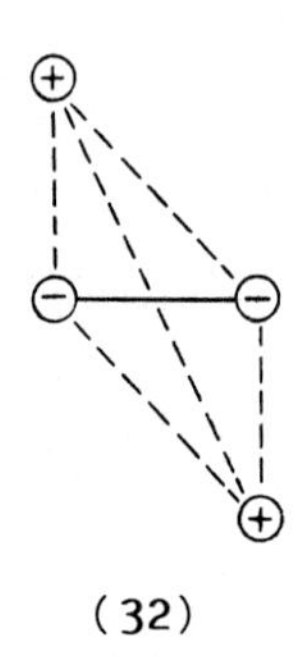

(**32**)

as in compounds (**5**), (**6**), (**7**), (**13**), (**21**), and (**22**). For this structure the electrostatic energy is given by equation (3),

$$E = \frac{1}{R_{--}} + \frac{1}{R_{++}} - \frac{4}{\sqrt{R_{++}^2 - R_{--}^2}} - \frac{4}{\sqrt{3R_{--}^2 + R_{++}^2}} \quad (3)$$

where the quantities are defined as in equation (2). For values of R_{--}/R_{++} from +0.146 up to 1 the proximity of charges provides stabilization; however as R_{--}/R_{++} approaches 1, E approaches negative infinity, (the positive charges coalesce with the negative charges). Obviously model (**32**) can provide greater stabilization than (**31**) at values of R_{--}/R_{++} near unity.

The change from model (**31**) to (**32**) can be understood by considering (**33**). When the parameter X is set equal to zero, R_{++} becomes $2 \cdot Z$ and the model corresponds to (**31**) (dotted position of the cations). When X is equal to S, the model corresponds to (**32**). Model (**33**) is drawn for an intermediate value of X. The

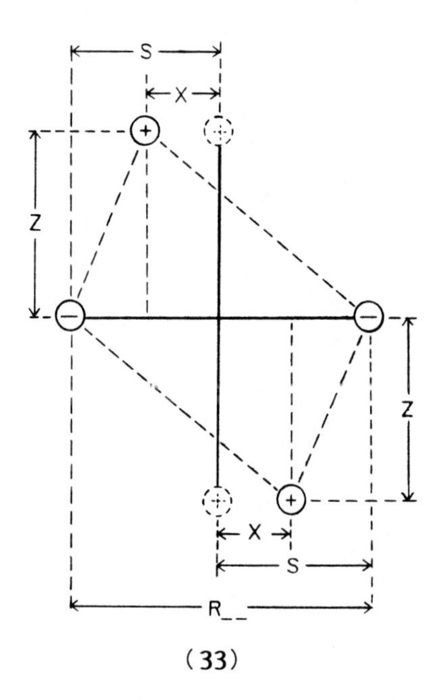

(**33**)

general equation[55] for the electrostatic energy (E) of (**33**) is (4):

$$E \cdot Z = \frac{1}{2\sqrt{1 + \left(\frac{X}{Z}\right)^2}} + \frac{1}{2\left(\frac{S}{Z}\right)} - \frac{2}{\sqrt{1 + \left(\frac{S-X}{Z}\right)^2}} - \frac{2}{\sqrt{1 + \left(\frac{S+X}{Z}\right)^2}} \quad (4)$$

In equation (4) all of the parameters are scaled to the value of Z, which corresponds to the distance of the two positive charges from the line passing through the negative charges (in the real three dimensional model, Z is the distance of the cations from the plane of the dianion and is considered to be essentially a fixed value corresponding to the nearest approach of cation to dianion as allowed by the van der Waals radii of the ions in a contact ion pair). For a fixed value of Z and a chosen value of

R_{--}/Z, equation (4) may be solved for the value of X/Z which gives the minimum value of E·Z (note that $R_{--} \equiv 2S$). The so calculated values of X/Z at various R_{--}/Z are displayed as the lower curve in FIGURE 1 (left ordinate). Note that as the negative charges are separated (at constant Z), X/Z is at first zero, a result corresponding to model (**31**), and remains zero

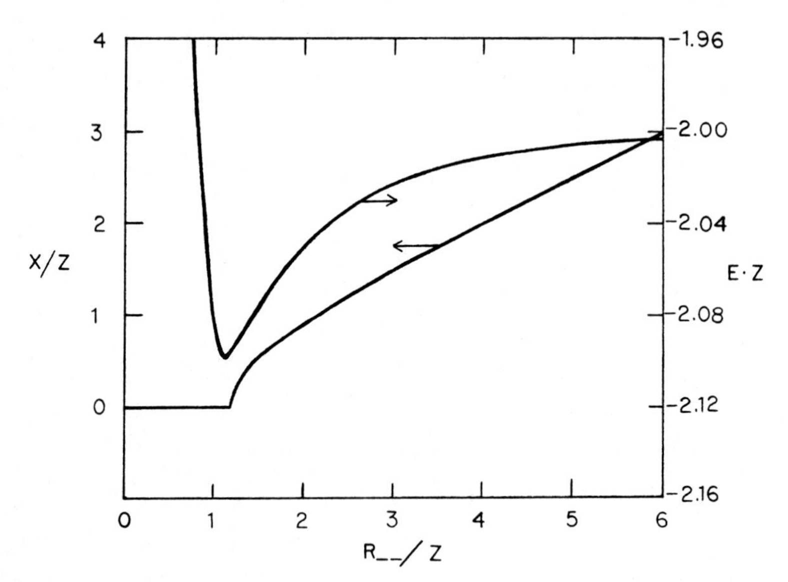

FIGURE 1. Plot of values of X/Z (lower curve) calculated for minimum values of E·Z (upper curve) for a given value of R_{--}/Z from solution of equation (4) for model (**33**).

through R_{--}/Z equal to 1.207. From this point X/Z begins to increase until, at large values of R_{--}/Z, X equals $(1/2)R_{--}$, a result corresponding to model (**32**). The change from model (**31**) to (**32**) after the transition point of 1.207 occurs rather rapidly. Thus at R_{--}/Z equal to 2, X equals 90% of $(1/2)R_{--}$; at R_{--}/Z equal to 3, X equals 98% of $(1/2)R_{--}$. The corresponding calculated minimum values of E·Z are shown in the upper curve of FIGURE 1 (right ordinate). It is of interest that the minimum in the plot of E·Z versus R_{--}/Z occurs at R_{--}/Z equal to 1.154 or in the region of model (**31**) just before the transition toward model (**32**) begins to occur.

According to FIGURE 1, the electrostatic stabilization energy, E·Z, rapidly decreases as R_{--}/Z goes from 1.0 to smaller values [this corresponds to moving the negative charges closer together in (**31**)]. For example at R_{--}/Z equal to 0.80, $E \cdot Z$ is now -1.96 or the system is less stable than two ion pairs separated at large distance [for model (**33**) as $R_{--}/Z \rightarrow \infty$, $E \cdot Z \rightarrow -2.00$]. Indeed the calculated additional electrostatic stability of the ion triplet over two well separated ion pairs (at the same value of Z) is at most only 4.9% at the minimum in the curve of $E \cdot Z$ versus E_{--}/Z. Moreover this minimum in the plot of $E \cdot Z$ is very shallow (the $E \cdot Z$ scale is much expanded). Other calculations show that models (**31**) and (**32**) do not differ more than 5% in electrostatic energy from R_{--}/Z equal to 1.0 to 1.7 and have identical energies at R_{--}/Z equal 1.47. Hence in this chemically interesting region factors other than those included in the simple electrostatic calculation may determine whether the observed geometry is similar to that shown in (**31**), (**32**), or (**33**). Thus dilithium pentalenide (**5**) and dilithium naphthalenide (**6**) (if one can assume that the negative charge is centered in the middle of the five- and six-membered rings) have R_{--}/Z of about 1.0 and 1.3 respectively and yet both resemble model (**32**). As might be expected from FIGURE 1, the stilbene dilithium adduct (**9**) with R_{--}/Z of 0.68 (if the negative charge is considered to be located on the two ethylenic carbon atoms) has the general structure of model (**31**). Bis(tetramethyl-ethylenediaminelithium) trans, trans-1,4-diphenylbutadienide has the unsymmetrically bridged structure (**14**) similar to that of model **33**. Finally it should be pointed out that the assumption of a constant value of Z in all of the present systems is not entirely justified; thus in dilithium naphthalenide (**6**) the distance from lithium to an average plane of the naphthalene ring is 1.96 Å whereas in bis(lithium-tmeda) stilbenide (**9**) the distance from lithium to the plane of the stilbene moiety is 2.03 Å. A similar shortened distance of 1.95 Å applies to lithium above the five-membered ring of acenaphthylene in (**8**). This shortened distance of lithium to the plane of five- and six-membered aromatic rings corresponds to lithium "falling into the hole" of the middle of such ring systems and may be a determining factor in the structure of compounds such as (**5**) and (**6**).

Bushby and Tytko[56] have performed calculations for benzyllithium and related compounds with a more sophisticated hard sphere electrostatic model in which a computer is programed to effectively "roll" the lithium cation over the surface of the anion and calculate the electrostatic binding energy at each position from the sum of the electrostatic interactions between the cation (as a point charge) and each nucleus of the anion (whose charge distribution is calculated *a priori* by some quantum mechanical method and then treated as a point charge centered on each nucleus). From such calculations the position of minimum electrostatic energy may be calculated. These calculations suggested that the lithium cation was either approximately centered above the benzene ring or bridged between the *ortho*- and *alpha*-carbon atoms of the benzyl anion. The latter is the X-ray result[57]. Charges from an STO-3G calculation gave the former result while HMO and CNDO charges gave the latter. These calculations emphasize the importance of the charge distribution in the anion. Since, however, the charge distribution in an anion is expected to vary with the location of the cation, correct quantum mechanical calculations must include the cation and should also lead to the optimum location of the cation. Hence simple electrostatic calculations are at most only an aid to our understanding the structures of organodialkali metal compounds. This is not a trivial utility since the more accurate the quantum mechanical calculation, the more difficult it is to interpret the results of the calculation, to paraphrase Mulliken.[58]

B. Quantum Mechanical Calculations

Naively it might have been thought that a molecular orbital calculation which was done with sufficient care and precision to give the correct dipole moment and correct spatial electron density distribution for a simple molecule such as monomeric methyllithium would allow one to answer with precision the question, raised by Pauling's valence bond approach, of the amount of charge separation in the carbon-lithium bond. However this is NOT the case; even imprecise answers to this question are difficult to defend. Recent answers[59,60] range from 0.44 to 0.8 electron for the charge separation along the C-Li bond. While electrons can be assigned with confidence to isolated atoms, no generally agreed upon method of assigning electrons to atomic centers in molecules exists; all such assignments are necessarily arbitrary. Thus the usual Mulliken population analysis[61] is basis-set dependent and gives values of +0.56 to +0.44 for the charge on lithium in methyllithium monomer, yet the corresponding dipole moments from these molecular orbital calculations are 5.6 and 5.40 or essentially the same.[60]. A different method of allocating electrons to atomic centers is to slice the electron density projection diagram at the minimum between the connecting atoms. On this basis the charge on lithium in methyllithium monomer is estimated at +0.8 and the bonding is described as "largely ionic" by Streitwieser and co-workers.[59] A similar but better mathematically defined method of allocating coordinate space to atoms in molecules is given by the topology of the molecular charge distribution with the condition that the surface bounding the atom not be crossed by any gradient vector of the charge density.[62] This procedure gives the charge on lithium in methyllithium monomer as +0.90.[63] From the viewpoint of modern quantum mechanics, the definitions of "covalency" and "ionicity" are so imprecise that these terms are useful only in qualitative discussions.

(1) Calculations upon Compounds of Known Crystal Structure

A triumph of molecular orbital theory is the calculation of structures of organolithium compounds frequently in agreement with the results of x-ray diffraction.[64] Indeed computer time-efficient MNDO calculations[65] (semi-empirical calculations which require parameterization) often give good geometries. It is thought that the high ionic character of organolithium compounds makes them favorable subjects for these calculations.[64]

For example Kos and Schleyer[66] predicted in 1980, on the basis of MNDO calculations, that 2,2'-dilithiobiphenyl would have the doubly bridged structure (**34**) which is considerably more stable

than any of the conformers[67] of the classical structure (**35**). This prediction was subsequently confirmed by x-ray structure determination[19] of the tmeda complex [see structure (**4**)]. The doubly bridged structure (**34**) is the intramolecular analog of phenyllithium association to the similar $(PhLi{\cdot}tmeda)_2$ complex (**3**).

In another example MNDO calculations[68] predicted that dilithium pentalenide should have the structure (**36**) which was

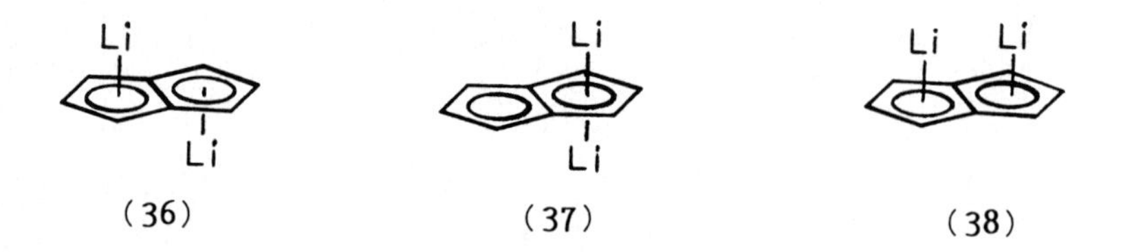

(**36**) (**37**) (**38**)

some 5 kcal/mole more stable than the arrangement (**37**) which in turn was more stable than (**38**).[20] The x-ray diffraction structure of the dimethoxyethane complex was that given earlier as (**5**) which is (**36**) without the two dimethoxyethane ligands. It seems unlikely that MNDO calculations which neglect the ligands about lithium can always be relied upon to give the correct structure. Thus calculations[30] show that the dilithio-o-xylene (**39**) is 0.3 kcal/mole more stable than (**40**) while, if four

(**39**) (**40**) (**41**) (**42**)

molecules of water are added to increase the coordination of lithium to its most common value of four, the relative stabilities are reversed with (**41**) being 6 kcal/mole more stable than (**42**). The calculated values now agree with the x-ray structure (**16**) for the tmeda complex of the diphenyl derivative.

MNDO calculations by Schleyer and coworkers[28] indicate that the unsolvated, cation-free 1,4-diphenyl-1,3-butadienide dianion is 3.1 kcal/mole more stable in the _trans_ geometry (**43**) than in the _cis_ form (**44**). When, however, two lithium cations are included in the calculations, the _cis_ structure is now 8.2 kcal/mole more stable than the _trans_ and has a structure like that observed in x-ray diffraction of the bis(tmeda) complex (**14**). These results illustrate the important stabilization afforded by two bridging lithium cations. It was noted that the

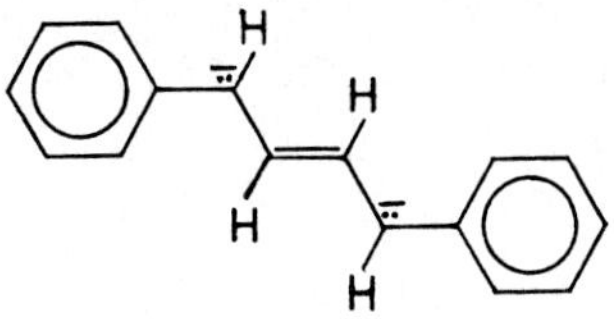

(43) 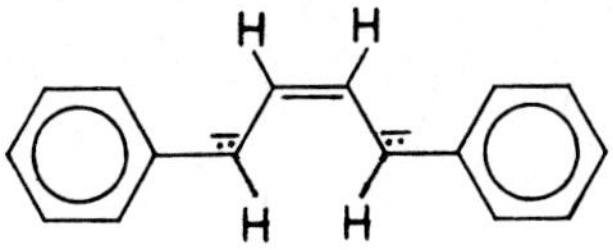 (44)

experimental Li-C bond lengths were all longer than those calculated; this result was said[28] to be due to a "systematic error in the MNDO parametization." A similar result was obtained for MNDO calculations[36] upon 1,3-dilithiodibenzyl ketone whose experimental Li-C bond lengths for the bis(tmeda) complex averaged 0.2 Å longer than those calculated by MNDO. Note that the calculated results for these systems did not include solvent or other ligands added to lithium.

Stucky[57] noted that the structures of organolithium compounds are consonant with some covalent bonding of the metal to carbon and that this interaction can be qualitatively modeled simply as an interaction of the HOMO of the carbanion with a suitable vacant s or p orbital of the lithium ion. In the case of the doubly bridged systems (**14**) and (**15**), the interaction with lithium may serve to complete a conjugated aromatic ring[64] of the 6 π electron system as shown by the Huckel MO (**45**) of the HOMO

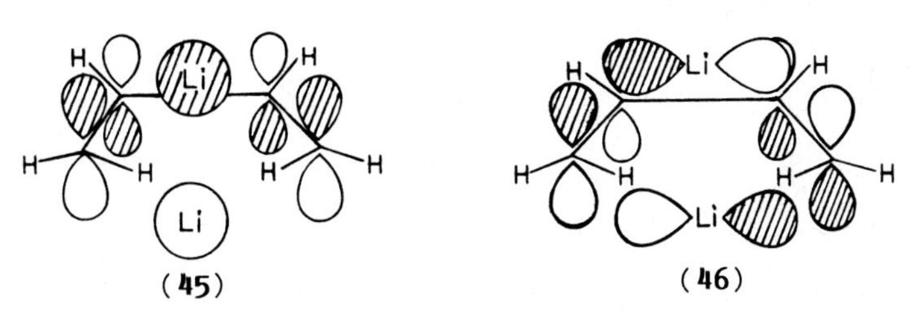

(**45**) (**46**)

and (**46**) for the subadjacent HMO which have proper symmetry for interaction with the lithium 2s and 2p orbitals respectively.

MNDO calculations,[33] both with and without water as a ligand for each lithium, model rather well the x-ray crystal structure of the dilithium adduct of 1,4-di-t-butyldiacetylene (**17**). This dimeric structure can be regarded as formed from two cis, planar 1,4-butatriene dianions which are bridged with lithiums as in (**47**). Structure (**47**) has two orthogonal π-systems, one perpendicular to the plane of the drawing with 8 π electrons and

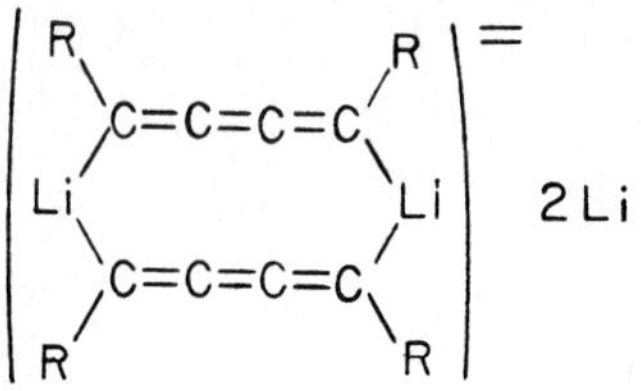

R R = C=C=C=C Li Li 2Li C=C=C=C R R

(47)

R R C=C—C=C Li Li C=C—C=C R R

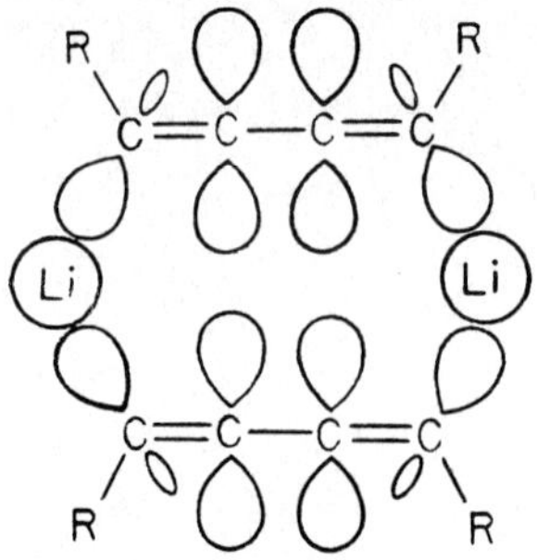

(48)

the other in the plane as drawn* in (48) with 12 electrons. While the overlap of orbitals is diminished in (48) especially because of the sp^2 hybridization of the terminal carbon atoms of the triene moiety, this set of 12 electrons constitutes a Huckel system of 4n electrons and would be expected to be somewhat anti-aromatic. Likely to avoid this anti-aromaticity and steric interaction between the _t_-butyl groups and the ligands on lithium, the molecule undergoes out of plane bending to give (49).

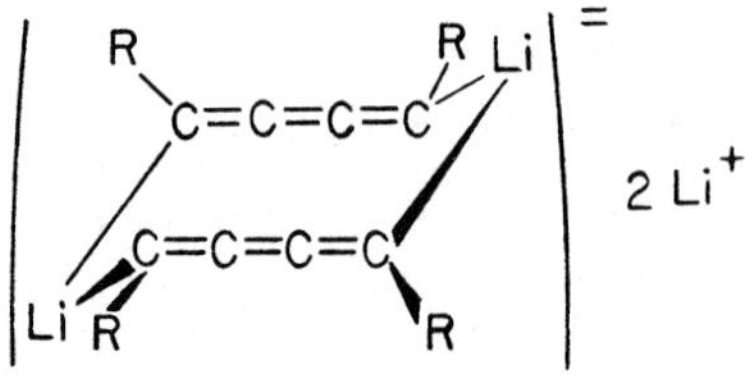

R R Li = C=C=C=C 2 Li^+ C=C=C=C Li R R

(49)

In the distorted structure the Li-C bonds are no longer orthogonal to the 8 electron π system and hence mixing will occur with transfer of some of the negative charge thereto. The two remaining lithium cations now bind to the more highly delocalized

*Also 2_p_ orbitals on lithium may be employed in the basis set.

dianion as in our electrostatic model (**32**) to give the observed structure (**17**).

Finally it should be noted that x-ray crystal structures do not always agree with MNDO or ab initio structure calculations for organolithium compounds. Thus while both MNDO[69] and ab initio[70] calculations upon allyllithium (both with and without water of solvation for MNDO) gave a symmetrically bridged structure (**50**), x-ray diffraction[71] indicated that the terminal

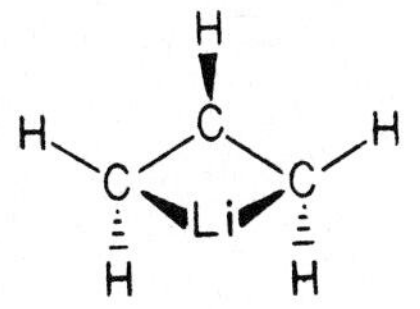

(**50**) 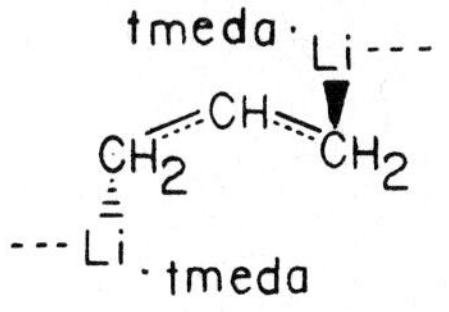 (**51**)

carbons of the allyl group in the crystal are linked to lithium atoms forming a polymeric chain (**51**). Of course it must be noted that all the quantum mechanical calculations assumed monomeric allyllithium; this assumption was incorrect for the crystalline phase which has been isolated. While it is to be hoped that x-ray structures upon organolithium compounds with added ligands will model solutions and the gas phase, unfortunately this will not always be the case.

(2) Calculations upon Compounds of Unknown Structure

The success of molecular orbital calculations in replicating structures obtained by x-ray diffraction encourages giving serious consideration to structures obtained so far only by molecular orbital calculations. For example Kos and Schleyer[66] have investigated many possible structures for $C_4H_4Li_2$ first by MNDO and then by more refined ab initio techniques. The doubly lithium-bridged structure (**52**) was found to be more stable than

(**52**) 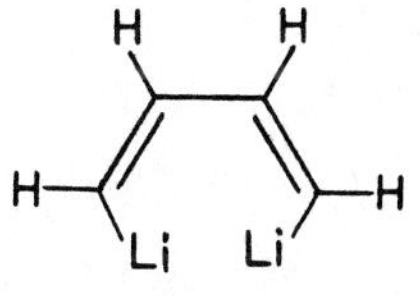 (**53**) 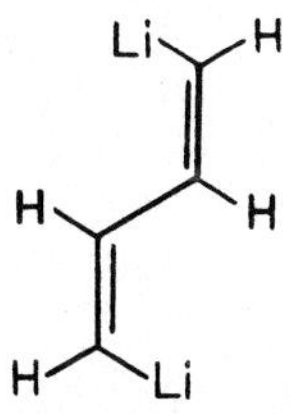 (**54**) 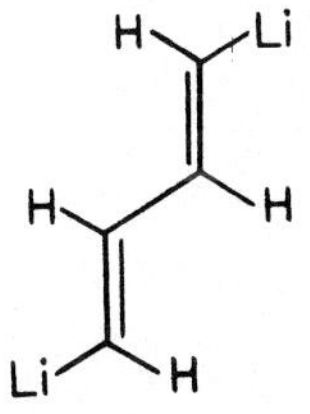 (**55**) 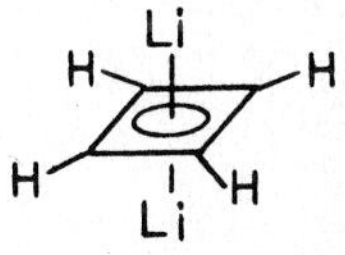 (**56**)

any other structures examined. It was 50 kcal/mole more stable than the classical structure (53), 35 kcal/mole more stable than (54), 44 kcal/mole more stable than (55), and even 26 kcal/mole more stable than the cyclic "aromatic dianion" (56). While it was first suggested that (52) is stabilized by "Mobius-Huckel" aromaticity,[66] i.e., contributions from structures such as (57)

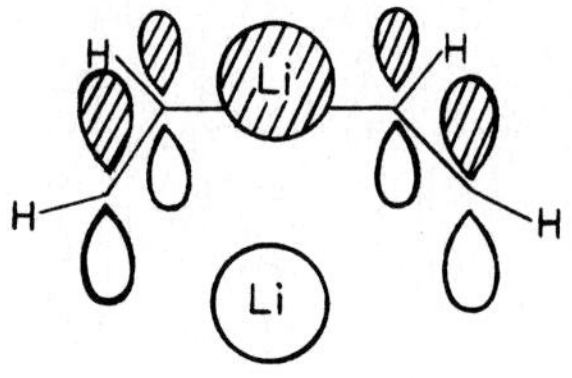

(57)

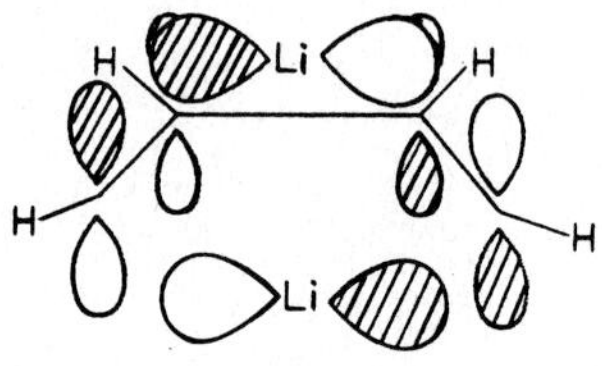

(58)

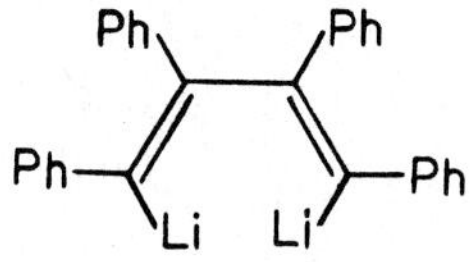

(59)

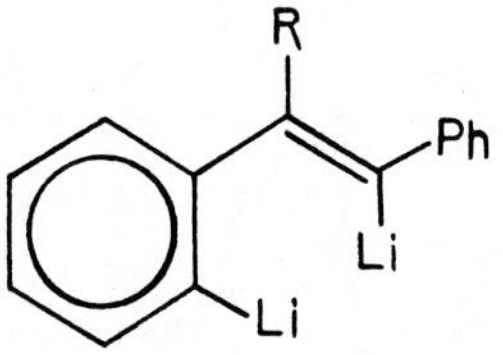

(60)

and (58), it is now thought[13] that electrostatic interaction [cf model (31)] are of "comparable if not of greater importance." It has been proposed[66] that the dimer of diphenylacetylene with lithium[72] has a structure analogous to (52) rather than the classical structure (59) and that the reaction product of alkyllithiums (RLi) with diphenylacetylene[73] also has a doubly lithium-bridged structure rather than the classical structure (60). These suggestions can be regarded as confirmed on the basis of the crystal structure determination[19] of 2,2'-di(lithium-tmeda)-biphenyl as the doubly lithium-bridged structure (4) rather than the classical structure (61).

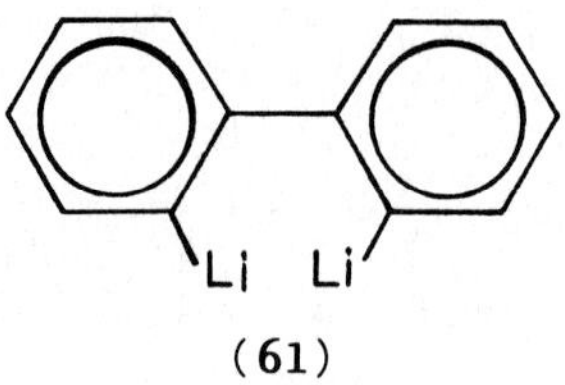

(61)

Reaction of propylene (or allylbenzene) with n-butyllithium in presence of tetramethylethylenediamine gives first allyllithium (or phenylallyllithium) and then a compound $CH_2CHCHLi_2$ ($PhCHCHCHLi_2$) which was thought to have the structure (**62**) in

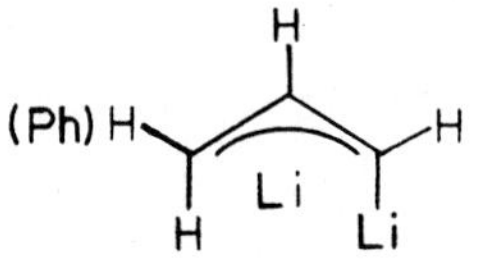

(**62**)

which one lithium is associated with a vinyl anion and the second with an allyl anion.[74] Calculations[75,76] for the monomeric species at the ab initio level suggest, however, that the structure is the doubly bridged structure (**63**), which is only

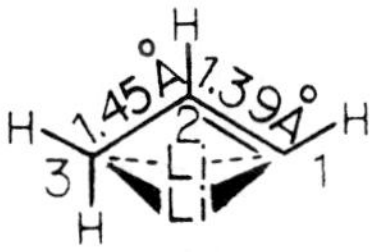

(**63**) 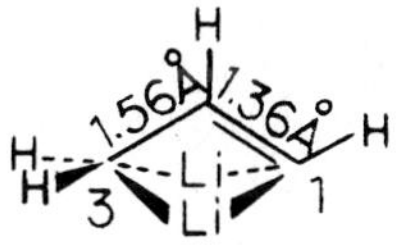 (**64**)

some 0.7 kcal/mole more stable than the alternative bridged structure (**64**). That (**63**) is only slightly more stable than (**64**) may be understood[75] on the basis that the usual allylic delocalization of negative charge in (**63**) is impeded by the negative charge on C(1); hence about one unit of negative charge resides on both C(1) and C(3) in both (**63**) and (**64**) and the C(1)-C(2) bond in each structure is shorter than the C(2)-C(3) bond. It is interesting that the reaction of methyllithium with allyllithium to give (**63**) is calculated[75] to be almost as exothermal (-15.4 kcal/mole) as the reaction of methyllithium with propylene to give allyllithium (-18 kcal/mole). This result emphasizes the important stabilization of the dianion provided by double lithium bridging; this stabilization is likely largely electrostatic[76] [cf. model (**31**)].

Calculations[77] upon 1,2-dilithioethylenes indicate similarly that the doubly bridged structure (**65**) has the most stable structure for the cis isomer and is about the same stability as

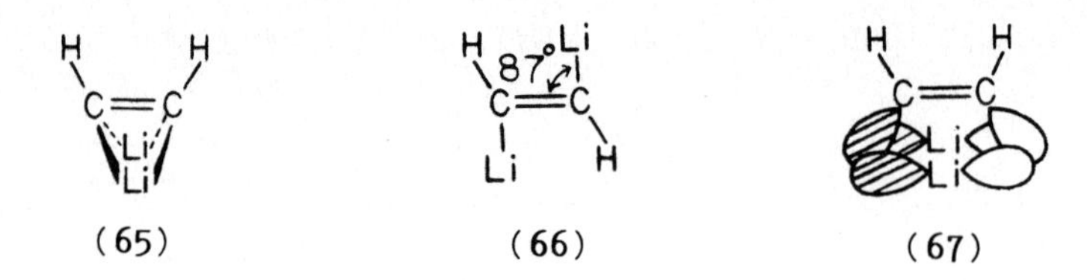

(65) (66) (67)

the "partially bridged" *trans* isomer (**66**). Attempts to prepare 1,2-dilithioethylene have so far been unsuccessful[78] although 1,2-dilithiobenzene is known[79] and could have a structure analogous to (**65**).[80] In addition to electrostatic stabilization [*cf*. models (**32**) and (**33**)], it has been suggested that *p*-orbitals of lithium help to stablize the *cis* dianion as shown in (**67**) and that Li-H bonding helps to stabilize the trans structure (**66**).[77] The structure of 1,1-dilithioethylene is calculated[77] to have the perpendicular geometry (**68**) rather than the classical geometry

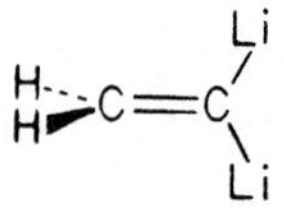

(68) 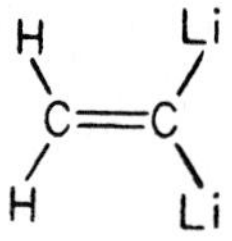 (69) 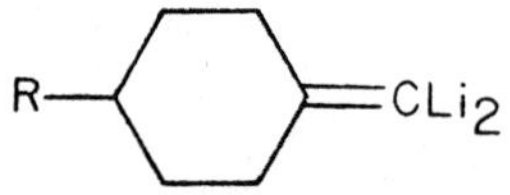 (70)

(**69**); however, the molecule is calculated to have a preference for the perpendicular triplet rather than the perpendicular singlet state.[81] While 1,1-dilithioethylene is likely unstable at room temperature because of ready loss of LiH, the compounds[82] (**70**), R=H and *t*-Bu, have recently been prepared.[80] It is notable that the geometry of structure (**68**) is just what would have been expected from the electrostatic model (**31**).

Dilithioacetylene is estimated by *ab initio* calculations[83] to have the planar doubly bridged singlet structure (**71**) rather than

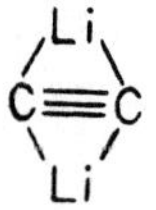

Li−C≡C−Li

(71) (72)

the classical linear structure (**72**). Of course the calculations pertain to the gaseous monomeric species rather than crystalline lithium carbide which is polymeric.[84] Structure (**71**) provides a simple example of the electrostatic model (**31**); however, orbital symmetries permit some multicenter covalent bonding of lithium.

An early attempt to make 1,2-dilithioethane was unsuccessful.[85] More recently this substance has been claimed as a poorly characterized powder[86] and as a reactive intermediate in dimerization of ethylene by lithium powder in dimethoxyethane at -10°C.[87] Monomeric, unsolvated 1,2-dilithioethane is calculated[88] to possess the partially bridged structure (**73**) which, however is only 1.9 kcal/mole more stable than the symmetrically bridged structure (**74**). Structure (**73**) is similar in geometry to the electrostatic model (**33**) while (**74**)

(**73**) (**74**)

is similar to (**31**). These favorable electrostatic geometries are thought to be stabilized by interaction with the p-orbitals of lithium as shown schematically in (**75**) and (**76**) for the HOMO's.

(**75**) (**76**)

The slightly lower energy of (**75**) likely results from the reduction in antibonding character of the π^* orbital by tilting of the p-orbitals through hybridization with carbon 2s orbitals. The gaseous compound 1,2-dilithioethane is calculated to be slightly unstable (1.5 kcal/mole) with respect to dissociation into ethylene and gaseous Li_2. This compound, however, likely can be stabilized by association and solvation.

While 1,2-dilithioethane is questionable as a stable organolithium compound, 1,3-dilithiopropane (**77**) and 1,3-dilithio-2,2-dimethylpropane (**78**) have recently been prepared.[89] Compound (**77**) readily lost LiH to give allyllithium whereas (**78**), with no

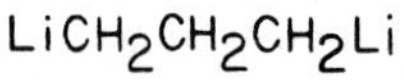

$LiCH_2CH_2CH_2Li$

(77)

$LiCH_2C(CH_3)_2CH_2Li$

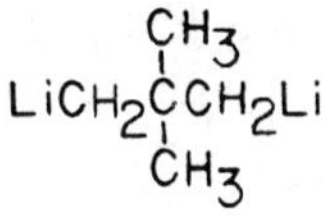

(78)

hydrogens available for 1,2-elimination of LiH, was more stable in diethyl ether solution than ordinary primary organolithium compounds.[89] *Ab initio* calculations[90] indicate that monomeric, unsolvated 1,3-dilithiopropane in the classical W conformation (**79**) is 24.6 kcal/mole less stable than the symmetrically bridged

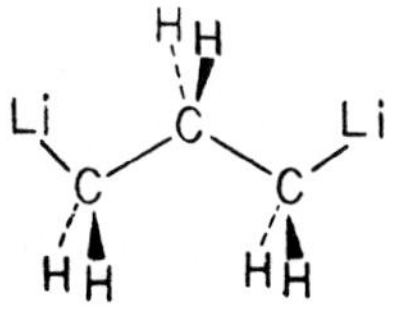

(**79**) 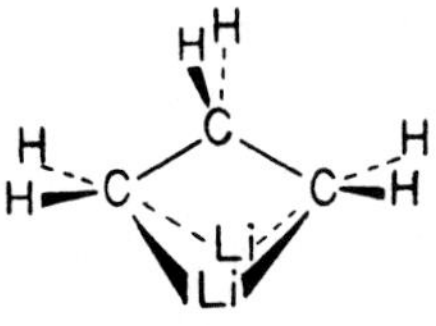 (**80**) 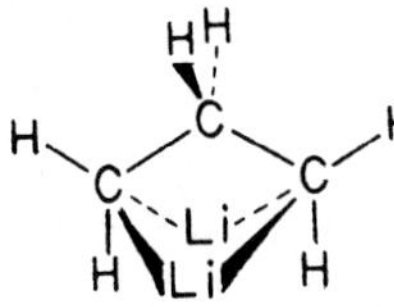 (**81**)

structure (**80**). Structure (**80**) is 16.7 kcal/mole more stable than the alternative bridged structure (**81**), likely because the sp^3 orbitals of the carbanion lone pairs of (**80**) locate the centers of negative charge closer to the lithiums than the *p* orbitals of the carbanion lone pairs of (**81**). Also negative charge is more stable in sp^3 than in *p* orbitals. Schleyer and co-workers[90] believe that both multicenter covalent bonding and electrostatic interactions of lithium are important in stabilizing (**80**).

The cyclization of 1,4-dilithiobutane from the classical structure (**82**) to the doubly bridged structure (**83**), according to

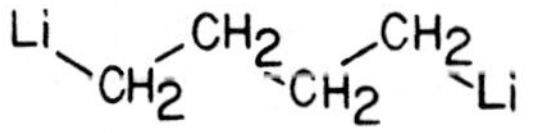

$LiCH_2CH_2CH_2CH_2Li$

(**82**) 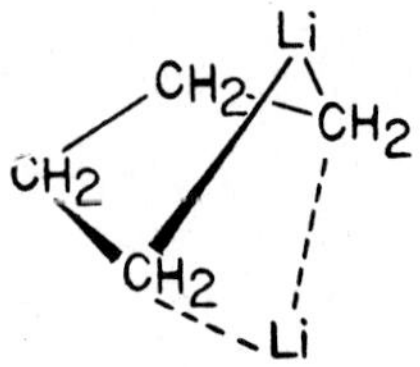 (**83**)

ab initio calculations,[90] liberates 34.6 kcal/mole of energy or 10 kcal/mole more energy than the cyclization of (**79**) to (**80**).

For comparison the dimerization of methyllithium to (**84**) evolves 46.3 kcal/mole. Note that double bridging of (**79**) to (**80**) and (**82**) to (**83**) are the intramolecular analogs of dimerization; however, double bridging is subject to geometric constraints which are larger for (**80**) than for (**83**) and absent in the dimer (**84**).

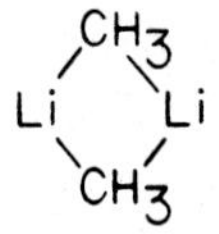

CH_3 Li Li CH_3

(**84**)

The structure of dilithiomethane has been examined on a theoretical basis by several groups of workers. Schleyer, Pople, and co-workers[90] and, independently, Nilssen and Skancke[91] noted that the tetrahedral and planar forms of CH_2Li_2 were surprisingly close in stability as estimated by <u>ab initio</u> methods. Laidig and Schaefer,[92] in more complete calculations including configuration interaction, concluded that the tetrahedral singlet structure (**85**) was the most stable structure but was only 2.1 kcal/mole

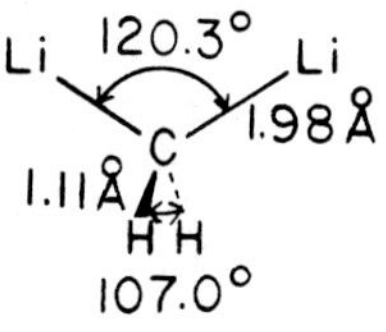

Li 120.3° Li 1.98 Å C 1.11 Å H H 107.0°

(**85**)

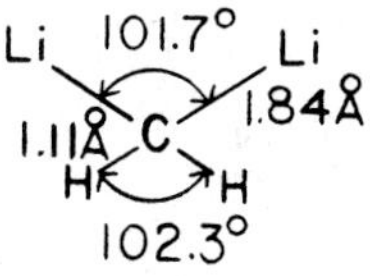

Li 101.7° Li 1.84 Å 1.11 Å C H H 102.3°

(**86**)

more stable than the tetrahedral triplet state or 8.3 kcal/mole more stable than the planar singlet structure (**86**). Backrach and Streitwieser[93] have concluded from an electron density analysis that both the tetrahedral and planar singlet structures of dilithiomethane are largely ionic in character like methyllithium. The charge distributions may be approximated (if we assume the usual tetrahedral distribution of electron density about carbon) by the point charge diagrams (**87**) and (**88**) for the tetrahedral and planar structures respectively. In figure (**87**) all the charges are coplanar; in (**88**) the negative charges are below and above the plane of the nuclei. Because the negative charges of the "dianion" are located on one carbon atom

and hence are close together in space as well as close to two C-H bonds, none of the earlier electrostatic models (**31**), (**32**), or (**33**) are applicable. The similarity of energy of (**87**) to (**88**) can be understood on the basis that in (**87**) one cation is close to each negative charge but further from the other negative charge whereas in (**88**) the cations are equidistant from the two negative charges at a distance intermediate between those of (**87**).

Dilithiomethane was first prepared by Ziegler and co-workers[94] from pyrolysis of methyllithium at 230-240°C; the methyllithium must be free of lithium halide and the optimum temperature for the preparation is 223-226°C.[95] Somewhat earlier West and Rochow[85] obtained evidence for dilithiomethane from trapping experiments with trimethylchlorosilane during reaction of dibromomethane with lithium. Dilithiomethane has also been obtained[96] by reaction of bis(iodomercurio)methane with lithium powder or excess _t_-butyllithium; lithium powder with 1,1-bis(chloromercurio)ethane similarly gives 1,1-dilithioethane.* As ordinarily prepared dilithiomethane is a highly associated substance of negligible vapor pressure below its decomposition temperature of about 225°C.[95] The detailed structure of this polymeric substance is unknown** but the structure of the dimer by _ab initio_ calculations[97] is (**89**) in which the LiCLi is

*This synthesis of 1,1-dilithioethane should lend encouragement to synthesis of 1,2-dilithioethane which is calculated to be 17.6 kcal/mole more stable than 1,1-dilithioethane as monomeric species.[96]

**R. J. Lagow (Symposium on Advances in Carbanion Chemistry, Am. Chem. Soc. National Meeting, Chicago, Sept. 8-13, 1985) has reported that a study of the crystal structure of dilithiomethane revealed a high polymer in which each carbon is coordinated with eight lithiums and two hydrogens for an unusual coordination number of ten for carbon, likely the highest coordination number ever observed for carbon.

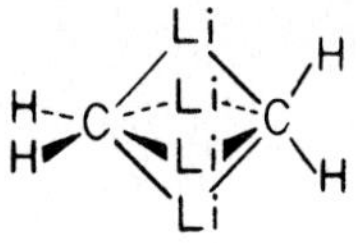

(89)

near 90° and the Li-C bonds are 2.04 and 2.17 Å with the four lithiums almost in a plane. This is a beautiful example of multiple lithium bridging.

Finally, while the topic is beyond the scope of this review, more than two alkali metals can be bound in a single molecule to give so-called "polyanions". A few examples must suffice. Reaction of 1,4-cycloheptadiene, 1,3,6-heptatriene, and 1,6-heptadiene with excess of *n*-butyllithium-tetramethylethylenediamine in hexane gives a black trilithium derivative which has been characterized as cycloheptatrienyl trianion.[98] The same trianion is also formed from cycloheptene with *n*-butyllithium-potassium *t*-amyloxide[99] and from 3-chloroquadricyclane with lithium p,p'-di-*t*-butylbiphenylide.[100] Schleyer and co-workers[101] emphasize that the product is certainly not a "trianion" but must be a trialkali metal compound. The monomeric trilithium compound is calculated[101] by MNDO to have the non-planar structure (**90**) in which two lithium ions are bound to the

(**90**)

convex side of the trianion and one to the concave side. This structure seems designed to provide maximum electrostatic stabilization of the substance.

In an earlier example West and co-workers[102] discovered that propyne reacts with excess *n*-butyllithium in hexane at reflux to give finally a compound C_3Li_4, which according to chemical derivatization appeared to have the structure (**91**) or (**92**).

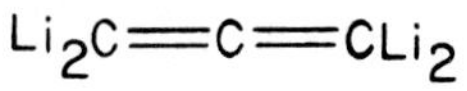

$Li_2C{=}C{=}CLi_2$

(**91**)

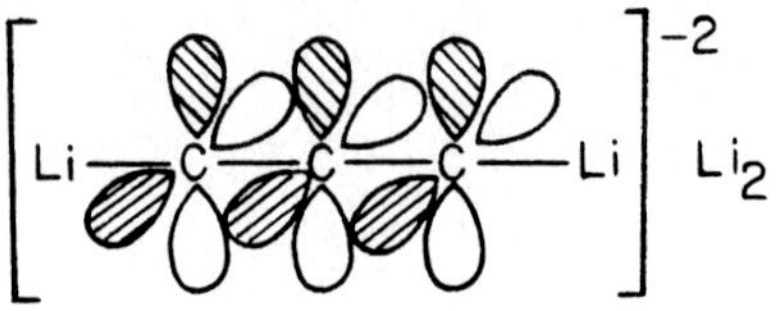

$[Li-C-C-C-Li]^{-2}$ Li_2

(**92**)

The same compound was obtained as the principal product in the reaction of carbon vapor with lithium atoms.[103] The stability of the substance may be rationalized on the basis that the anion C_3^{-4} is isoelectronic with carbon dioxide and has two lithiums attached at the terminal carbons by lithium acetylide-like bonds, while the remaining lithiums are attached to two mutually orthogonal allylic systems [see (**92**)]. Ab initio calculations[13,104] indicate that the structure of the monomeric compound is (**93**), where the angle LiC(2)Li is 102.5°. This

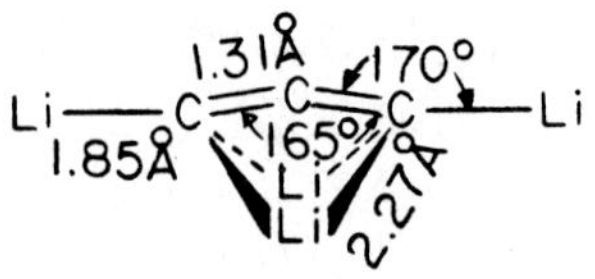

(**93**)

structure is analogous to (**92**) except that the carbon skeleton is bent 15° from linearity evidently to facilitate bridging by lithium. The bending would enhance both electrostatic interaction of lithium with the negative charge at the terminal carbon atoms and overlap of lithium 2p orbitals with the p-orbitals of the terminal carbon atoms. Metalation of 1,3-pentadiyne and 2,4-hexadiyne by butyllithium similarly gives tri- and even tetralithium compounds.[105]

More surprising is the discovery that trilithiomethane[106,107] and tetralithiomethane[107,108] can be prepared by reaction of lithium vapor with chloroform and carbon tetrachloride respectively. Tetralithiomethane is also preparable by reaction of excess t-butyllithium with tetrakis(chloromercurio)methane or tetrakis(ethylmercurio)methane.[109] According to ab initio calculations monomeric trilithiomethane and tetralithiomethane are 7.4 and 16.5 kcal/mole more stable as the classical tetrahedral structures (**94**) and (**95**) respectively than in

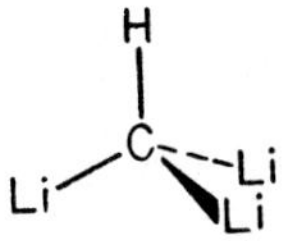

(**94**) 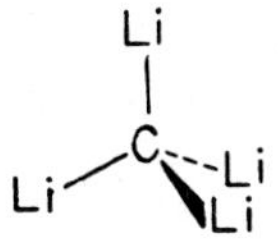 (**95**) 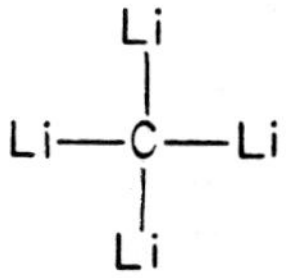 (**96**)

alternative planar structures, e.g. (**96**). Structure (**95**) would be anticipated based either on the covalent model of methane or the ionic model of a spherical negative charge interacting with four positive charges. Trilithiomethane and tetralithiomethane as usually prepared are high aggregates or are complexes such as tetralithiomethane associated with di-*t*-butylmercury.[109]

Flash vaporization of $(CLi_4)_n$ in a mass spectrometric probe gives many species[110] including CLi_5^+. The formation of CLi_5^+ from CLi_4 plus Li^+ or from CLi_3^+ plus Li_2 is calculated to be a highly exothermic process. The predicted structure of CLi_5^+ is (**97**).[110] The corresponding electroneutral species CLi_5 is

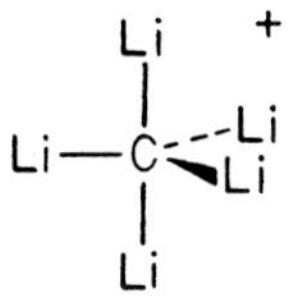

(**97**) 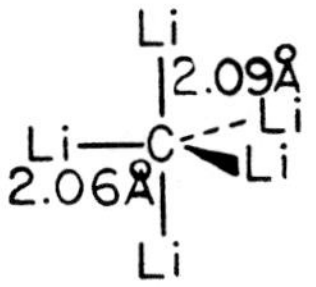 (**98**) 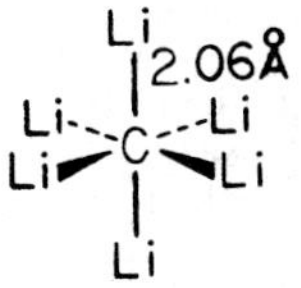 (**99**)

calculated[111] to have the trigonal-bipyramidal geometry D_{3h} (**98**) and to be highly stable with respect to loss of a lithium atom. Similarly CLi_6 is calculated to be octahedral, O_h, as shown in (**99**), and is stable toward loss of Li_2. Compounds such as (**98**) and (**99**) have "extra" electrons and a high coordination about carbon and are called "hypervalent". The extra electrons, however, are calculated to contribute to Li-Li bonding; the central carbon thus is "encaged" in lithium.[111]

IV. CONCLUSIONS

Organodialkali metal compounds are now in many instances well characterized substances. This is especially true of crystalline compounds whose structures are known in detail from x-ray diffraction upon single crystals. While organoalkali metal compounds are generally highly associated substances, they can frequently be dissociated into monomeric or dimeric units by solvation, especially by bidentate and tridentate ligands such as glyme (dimethoxyethane), diglyme, tetramethylethylenediamine, and pentamethyldiethylenetriamine. The structures of organodialkali metal compounds can frequently be understood in terms of ionic structures in which the ions are arranged to maximize electrostatic attraction between cations and anion and minimize repulsion between cations.

Organodilithium compounds are the most thoroughly studied of the dialkali metal compounds. While their structures are generally in qualitative agreement with simple ionic models, frequently their structures are very different from classical structures modeled after compounds of carbon and hydrogen. While the ionic model may suffice, a more complete explanation may be that lithium, unlike carbon or hydrogen, may utilize _empty_ low-lying 2_s_ and 2_p_ orbitals to enhance bonding. Arguments over the degree of ionicity of carbon-metal bonds become entrapped in the dilemma that there is no generally agreed upon method of deciding which atom "owns" which electron (or which part of coordinate space). Yet the structures of organolithium compounds can be said to be _better understood_ than many other classes of compounds in the sense that structures can now frequently be predicted in advance by _ab initio_ or even simple MNDO calculations.

Chemical evidence as well as calculations show that, far from being unstable, organodialkali metal compounds have a stability that can rival that of monoalkali metal compounds. Not only can two lithiums be attached to one carbon but even three, four and likely five or six!

The compounds of the heavier alkali metals are much less well studied than lithium. The organometallic chemistry of sodium would be expected to parallel that of lithium fairly closely. On the other hand, the largest metal readily available for study by chemists, namely cesium, whose cation is about the diameter of a benzene ring, has subtleties in its behavior not yet fully understood. For example cesium is the only alkali metal which reduces benzene to benzenide ion in appreciable yield.[5] Hopefully the next fifteen years may see development of the organometallic chemistry of the heavier alkali metals to rival that of lithium for the past fifteen years.

Acknowledgment. The author wishes to acknowledge many helpful discussions with Professors William H. Eberhardt and Donald J. Royer concerning the structures of organoalkali metal compounds.

REFERENCES

1. A. A. Morton, W. J. LeFevre, and I. Hechenbleikner, J. Am. Chem. Soc., 58 (1936) 754; A. A. Morton and I. Hechenbleikner, ibid., 58 (1936) 1697.

2. H. Gilman and H. A. Pacevitz, J. Am. Chem. Soc., 62 (1940) 1301; A. A. Morton, J. B. Davidson, and H. A. Newey, ibid., 64 (1942) 2240.

3. L. Hackspill, Proc. Int. Congr. Appl. Chem., 8th, 2 (1912) 113; Ann. Chim. Phys. (Paris., 28 (1913) 653; Helv. Chim. Acta, 11 (1928) 1026.

4. J. de Postis, Proc. Int. Cong. Pure Appl. Chem., 11th, 5 (1947) 867; see also L. Hackspill,"Nouveau Traite de Chemie Minerale", Vol. 3, P. Pascal, Ed., Masson, Paris, 1956, p 124.

5. E. Grovenstein, Jr., T. H. Longfield, and D. E. Quest, J. Am. Chem. Soc., 99 (1977) 2800.

6. A. Minsky, A. Y. Meyer, and M. Rabinovitz, J. Am. Chem. Soc., 104 (1982) 2475; see also M. Rabinovitz, I. Willner, and A. Minsky, Acc. Chem. Res., 16 (1983) 298.

7. B. Eliasson, T. Lejon, and U. Edlund, J. Chem. Soc., Chem Commun., (1984) 591.

8. C. G. Screttas and M. Micha-Screttas, J. Org. Chem., 48 (1983) 153.

9. K. D. Jordon and P. D. Burrow, Acc. Chem. Res., 11 (1978) 341.

10. E. Weiss and G. Hencken, J. Organomet. Chem., 21 (1970) 265.

11. H. Köster,D. Thoennes and E. Weiss, J. Organomet. Chem., 160 (1978) 1. See also the crystal structure of tetrameric 1-(dimethylamino)-3-lithiopropane. [G. W. Klumpp, M. Vos, and F. J. J. de Kanter, J. Am. Chem. Soc., 107 (1985) 8292].

12. P. West and R. Wack, J. Am. Chem. Soc., 89 (1967) 4395.

13. P. v. R. Schleyer, Pure. and Appl. Chem., 56 (1984) 151. See also G. Graham, S. Richtsmeier, and D. A. Dixon, J. Am. Chem. Soc., 102 (1980) 5759.

14. H. Hopp and P. P. Power, J. Am. Chem. Soc., 105 (1983) 5320.

15. D. Thoennes and E. Weiss, Chem. Ber., 111 (1978) 3157.

16. L. M. Jackman and L. M. Scarmoutzos, J. Am. Chem. Soc., 106 (1984) 4627.

17. G. Fraenkel, H. Hsu, and B. M. Su, Lithium Chemistry Symposium ACS S.E. Regional Meeting, Charlotte, N. C., Nov. 1983.

18. U. Schümann,J. Kopf, and E. Weiss, Angew. Chem. Int. Ed. Eng., 24 (1985) 215.

19. U. Schubert, W. Neugebauer, and P. v. R. Schleyer, J. Chem. Soc., Chem. Commun., (1982) 1184.

20. J. J. Stezowski, H. Hoier, D. Wilhelm, T. Clark, and P. v. R. Schleyer, J. Chem. Soc., Chem. Commun., (1985) 1263.

21. J. J. Brooks, W. Rhine, and G. D. Stucky, J. Am. Chem. Soc., 94 (1972) 7346.

22. W. E. Rhine, J. Davis, and G. D. Stucky, J. Am. Chem. Soc., 97 (1975) 2079.

23. W. E. Rhine, J. H. Davis, and G. Stucky, J. Organomet. Chem., 134 (1977) 139.

24. M. Walczak and G. Stucky, J. Am. Chem. Soc., 98 (1976) 5531.

25. M. Walczak and G. D. Stucky, J. Organomet. Chem., 97 (1975) 313.

26. G. Boche, H. Etzrodt, W. Massa, and G. Baum, Angew. Chem., 97 (1985) 858; Angew. Chem. Int. Ed. Engl., 24 (1985) 863. The author is indebted to Professor Boche for a copy of this article prior to publication.

27. S. K. Arara, R. B. Bates, W. A. Beavers, and R. S. Cutler, J. Am. Chem. Soc., 97 (1975) 6271.

28. D. Wilhelm, T. Clark, P. v. R. Schleyer, H. Dietrich, and W. Mahdi, J. Organometal. Chem., 280 (1985) C6.

29. M. F. Lappert, C. L. Raston, B. W. Skelton, and A. H. White, J. Chem. Soc., Chem. Commun., (1982) 14.

30. G. Boche, G. Decher, H. Etzrodt, H. Dietrich, W. Mahdi, A. J. Kos, and P. v. R. Schleyer, J. Chem. Soc., Chem. Commun., (1984) 1493.

31. S. P. Patterman, I. L. Karle, and G. D. Stucky, J. Am.

Chem. Soc., 92 (1969) 1150.

32. P. v. R. Schleyer, A. J. Kos, D. Wilhelm, T. Clark, G. Boche, G. Decher, H. Etzrodt, H. Dietrich, and W. Mahdi, J. Chem. Soc., Chem. Commun., (1984) 1495.

33. W. Neugebauer, G. A. P. Geiger, A. J. Kos, J. J. Stezowski, and P. v. R. Schleyer, Chem. Ber., 118 (1985) 1504.

34. W. N. Setzer and P. v. R. Schleyer, Adv. Organomet. Chem., 24 (1985) 353.

35. D. Wilhelm, H. Dietrich, T. Clark, W. Mahdi, A. J. Kos, and P. v. R. Schleyer, J. Am. Chem. Soc., 106 (1984) 7279.

36. H. Dietrich, W. Mahdi, D. Wilhelm, T. Clark, and P. v. R. Schleyer, Angew. Chem. Int. Ed. Engl., 23 (1984) 621.

37. B. Bogdanovic, C. Krüger,and B. Wermeches, Angew. Chem. Int. Ed. Engl., 19 (1980) 817.

38. M. Walczak, K. Walczak, R. Mink, M. D. Rausch, G. Stucky, J. Am. Chem. Soc., 100 (1978) 6382.

39. R. Zorger, W. Rhine, and G. D. Stucky, J. Am. Chem. Soc., 96 (1974) 6048.

40. E. Weiss, G. Sauermann, and G. Thirase, Chem. Ber., 116 (1983) 74.

41. E. Weiss and G. Sauermann, Angew. Chem. Int. Ed. Engl., 7 (1968) 133; Chem. Ber., 103 (1970) 265.

42. E. Weiss and H. Köster,Chem. Ber., 110 (1977) 717.

43. M. Atoji, J. Chem. Phys., 60 (1974) 3324.

44. M. Atoji, J. Chem. Phys., 56 (1972) 4947.

45. S. Z. Goldberg, K. N. Raymond, C. A. Harmon, and D. H. Templeton, J. Am. Chem. Soc., 96 (1974) 1348.

46. J. H. Noordik, Th. E. M. van den Hark, J. J. Mooij, and A. A. K. Klassen, Acta Cryst., B30 (1974) 833.

47. J. H. Noordik, H. M. L. Degens, and J. J. Mooij, Acta Cryst., B31 (1975) 2144.

48. E. de Boer and A. A. K. Klassen, Kem.-Kemi, 7 (1980) 257.

49. L. Pauling, "The Nature of thc Chemical Bond", Cornell University Press, Ithaca, N. Y., 3rd ed. 1960, pp 91-102.

50. D. J. Royer, "Bonding Theory", McGraw Hill, N. Y., 1968, p 193.

51. A. Streitwieser, Jr., J. Organomet. Chem., 156 (1978) 1.

52. For a further discussion of this electrostatic model and its application to $(LiF)_4$, $(LiOH)_4$, $(LiH)_4$, and $(LiNH_2)_4$ see: A.-M. Sapse, K. Raghavachari, P. v. R. Schleyer, and E. Kaufmann, J. Am. Chem. Soc., 107 (1985) 6483.

53. A. Streitwieser, Jr., and J. T. Swanson, J. Am. Chem. Soc., 105 (1983) 2502.

54. A. Streitwieser, Jr., Acc. Chem. Res., 17 (1984) 353.

55. We are greatly indebted to Prof. Donald J. Royer for derivation and solution of this equation.

56. R. J. Busby and M. P. Tytko, J. Organomet. Chem., 270 (1984) 265.

57. G. Stucky, Advan. Chem. Ser., 130 (1974) 56.

58. R. S. Mulliken, J. Chem. Phys., 43 (1965) S2.

59. A. Streitwieser, Jr., J. E. Williams, Jr., S. Alexandros, and J. M. McKelvey, J. Am. Chem. Soc., 98 (1976) 4778.

60. G. D. Graham, D. S. Marynick, and W. N. Lipscomb, J. Am. Chem. Soc., 102 (1980) 4572.

61. R. S. Mulliken, J. Chem. Phys., 23 (1955) 1833.

62. R. F. W. Bader, Acc. Chem. Res., 18 (1985) 9.

63. R. F. W. Bader and P. J. MacDougall, J. Am. Chem. Soc., 107 (1985) 6788.

64. P. v. R. Schleyer, Pure Appl. Chem., 55 (1983) 355; 56 (1984) 151.

65. M. J. S. Dewar and M. Thiel, J. Am. Chem. Soc., 99 (1977) 4899, 4907.

66. A. J. Kos, P. v. R. Schleyer, J. Am. Chem. Soc., 102 (1980) 7928; see also A. J. Kos, P. Stein, and P. v. R. Schleyer, J. Organomet. Chem., 280 (1985) C1.

67. W. Neugebauer, A. J. Kos, and P. v. R. Schleyer, J. Organomet. Chem., 228 (1982) 107.

68. D. Wilhelm, J. L. Courtneidge, T. Clark, and A. G. Davies, J. Chem. Soc., Chem. Commun., (1984) 810.

69. G. Decher and G. Boche, J. Organomet. Chem., 259 (1963) 31.

70. T. Clark, E. D. Jemmis, P. v. R. Schleyer, J. S. Binkley, and J. A. Pople, J. Organomet. Chem., 150 (1978) 1; T. Clark, C. Rohde, and P. v. R. Schleyer, Organometallics, 2 (1983) 1344.

71. H. Köster and E. Weiss, Chem. Ber., 115 (1982) 3422.

72. W. Schlenk and E. Bergmann, Liebigs Ann. Chem., 463 (1928) 71; L. I. Smith and H. J. Hoehn, J. Am. Chem. Soc., 63 (1941) 1184.

73. J. E. Mulvaney, Z. G. Gardlund, S. L. Gardlund, and D. J. Newton, J. Am. Chem. Soc., 88 (1966) 476; J. E. Mulvaney, S. Groen, L. J. Carr, Z. G. Gardlund, and S. L. Gardlund, ibid, 91 (1969) 388; J. E. Mulvaney, and D. J. Newton, J. Org. Chem., 34 (1969) 1936.

74. J. Klein and A. Medlik-Balan, J. Chem. Soc., Chem. Commun., (1975) 877.

75. P. v. R. Schleyer and A. J. Kos, J. Chem. Soc., Chem. Commun., (1982) 448.

76. D. Kost, J. Klein, A. Streitwieser, Jr., and G. W. Schriver, Proc. Natl. Acad. Sci. USA, 79 (1982) 3922.

77. Y. Apeloig, T. Clark, A. J. Kos, E. D. Jemmis, and P. v. R. Schelyer, Israel J. Chem., 20 (1980), 43; see also Y. Apeloig, P. v. R. Schleyer, J. S. Binkley, and J. A. Pople, J. Am. Chem. Soc., 98 (1976) 4332.

78. D. Seyferth and S. C. Vick, J. Organomet. Chem., 144 (1978) 1.

79. G. Wittig and F. Bickelhaupt, Chem. Ber., 91 (1958) 883; H. J. S. Winkler and G. Wittig, J. Org. Chem., 28 (1963) 1733.

80. For preparation of $Li_2C=C(CH_3)_2$, $CH_3C(Li)=C(Li)CH_3$ and related compounds see: J. A. Morrison, C. Chung, and R. J. Lagow, J. Am. Chem. Soc., 97 (1975) 5015.

81. W. D. Laidig and H. F. Schaefer, III, J. Am. Chem. Soc., 101 (1979) 7184.

82. A. Maercker and R. Dujardin, Angew. Chem. Int. Ed. Eng., 23 (1984) 224.

83. Y. Apeloig, P. v. R. Schleyer, J. S. Binkley, J. A. Pople, and W. L. Jorgensen, Tetrahedron Lett., (1976) 3923.

84. V. R. Juza, W. Wehle, and H. V. Schuster, Zeits. für Anorg. und Allgem. Chem., 352 (1967) 252.

85. R. West and E. G. Rochow, J. Org. Chem., 18 (1953) 1739; however polylithiated alkanes, e. g. , $CLi_3CLi_2CLi_3$, have been reported: L. G. Sneddon and R. J. Lagow, J. Chem. Soc., Chem. Commun., (1975), 302; L. A. Shimp and R. J. Lagow, J. Org. Chem., 44 (1979) 2311.

86. H. Kuus, Chem. Abstr., 69 (1968) 67443; 71 (1969) 49155.

87. V. Rautenstrauch, Angew. Chem., Int. Ed. Engl., 14 (1975) 259.

88. A. J. Kos, E. D. Jemmis, P. v. R. Schelyer, R. Gleiter, V. Fischbach, and J. A. Pople, J. Am. Chem. Soc., 103 (1981)

4996.

89. J. W. F. L. Seetz, G. Schat, O. S. Akerman, and F. Bickelhaupt, J. Am. Chem. Soc., 104 (1982) 6848.

90. J. B. Collins, J. D. Dill, E. D. Jemmis, Y. Apeloig, P. v. R. Schelyer, and J. A. Pople, J. Am. Chem. Soc., 98 (1976) 5419.

91. E. W. Nilssen and A. Skancke, J. Organomet. Chem., 116 (1976) 251.

92. W. D. Laidig and H. F. Schaefer III, J. Am. Chem. Soc., 100 (1978) 5972.

93. S. M. Bachrach and A. Streitwieser, Jr., J. Am. Chem. Soc., 106 (1984) 5818.

94. K. Ziegler, K. Nagel, and M. Patheiger, Z. Anorg. Allgem. Chem., 282 (1955) 345.

95. J. A. Gurak, J. W. Chinn, Jr., and R. J. Lagow, J. Am. Chem. Soc., 104 (1982) 2637.

96. A. Maercker, M. Theis, A. J. Kos, and P. v. R. Schleyer, Angew. Chem. Int. Ed. Engl., 22 (1983) 733.

97. E. D. Jemmis, P. v. R. Schleyer, and J. A. Pople, J. Organomet. Chem., 154 (1978) 327.

98. J. J. Bahl, R. W. Bates, W. A. Beavers, and C. R. Launer, J. Am. Chem. Soc., 99 (1977) 6126.

99. D. Wilhelm, T. Clark, P. v. R. Schleyer, J. L. Courtneidge, and A. G. Davies, J. Organomet. Chem., 273 (1984) C1.

100. J. Stapersma, P. Kuipers, and G. W. Klumpp, J. Roy. Neth. Chem. Soc., 101 (1982) 213.

101. P. v. R. Schleyer, D. Wilhelm, and T. Clark, J. Organomet. Chem., 281 (1985) C17.

102. R. West, P. A. Carney, and I. C. Mineo, J. Am. Chem. Soc., 87 (1965) 3788; R. West and P. C. Jones, J. Am. Chem. Soc., 91 (1969) 6156; W. Priester, R. West, and T. L. Chwang, J. Am. Chem. Soc., 98 (1976) 8413; W. Priester and R. West, J. Am. Chem. Soc., 98 (1976) 8421.

103. L. A. Shimp and R. Lagow, J. Am. Chem. Soc., 95 (1973) 1343.

104. E. D. Jemmis, D. Poppinger, P. v. R. Schleyer, and J. A. Pople, J. Am. Chem. Soc., 99 (1977) 5796; E. D. Jemmis, J. Chandrasekhar, and P. v. R. Schleyer, J. Am. Chem. Soc., 101 91979) 2848.

105. J. Klein and J. Y. Becker, J. C. S. Perkin II, (1973) 599; T. L. Chwang and R. West, J. Am. Chem. Soc., 95 (1973) 3324; W. Priester and R. West, J. Am. Chem Soc., 98 (1976)

8426.

106. F. J. Landro, J. A. Gurak, J. W. Chinn, Jr., R. M. Newman, and R. J. Lagow, J. Am. Chem. Soc., 104 (1982) 7345.

107. F. J. Landro, J. A. Gurak, J. W. Chinn, Jr., and R. J. Lagow, J. Organomet. Chem., 249 (1983) 1.

108. C. Chung and R. J. Lagow, J. C. S. Chem. Comm., (1972) 1078; L. A. Shimp, J. A. Morrison, J. A. Gurak, J. W. Chinn, Jr. and R. J. Lagow, J. Am. Chem. Soc., 103 (1981) 5951.

109. A. Maercker and M. Theis, Angew. Chem. Int. Ed. Engl., 23 (1984) 995.

110. E. D. Jemmis, J. Chandrasekhar, E.-U. Würthwein,P. v. R. Schleyer, J. W. Chinn, Jr., F. J. Landro, R. J. Lagow, B. Luke, and J. A. Pople, J. Am. Chem. Soc., 104 (1982) 4275; J. W. Chinn, Jr., and R. J. Lagow, J. Am. Chem. Soc., 106 (1984) 3694.

111. P. v. R. Schleyer, E.-U. Würthwein,E. Kaufmann, and T. Clark, J. Am. Chem. Soc., 105 (1983) 5930.

CHAPTER 4

THE PHOTOCHEMISTRY OF RESONANCE-STABILIZED ANIONS.

Laren M. Tolbert
School of Chemistry, Georgia Institute of Technology, Atlanta, Georgia 30332

CONTENTS

I.	INTRODUCTION	224
II.	THEORY	224
III.	PHOTOPHYSICAL PROPERTIES	226
	A. Absorption	226
	B. Emission	228
	C. Acid-Base Properties	228
	D. Ion-Pair Phenomena	230
	E. Photoejection	230
	F. Intersystem Crossing	232
	G. Isomerization	233
IV.	SIMPLE CARBANIONS	233
	A. Triarylmethyl Anions	234
V.	ALLYL ANIONS	235
	A. 1,3-Diphenylpropenyl Anions	236
	B. Phenalenyl Anion	240
	C. 2-Substituted Allyl Anions	246
	D. Allyl Anions from Cyclopropyl Anions	246
VI.	OTHER POLYENYL ANIONS	247
VII.	CYCLIC ANIONS	249
	A. Cyclopentadienyl, Indenyl, and Fluorenyl Anions	249
	B. 2-Haloindenyl Anions	252
	C. Cyclooctatetraenyl Dianion	253
	D. Cyclononatetraenyl Anion	254
VIII.	OXYGEN-CENTERED ANIONS	255
	A. Enolates	255
	B. Polyenolates	259
	C. Phenolates	260
IX.	CONCLUSION	264
X.	REFERENCES	265

I. INTRODUCTION

The excited states of organic anions are playing an increasingly important role in chemistry, not only as photoinitiators for $S_{RN}1$ reactions and other processes involving electron transfer, but also as convenient sources of photogenerated electrons and as intermediates for investigating ion pairing in organic solvents.

Recent reports have appeared in which photoexcited anions themselves, in addition to being potent photoinitiators, have a rich and varied chemistry of their own when electron-transfer and other photooxidative phenomena are circumvented. Thus an understanding of the nature of the anionic excited state will lead to an increased sophistication in the design of experiments using their unique properties. This report will attempt to review the literature in the area,[1] paying attention not only to theoretical and photophysical constraints, but also the chemistry which results from those constraints.

II. THEORY

Qualitative organic molecular orbital theory has been enormously useful in predicting and understanding the photochemical reactivity of neutral molecules. It is natural to extrapolate such predictions to the photochemistry of anions. The disrotatory ring closure of allyl anion to cyclopropyl anion, for instance, is photochemically allowed in much the same way that the disrotatory ring closure of butadiene to cyclobutene is photochemically allowed (see Fig. 1).[2] It is thus tempting to consider the nonbonded pair of electrons as isoelectronic to a pi-bond and generalize from there. In practice, however, the presence of negative charge has enormous consequences for the photochemistry of anions. The increased electron-electron repulsion over that of neutrals results in decay pathways which minimize that repulsion, particularly through electron transfer or acid-base chemistry. Moreover, since the degree of solvation affects the reaction pathway, solvent becomes involved in a way generally uncharacteristic of neutral photochemistry. All of these properties are a function of the way in which the negative charge is distributed in the excited state as opposed to the ground state, and it is useful to consider some elementary quantum mechanical calculations bearing on this phenomenon.

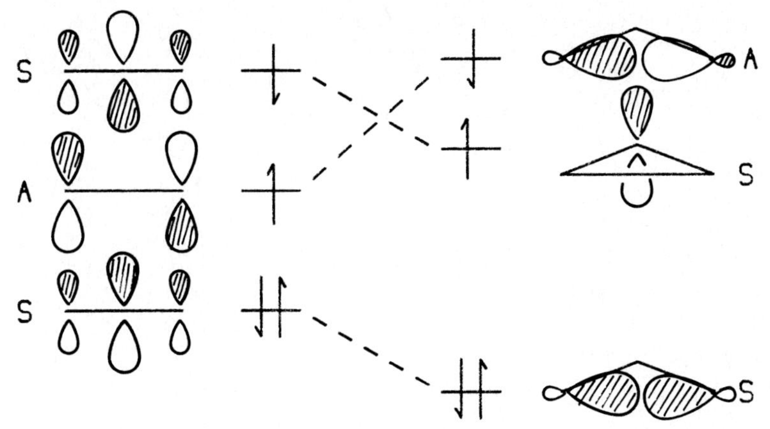

Figure 1. Correlation diagram for conrotatory allyl to cyclopropyl anion photoconversion (C_s symmetry).

The most useful carbanions to treat using elementary Hückel theory are odd alternant hydrocarbon anions, of which the allyl anion (Fig. 1) is the prototypical example. For odd alternants, the non-bonding molecular orbital (NBMO) places negative charge density at all odd-numbered ("starred") carbon atoms,[3] in accord with the simple resonance picture. Thus an immediate difference between the photochemistry of hydrocarbon anions and neutrals might be expected to result from the relatively even distribution of charge in the latter as opposed to the former. More importantly, this distribution of charge, although altered in degree, is not altered in substance upon substitution by, say, aromatic groups. Thus 1,2,3-triphenylallyl anion retains 44% of the negative charge density at C-1,3 and 42% at the remaining o,p aromatic positions in the ground state (see Fig. 2).[3]

To a first approximation, photoexcitation removes an electron from the NBMO and places it in an antibonding MO. In general, this antibonding MO will have coefficients which are smaller on the "starred" positions and larger on the "unstarred" positions. The necessary result, then, of photoexcitation is an "intramolecular charge transfer" resulting in a redistribution of charge. This charge redistribution has important consequences for absorption-emission behavior, for ion-pair formation, and for acid-base properties, all of which are intrinsically related through the Förster equation (see below).

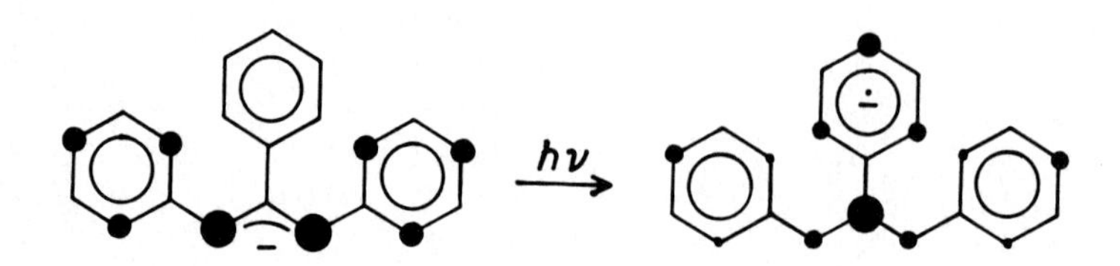

Figure 2. HMO Charge distribution in ground and excited-state 1,2,3-triphenylallyl anion.

III. PHOTOPHYSICAL PROPERTIES

Carbanions, of course, have the same excited state properties available to them as do neutrals. These include absorption, emission, photorearrangement, and electron ejection. Additionally, the chemical properties associated with ground state carbanions must now be dealt with in the excited state. These include acid-base and ion-pair phenomena. Since the high reactivity already associated with carbanions in the ground state is often greatly enhanced in the excited state, the photophysical properties of carbanions present unique challenges to the spectroscopist and photochemist, and the various excited-state decay properties are often poorly defined.

A. Absorption. In general, the removal of a proton from a carbon acid is associated with a marked bathochromic shift of the absorption (see Table 1). For instance, triphenylmethane has λ_{max} of 254 nm, while the conjugate base has λ_{max} of 496 nm,[5] a shift of 242 nm! Although most organic acids do not exhibit shifts of this magnitude, the general rule is that some shift to longer wavelengths ranging from 20 to 300 nm is generally observed (see Table 1). Exceptions to this general trend are carbon acids with highly aromatic conjugate bases and benzoic acids. As we shall see, this shift is intimately associated with changes in excited state basicities as well.

Table 1. Effect of resonance-stabilization on absorption spectra of ion pairs.

Anion	Solvent	Counterion	Form[a]	λ_{max}[b]	Ref.
		Arylmethyl anions			
$Ph\ddot{C}H_2^-$	THF	Na	SSIP	355	5f
	THF	Li	SSIP	330	5g
	THF	-	FI	362	5h
$Ph_2\ddot{C}H^-$	DME	Li	SSIP	448	5b
$PhC_6H_4\ddot{C}H_2^-$	THF	Li	CIP	430	5i
	THF	K	CIP	470	5j
$Ph_3\ddot{C}^-$	THF	Li	SSIP	500	5a
	THF	K	SSIP	486	5a
		Polyenyl Anions			
$PhCH{=}CH{-}\ddot{C}HPh^-$	Ether	Na		535	6b
$Ph(CH{=}CH)_2{-}\ddot{C}HPh^-$	Ether	Na		568	6b
$Ph(CH{=}CH)_3{-}\ddot{C}HPh^-$	Ether	Na		600	6b
$Ph(CH{=}CH)_4\ddot{C}HPh^-$	Ether	Na		635	6b
		Cyclopentadienyl anions			
Indenyl	DME	Li	SSIP	344	51d
Fluorenyl	THF	Li	SSIP	372	51c
9-Phenylfluorenyl	THF	Cs	SSIP	397	5k

a. SSIP, solvent-separated ion pair; CIP, contact ion pair; FI, free ion. b. nm.

Another characteristic of anion absorption spectra is a pronounced bathochromic shift with increasing resonance stabilization. For instance, substitution of benzyl anion by successive phenyl groups produces shifts of 60-90 nm, while anions of the form $Ph{-}(CH_2{=}CH_2)_n{-}\ddot{C}HPh^-$ exhibit 35 nm shifts per vinylene group (see Table 1).[6] When one considers the fact that similar homologation of diphenylpolyenes produces only a 26 nm shift per vinylene group, the effect on anions is quite remarkable. This observation leads to the conclusion that the excited states of anions are stabilized by resonance significantly more than those of their conjugate acids.

B. Emission. Fluorescence from anionic excited states has some similarities to and some important differences with fluorescence from neutral excited states. Just as in neutrals, linear flexible systems fluoresce less efficiently than cyclic ones. Anions such as 1,3-diphenylpropenyl anions fluoresce weakly,[7] and such fluorescence can only be observed at low temperature. Thus most studies involving determination of excited-state lifetimes have relied on cyclic systems such as fluorenyl anions. Aside from the distinct long wavelength emissions associated with low-lying excited states, emission from organic anions is characterized by a strong dependence upon counterion, with ion pair formation leading to shorter excited state lifetimes and, hence, weaker emission.

C. Acid-base properties. The spectral shifts associated with anion formation are intrinsically related to the basicities of the anion and its excited state through the Förster equation[8], Equation 1.

$$pK_a{*} = pK_a + \frac{E_{o,o}(R{:}^-) - E_{o,o}(RH)}{2.3\ RT} \qquad (1)$$

This can be illustrated through the use of a thermodynamic cycle represented in Fig. 3. Thus the pK_a of a photoexcited acid (pK_a*) can be calculated using the measured pK_a of the ground-state acid and the observed shifts in $E_{o,o}$, the excited state energy upon deprotonation. This calculation is only approximate, since it assumes that the solvation of the anion in both ground and excited states is comparable. To the extent that such errors are minor, we can predict that the thermodynamic pK_a*'s of triarylmethane hydrocarbons such as 9-phenylfluorene are below -10, and that their conjugate bases correspondingly are less basic.

The increased acidity in the excited state can be readily understood in terms of charge redistribution in the conjugate base. That is, charge is removed from a carbanion NMBO which is localized at the formal site of negative charge and placed in an antibonding MO which is, in general π* in character and delocalized around an aryl ring. Thus the basicity decreases, at least at the formal carbanionic center.

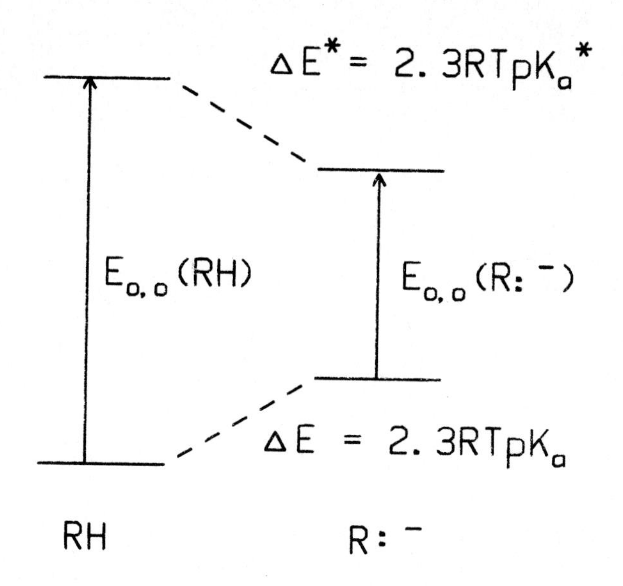

Figure 3. Acidities of photoexcited hydrocarbons.

Experimental verification for the decreased basicity in photoexcited carbanions or, equivalently, for increased acidity in the corresponding hydrocarbon acids, has not yet become available. The shorthand explanation for the strong thermodynamic acidity but weak kinetic acidity of hydrocarbons is attributed to the relatively slow rate of proton transfer to or from carbon, which must compete with excited state decay. If a heteroatom is involved instead, proton transfer can compete with excited state decay, and emission from both acid and conjugate base can be observed. Such effects are evident in systems such as phenolates or naphtholates, as well as aromatic amines, which are isoelectronic with odd-alternant hydrocarbon anions. For such species, the presence of enhanced acidities in the excited state has been observed for several years, particularly in the laboratories of Weller.[9] Due to the presence of non-bonding electrons in both anion and neutral, however, the pK_a changes upon photoexcitation are diminished relative to those of hydrocarbons. In any event, heteroatomic examples of carbanion chemistry will continue to serve as useful models for reactions which have not yet been observed for hydrocarbon anions.

D. Ion-pair Phenomena. Ion-pairing in the ground and excited states is intrinsically related to acid-base behavior. First, ion pairs typically absorb at shorter wavelengths than the free anions, a phenomenon which is related to the localization of negative charge in response to a point charge, the counterion. Because photoexcitation, in general, produces a more diffuse electron cloud, electrostatic terms are less effective at lowering the energies of ion-pairs in the excited state than they are in the ground state. This ground-state energy lowering leads to a higher excitation energy and the pronounced hypsochromic shift in absorption (see Fig. 4). If the excited state is sufficiently long lived, relaxation of the ion pair to free ions can occur. Thus, irradiation of an ion pair can produce fluorescence from the free anion.

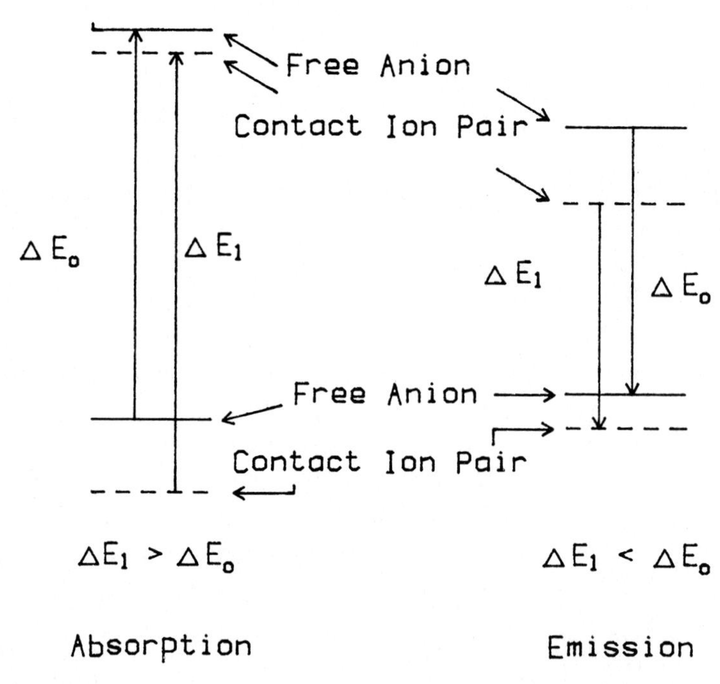

Figure 4. Effect of ion pairing on transition energies.

E. Photoejection. Since carbanions, by definition, have an excess of electron-electron repulsion terms relative to that for neutrals, the electrons in carbanions are less tightly bound. Placing a carbanion in the excited state thus greatly lowers the oxidation potential. For simple carbanions, the excited state will be virtual and autodetachment will occur. In the gas phase, photodetachment can be induced by irradiation at wavelengths below the excited state energy. The onset of photodetachment, corresponding to the minimum energy required to eject an electron, may be related to the electron affinity of the radical assuming negligible Franck-Condon factors. (see Fig. 5). At wavelengths corresponding to the excited state energy, i.e., at "resonance" (420 nm in Fig. 5), a marked increase in the photodetachment cross-section occurs.[10] In solution, the anion and a free electron will be differentially solvated, and the propensity toward photodetachment will depend upon the relative solvation energy of the two species.

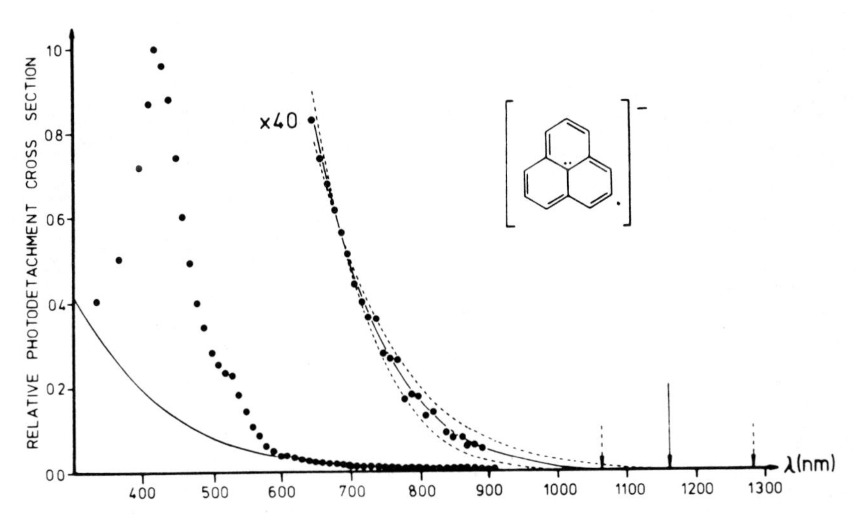

Relative experimental (●) and theoretical (—) photodetachment cross sections for the perinaphthenyl anion as a function of incident photon wavelength. The large arrow points to the threshold at 1.07 eV. Dashed lines and arrows indicate the theoretical curves with thresholds offset by ±0.1 eV from the best fit.

Figure 5. Photodetachment spectrum of phenalenyl anion. Reproduced from ref. 10, copyright 1979, American Chemical Society

When a carbanionic center is substituted by resonance-stabilizing groups, two important phenomena occur. One, a decrease in excited state energy, has already been noted. A second is that the electron affinity of the corresponding neutral increases. For instance, the electron affinity of methyl radical, as measured by photodetachment spectroscopy is 7 kcal/mole, while that of benzyl radical is 20 kcal/mole.[11] Thus the energy required to detach an electron from the carbanion increases and, if the carbanions is sufficiently stabilized, may exceed the excitation energy, in which case the carbanion excited state is bound with respect to photodetachment. This phenomenon is apparently related to the ability of resonance-stabilized molecules to accommodate more readily the greater electron-electron repulsion of the carbanion. While triphenylmethyl radical has 65% of the spin density at the α-carbon,[12] only 10% of the negative charge is localized on the corresponding carbon atom in triphenylmethyl anion.[13] For these reasons, resonance-stabilized carbanions have been the most useful substrates for investigating carbanion photochemistry unencumbered by photoejection leading to ground-state free radical pathways.

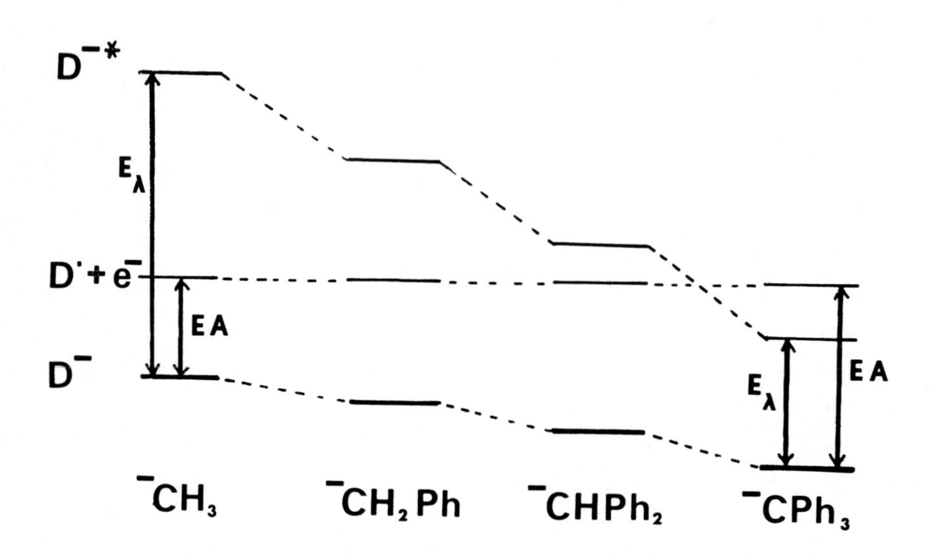

Figure 6. Effect of phenyl substitution on excitation energy and electron affinity.

F. Intersystem Crossing. No carbanionic photoreaction has yet been clearly identified as involving the triplet state. The apparent absence of triplet reactivity is not the result of any theoretical limitation, but rather reflects the experimental difficulties associated with triplet sensitization or, conversely, triplet quenching. On the one hand, the common sensitizers such as benzophenone are excellent oxidizing agents in the excited state and promote electron transfer rather than energy transfer. The effect of benzophenone in promoting the cyclization of phenalenyl anion (vide infra) is apparently not one of sensitization but of inhibition of polymerization. Even weak electron acceptors such as naphthalene are capable of acting as excited-state carbanion oxidants. On the other hand, common triplet quenchers such as perylene have triplet energies too high for efficient triplet energy transfer from what must be extremely low-lying excited states. Assuming such quenchers are available, their low-lying vacant orbitals again make them excellent oxidants of excited-state carbanions. Thus any conclusions about the absence or presence of triplet excited states involving anionic excited states must be met with a healthy skepticism.

G. Isomerization. To the extent that other excited state decay pathways do not compete, valence photoisomerization of a carbanion can result. Often a competing thermal retroisomerization can obscure the photochemistry, which then can be observed only at low temperature. Finally, a competing pathway involving first photoejection followed by isomerization of the resulting radical is very difficult to exclude.

IV. SIMPLE CARBANIONS.

Having dealt with the general photophysical phenomena which have been observed for carbanions, we now consider the various classes of carbanions and how their structure affects their photoreactivity. Carbanions which have been studied include carbon-centered anions in which charge resides mainly on a single carbon center, although delocalization through resonance may play a role. More complex systems are represented by polyenyl anions and cyclic anions. Finally, we include in this discussion enolates and related "carbanions" which are stabilized by electronegativity effects.

Since few isomerization pathways are available to simple carbon-centered anions, the photochemistry of these species is characterized mainly by electron ejection or other photophysical phenomena.

A. <u>Triarylmethyl Anions</u>. Triphenylmethyl anion fluoresces only weakly,[14] while the structurally related 9-phenylfluorenyl anion fluoresces strongly. The weak fluorescence in the former may be associated with decay mechanisms resulting from facile twisting of the phenyl groups in the excited state. More likely, however, is the intervention of photodetachment. In fact, triphenylmethyl anion undergoes photochemistry in the presence of very weak electron acceptors such as dimethyl sulfoxide through a mechanism which involves, at least transiently, formation of dimethyl sulfoxide radical anion.[15] The latter undergoes bond homolysis to yield methyl radical which can add to an additional triphenylmethyl anion to yield 1,1,1-triphenylethane in a mechanism analogous to the $S_{RN}1$ reaction (see Scheme 1). Intervention of the radical anion pathway also results in a significant yield of the para-methylated product, presumably the result of stabilization of the methylenecyclohexadiene radical anion intermediate.[15] Further evidence for a radical anion intermediate is the observation of quenching by electron acceptors.[15d]

$$Ph_3C:^- + Me_2SO \xrightarrow{h\nu} Ph_3C\cdot + Me_2SO^{\dot{-}}$$

$$Me_2SO^{\dot{-}} \longrightarrow Me\cdot \xrightarrow{Ph_3C:^-} Ph_3CMe^{\dot{-}} + Ph_2C{=}C_6H_5(Me)(H)^{\dot{-}}$$

$$Ph_3CMe^{\dot{-}} \xrightarrow{-e^-} Ph_3CMe$$

$$Ph_2C{=}C_6H_5(Me)(H)^{\dot{-}} \xrightarrow{-e^-} Ph_2CH{-}C_6H_4{-}CH_3$$

Scheme 1. Photomethylation of trityl anion in Me_2SO.

Photoalkylation of triphenylmethyl anion has been observed with other electron acceptors. With tert-butylmercury chloride, in which case electron transfer produces tert-butyl radical as the reactive intermediate, alkylation at the alpha position produces 1,1,1-triphenyl-2,2-dimethylpropane in 39% yield as well as the novel unrearranged product of p-tert-butylation, 6-tert-butyl-3-(diphenylmethylene)-1,4-cyclohexadiene, in 21% yield.[16]

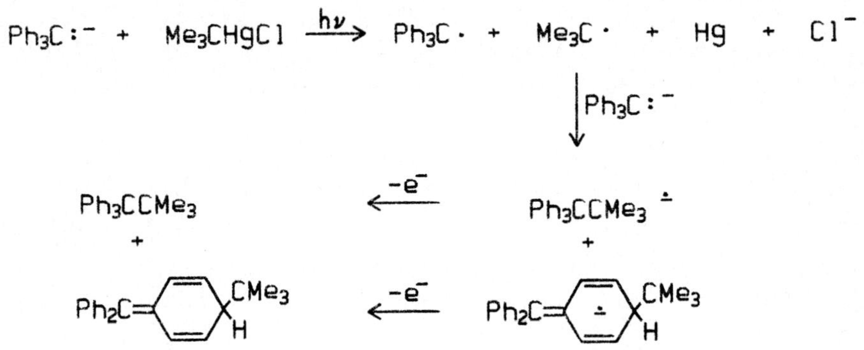

$$Ph_3C:^- + Me_3CHgCl \xrightarrow{h\nu} Ph_3C\cdot + Me_3C\cdot + Hg + Cl^-$$

$$Me_3C\cdot \xrightarrow{Ph_3C:^-} Ph_3CCMe_3^{\dot{-}} + Ph_2C{=}C_6H_5(H)(CMe_3)^{\dot{-}} \xrightarrow{-e^-} Ph_3CCMe_3 + Ph_2C{=}C_6H_5(H)(CMe_3)$$

Figure 7. Photoalkylation of trityl anion by tert-butylmercury chloride.

When an aryl halide is the electron acceptor, triphenylmethyl anion undergoes phenylation, again through a radical-chain $S_{RN}1$ mechanism.[17] In this case, the para-arylated product becomes the dominant product. Thus the role of radical-anions in controlling product regiochemistry remains an important mechanistic and synthetic consideration which remains to be exploited.

4-Biphenylyldiphenylmethyl anion undergoes similar electron transfer chemistry.[15] However, the efficiency is greatly reduced, perhaps reflecting the decreased ability of this lower energy excited state to photoeject.

V. ALLYL ANIONS

The Hückel molecular orbital description of the allyl anion places charge density at the terminal carbon atoms (C-1,3) and none at the central atom for the simple reason that the non-bonding molecular orbital (NBMO) is antisymmetric and thus possesses a node at the central carbon atom (see Fig. 1). Photoexcitation in the single electron description places an electron in the antibonding symmetric orbital and thus induces a shift in charge density to the central carbon atom. If we consider a prototypical allyl anion tagged with a stereochemical label R and an electronic label X we can predict several results of such charge redistribution which will be expressed as primary photochemical processes for such a system, as illustrated in

Figure 8. These are: (i) E-Z isomerization, (ii) protonation, (iii) cyclization to a cyclopropyl anion, (iv) bond cleavage, particularly if X is a suitable leaving group, and (v) electron ejection. These are in addition to the default decay pathways, (vi) fluorescence and (vii) radiationless decay. Of course, the excited state of allyl "anion" is too high in energy to undergo anything other than redox chemistry, and we must consider derivatives which are further stabilized through resonance. Although electron transfer remains the default primary photochemical process, in most cases it does not compete effectively and we are able to examine other pathways. Moreover, even with extensive substitution, the symmetries of the NBMO's are maintained and the basis for predicting the reactivity in such systems remains intact.

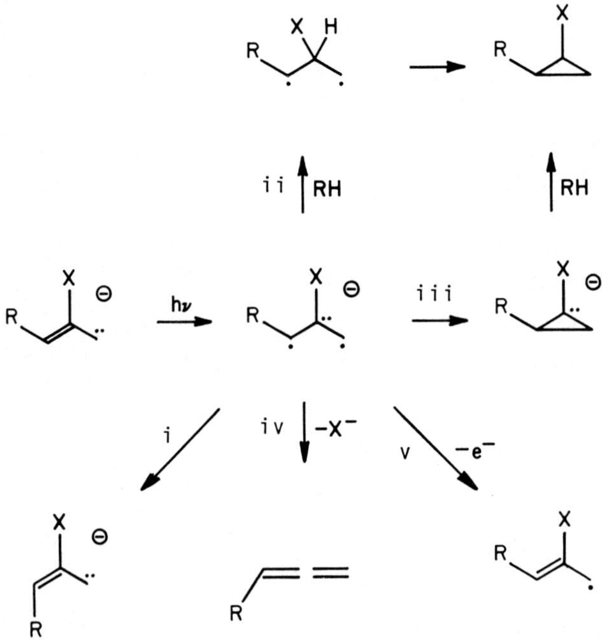

Figure 8. Excited-state decay pathways for photoexcited allyl anions.

A. 1,3-Diphenylpropenyl Anions. The photochemical interconversion of the E and Z ion pairs of 1,3-diphenylpropenyl anions has been reported by Parkes and Young[18] and confirmed by

Bushby[19] (Fig. 9). The observation of photochemistry in this system is complicated by the facile ground state interconversion of the E and Z isomers which has led to some controversy,[20] apparently as a result in solvent effects which affect the rate of this ground state process. Thus 1,3-diphenylallylsodium exhibits no photochemistry in liquid ammonia.[19] The barrier to rotation in 1,3-diphenylallyllithium in tetrahydrofuran is large enough (ca. 18 kcal)[21] that the interconversion can be minimized at low temperature. However, spectroscopic methods, particularly nmr, must be relied upon to sort out the species involved at subambient temperatures. The controversy surrounding the absence of photochemistry in liquid ammonia has apparently been resolved, and the photochemical properties may be summarized as follows: In tetrahydrofuran at low temperature, long wavelength (550 nm) excitation converts the more stable E,E form to the E,Z form, which reverts to the E,E form at higher temperatures. In contrast to the tetrahydrofuran results, irradiation of the anion in liquid ammonia produces no permanent photochemistry, either because electron ejection becomes the dominant pathway or because E,Z-isomerization is too fast for spectral observation.

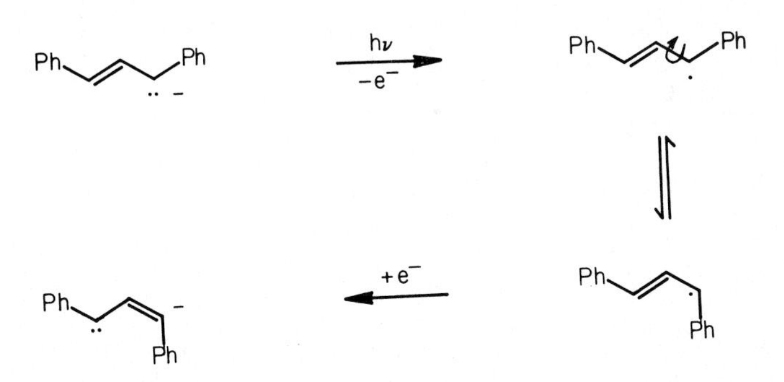

Figure 9. Photoisomerization of 1,3-diphenylallyl anion pairs.

Although the overall transformation is clearly established, the mechanistic details have not been delineated. The only mechanistic clues are the absence of photoisomerization in ammonia, a solvent known to produce solvent separated ions, and the presence of a solvent and counterion-sensitive room temperature fluorescence. The trans isomer is the more fluorescent for reasons which are obscure, although this may reflect a longer lifetime which results in a more competitive cis-trans conversion in the excited state. Some mechanistic possibilities can be discussed, however. In contrast to the ca. 18 kcal rotational barrier for the anion, the 1,3-diphenylallyl radical has a very low barrier to bond rotation, ca. 7 kcal.[22] Thus, photoejection followed by fast ground state E,Z isomerization of the radical and ultimate electron recapture (Path A, Fig. 8) provides a mechanism of simplicity and high appeal. A ground-state analogue of this reaction has been observed by Bushby in the electron-transfer catalyzed isomerization of 1,3-diphenylallyl ion pairs by anthracene.[23] The presence of fluorescence indicates that the excited anion possesses a true excited state of finite lifetime,[24] which excludes direct photodetachment, although not photoejection per se. Moreover, the fluorescence lifetime was shorter in the presence of more easily reducible counterions.[24] This suggests a modification of this mechanism, electron capture by alkali metal, as a possible mode for decay (and cis-trans isomerization). Alternatively, photoejection in solvents which stabilize this process by solvating electrons (i.e., ammonia) may divert the cis-trans isomerization by quenching rather than activating. Obviously, the mechanistic question is by no means resolved.

One attempt to resolve this mechanistic ambiguity for at least one system is represented by 1,3-diphenylallyl anions which are substituted at C-2 by an electronically perturbing substituent.[25] This approach is based upon the premise that substituents at C-2 will have no effect on either the ground state chemistry of the anion or of the radical because of the nodal properties of the NBMO. That is, the NBMO of allyl anion or radical and its derivatives has a zero coefficient at C-2 and the presence of a substituent will not perturb the energy of this MO in the Hückel approximation. Thus photochemistry which proceeds via prior photoejection to produce an allyl radical will be independent of C-2 substitution. Conversely, a bound excited state will have an electron in a symmetric antibonding orbital which will be strongly dependent upon the identity of the C-2 substituent. The variable chosen is the reduction potential of the substituent. The electronic probes phenyl, 4-biphenylyl, and 2-naphthyl have decreasingly negative reduction potentials (see Table 2). For these substituents, molecular models indicate that the steric effect of these substituents is identical insofar as the terminal C-1 and C-3 phenyl groups were concerned. In fact, both pmr and cmr spectra of all three anions in a variety of solvents are indistinguishable save for the resonances associated with the substituent. Moreover, the chemical shifts of the substituents are unperturbed by anion formation, indicating that the simple Hückel model placing no charge at the central atom is correct. In contrast to the ground state properties, the excited state properties are quite different for the three anions. The substituent which most stabilizes the LUMO, 2-naphthyl, is also the most effective for inducing E-Z-trans isomerization. When cis-stilbene is added as a probe for electron transfer--cis-stilbene is known to undergo efficient isomerization in the presence of a high potential electron donor--the 2-naphthyl anion is nearly ineffective (see Table 2). These results are consistent with the description of a bound excited state stabilized by a central substituent (see Fig. 10).

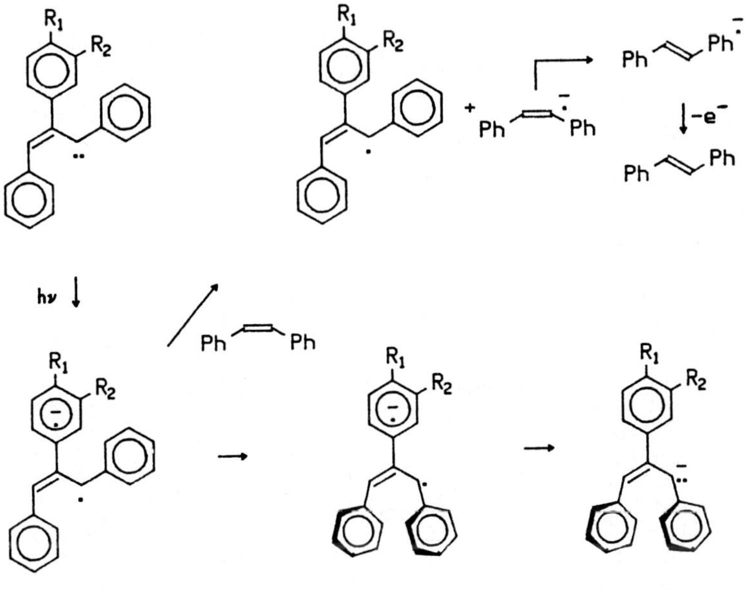

Figure 10. Photochemistry of 1,3-diphenyl-2-arylallyl anions.

Table 2. Photochemical Properties of 1,3-Diphenyl-2-arylallyl Anions.

property	Solvent	2-aryl group: Ph	Ph-C_6H_4-	$C_{10}H_7$
Z/E ratio after quenching				
before irrad	Me_2SO	0.37	0.30	0.86
4-h irrad	Me_2SO	0.38	0.56	0.69
10-h irrad	Me_2SO	0.40	0.82	0.36
% cis of added cis-stilbene	THF	0.86	0.08	0.11

Other substituted allyl anions have been shown to undergo photorearrangement. In particular, the 1,3-dicyanoallyl anion undergoes photoisomerization among all three E,E-, E,Z-, and Z,Z-geometric forms.[25]

B. Phenalenyl anion and the mechanism of electrocyclization. The cyclization of an allyl anion to a cyclopropyl anion, although formally allowed, has not been rigorously observed. Irradiation of 2-aryl-1,3-propenyl anions might have been expected to induce the formation of a cyclopropyl anion by stabilization of the incipient negative charge by the C-2 substituent. Such was not the case, even when, in addition to phenyl, 4-biphenyl, and 2-naphthyl, more strongly electron withdrawing substituents such as 4-chlorophenyl and 4-cyanophenyl were employed.[24b] In none of the photochemical mixtures were cyclopropane products observed, even when proton sources for fast protonation of intermediate cyclopropyl anions were available. Incorporation of the allyl anion into the antiaromatic 8-π-electron 7-phenyl-2:3,4:5-dibenzocycloheptatrienyl anion, for similar reasons, might have been expected to destabilize the open chain form relative to the closed form, but oxidative photodimers were the only observed products (see Fig. 11).[26] That is, photoejection became the default reaction pathway. In a similar attempt, cyclopropane formation is also avoided in a number of systems investigated by Fox.[27] Fortunately, "failure" also provides information, probably as a result of fundamental energetic constraints which prevent such cyclopropyl anion formation, as we shall see.

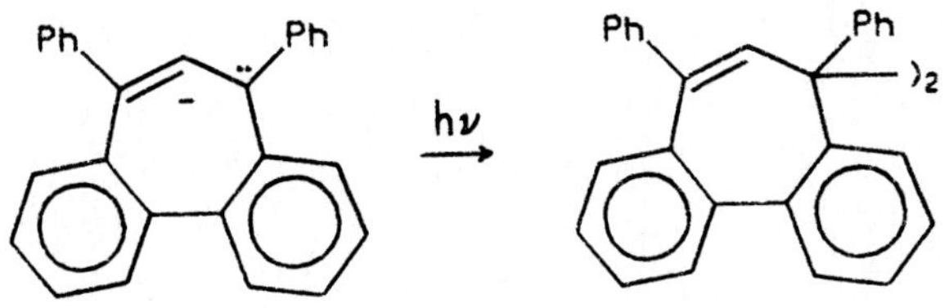

Figure 11. Photochemistry of phenyldibenzocycloheptatrienyl anion.

The one exception to the non-intervention of cyclization pathways in allyl anion photochemistry appears to be Hunter's discovery of an apparent allyl to cyclopropyl anion photoconversion. Irradiation of phenalenyl anion, formally a 1,3-diarylpropenyl anion, in t-butyl alcohol produces a cyclopropane, cyclopropa-acenaphthene,[28] a result which can be explained by invoking the intermediacy of the corresponding cyclopropyl anion which undergoes rapid protonation by solvent (see Fig. 12). Curiously, benzophenone is required for efficient reaction, although the role of this ubiquitous triplet sensitizer is, in this case, to suppress polymerization rather than to sensitize.

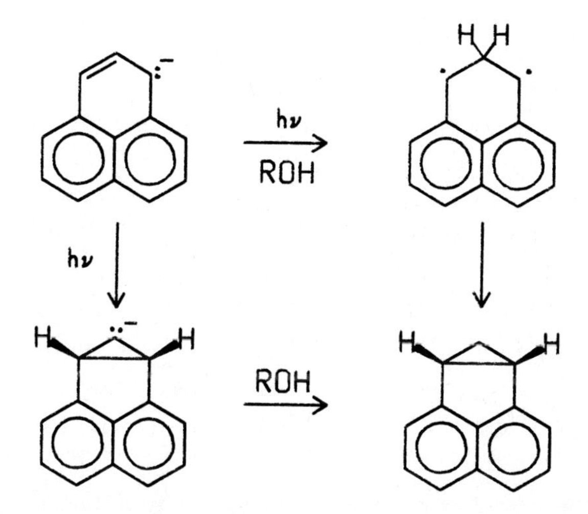

Figure 12. Cyclization pathways in phenalenyl anion.

Examination of the relative energies of the phenalenyl anion and the presumed cyclopropyl anion intermediate reveals that the excited state does not possess enough energy for its formation. For propenyl anion itself, MNDO calculations indicate a gas phase activation barrier of 50 kcal for the direct cyclization.[29] However, a delocalized anion such as the phenalenyl anion is more stable than propenyl anion, and a correction for these barrier estimates must be made by using the relative pK_a's. On the reactant side, the anion must be adjusted for the difference between pK_a's of propene (ca. 42)[30] and phenalene (18),[31] i.e., 2.3RT(42-18) = 34 kcal/mole. On the product side, since the carbanionic site is not stabilized in either case, the pK_a's will remain relatively unchanged, and the difference will be approximately zero. Since phenalenyl anion is stabilized by an additional 34 kcal over propenyl anion, the transition state will reflect some fraction (say, 50%) of that additional 34 kcal/mole required for charge localization. Assuming the difference in strain energy for propene to cyclopropane is comparable to phenalene to cyclopropa-acenaphthene, the (ground state) transition energy will be ca. 50 + 34/2, or 67 kcal/mole (see Fig. 13). The S_1 energy of phenalenyl anion itself is

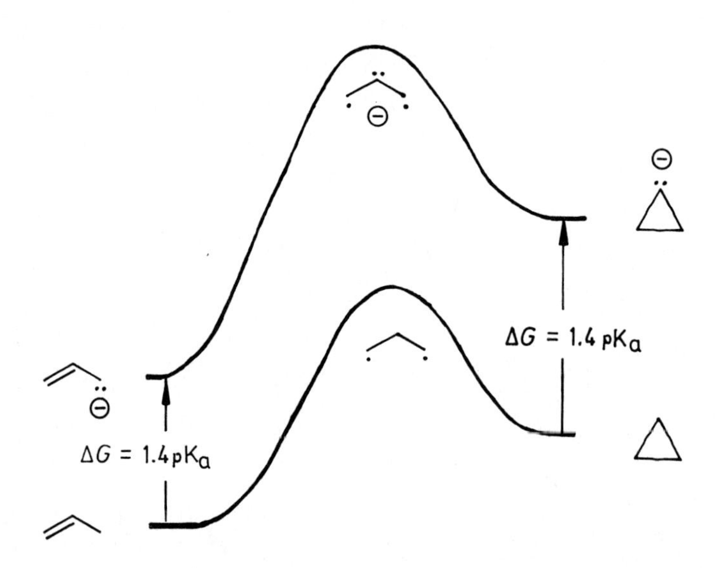

Figure 13. Energetics of allyl to cyclopropyl anion conversion.

approximately 500 nm (54 kcal). Thus there is not enough excitation energy to overcome the already high ground state barrier. This calculation, of course, ignores solvent effects. To the extent that a solvent can stabilize a localized charge better than a delocalized one, e.g., an ion-pairing solvent, the effect of charge localization can be minimized. As the anion-counterion bond becomes covalent, one can expect cyclization to occur. In fact, an allyl carbanionoid, cinnamyl magnesium bromide, is known to undergo exactly this cyclization (see Fig. 14).[32] In the limit of a completely covalent bond, e. g., carbon-hydrogen, such migration becomes a familiar feature of hydrocarbon photochemistry.[33] A possible theoretical model for such migrations is that they are proton migrations on a photoexcited allyl anion backbone. Thus the acid-base properties of hydrocarbons and their conjugate bases may provide insight into photochemical reactivity of neutral molecules.

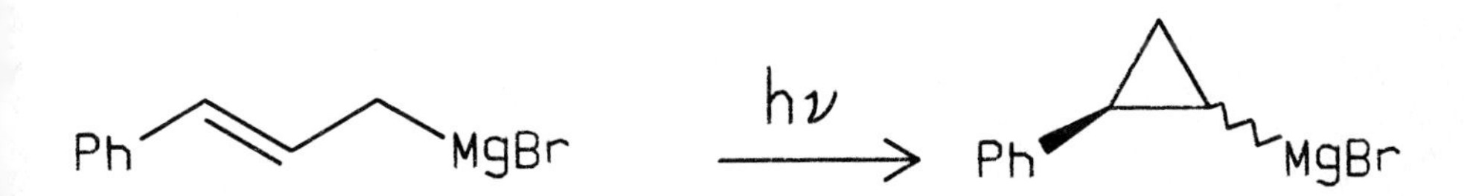

Figure 14. Photocyclization of cinnamylmagnesium bromide

Insofar as delocalized carbanions are concerned, since the product sp^3-hybridized cyclopropyl anion is only poorly stabilized by resonance, attempts to stabilize such pathways are apparently doomed. Thus the failure of delocalized allyl anions to undergo cyclization can be traced to a simple fact: The more efficient the delocalization, the lower the excited state energy and the higher the energy cost of charge localization. Attempts to destabilize the ground state, conversely, lead to excited states exceeding the photoejection threshhold and produce oxidative dimers.

Exceptions to this generalization which, in fact, prove the rule, are provided by hetereoatomic examples. Thus nitranions undergo photorearrangement through an oxazirine oxide intermediate,[34] a result which can be attributed to the facility of nitrogen to stabilize the lone pair of the product (see Fig. 15).

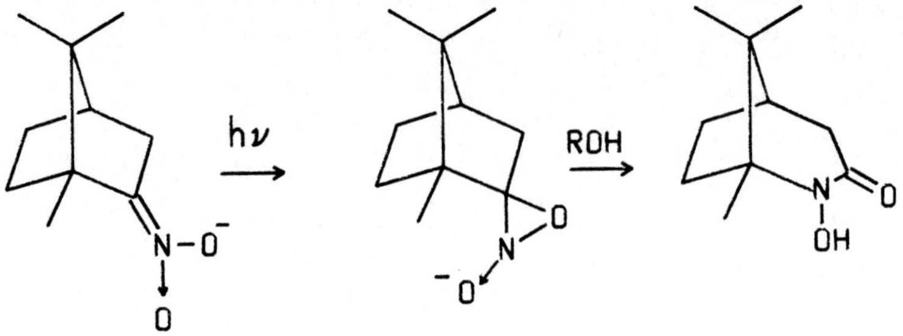

Figure 15. Photochemistry of nitronate anions.

The enolate of 3-pentanone, an "oxaallyl" anion, also produces a low yield of cyclic product, the corresponding epoxide (see Fig. 16).[35] Again, thermodynamic considerations suggest that the epoxide anion is not an intermediate, although this is a formally allowed cyclization pathway.

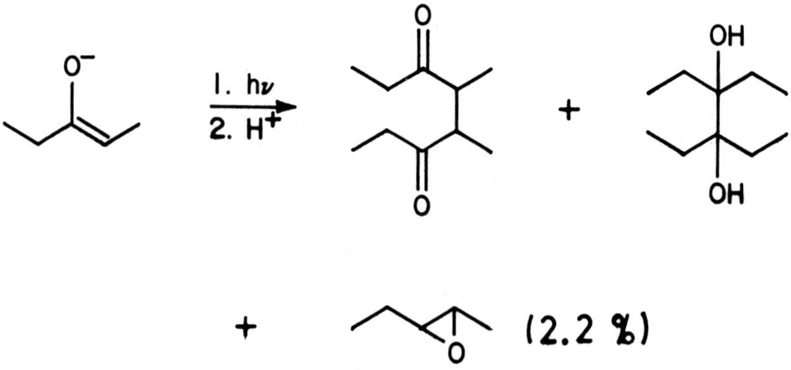

Figure 16. Photodimerization and photocyclization of 3-pentanone enolate.

Charge redistribution away from the formal site of charge in the ground state necessarily requires an increased basicity at alternate, formerly non-basic, sites. For instance, allyl anion might undergo protonation at C-2 to generate a biradical which can close to a cyclopropane. This possibility requires that protonation at C-2 be more favorable than at C-1 to (re)generate a propene. In fact, Bernardi and Schlegel have discovered an excited state energy minimum in the C_3H_6 surface at trimethylene,[36] a result exactly matching the Hückel prediction (see Fig. 17). Because of the inherently slow rates of protonation at carbon, however, such behavior has only rarely been realized. The 1,3-diphenylindenyl anion, although only formally an allyl anion, undergoes sluggish hydrogen-deuterium exchange in the presence of tert-BuOD.[37a] 1,1,3,3-Tetraphenylpropenyl anion undergoes hydrogen-deuterium exchange at C-2 without cyclopropane formation.[37b] Presumably, proton loss from the resulting biradical is faster than carbon-carbon bond formation. In the case of phenalenyl anion, protonation followed by ring closure provides a mechanism of extraordinary simplicity which, furthermore, is in accord with the known propensity for ring closure of the known 1,8-naphthoquinodimethane diradical.[38]

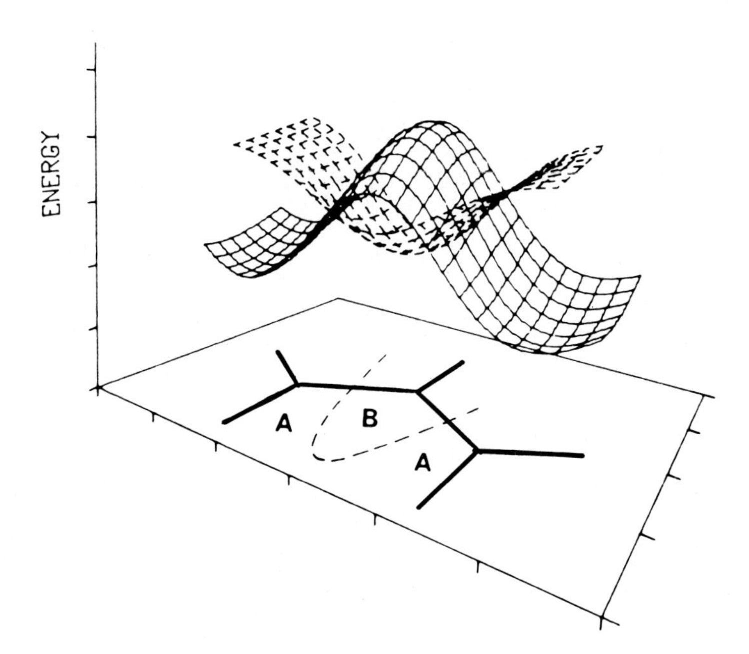

Figure 17. Intersection of diabatic energy surfaces for hydrogen moving 1.1Å above a rigid allyl radical framework. The solid grid corresponds to the closed shell (packet A) and the dashed grid to the diradical surface (packet B) (from ref. 36, copyright 1984, American Chemical Society).

C. <u>2-Substituted allyl anions and photochemical bond cleavage</u>. If C-2 of an allyl anion contains a nucleofuge, an "alpha-cleavage" can result. In the simple acyclic case, the product isolated is an allene. Thus both 1,3-diphenyl-2-phenylthiopropenyl anion and 1,3-diphenyl-2-phenylsulfonylpropenyl anion undergo photoelimination to produce 1,3-diphenylallene.[37b] Given that these eliminations occur thermally, this result is unexceptional, although this technique is useful in the generation of strained allenes (see below).

D. <u>Allyl anions from cyclopropyl anions</u>. Although cyclization of an allyl anion to a cyclopropyl anion has not been demonstrated except in the specialized case of cinnamylmagnesium bromide, the reverse reaction, photopromoted retrocyclization of a 2,3-diphenylcyclopropyl anion, has been demonstrated (see Fig. 18).[39,40] Curiously, the success of this rearrangement apparently depends on the ionicity of the cyclopropyl anion. When the carbanionic center is stabilized by a strongly electron-withdrawing group such as cyano or methoxycarbonyl, facile disrotatory ring-opening is observed. In contrast, when the anionic center is unstabilized or weakly stabilized by vinyl or bromo, the anion is photochemically inert.[27] This effect has been attributed to a strengthening of the carbon-carbon bond being broken when the carbon-alkali metal bond is made more ionic. Alternatively, MNDO calculations indicate a closer approach of the ground and excited state surfaces near the transition state, as well as more modest activation energies, when the carbon-metal bond is more ionic.[27]

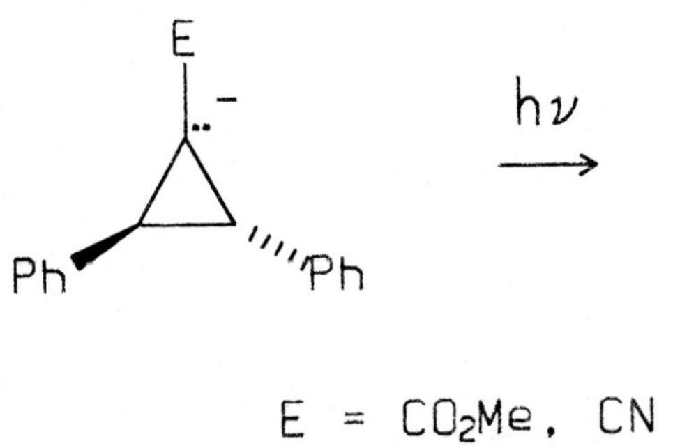

Figure 18. Retrocyclization of cyclopropyl anions.

VI. OTHER POLYENYL ANIONS

In contrast to allyl anions, the higher polyenyl anions have seen little effort, either at the ground state level or excited state level. Parkes and Young have carried out a study on ion pairing in 1,5-diphenylpentadienyl, 1,7-diphenylheptadienyl, and 1,9-diphenylnonatetraenyl anions[41] and have discovered two related effects associated with increasing chain length. First, there is an increasingly larger bathochromic shift on going from tight to loose ion pair. Second, the temperature for the transition from tight to loose ion pair increases. Both these effects are associated with an increasingly delocalized excited state compared to the ground state as chain length increases. That is, the ground state is polarizable and better able to accomodate a point charge and the excited state less so. These effects presumably mirror excited state pK_a effects, but these have not been calculated.

Electrocyclic reactions of certain polyenyl anions have been observed. For instance, 8,8-dimethyl-2,4,6-cyclooctatrienyl anion forms a photoequilibrium mixture with 8,8-dimethylbicyclo-[5.1.0]octa-3,4-dienyl anion (see Fig. 19).[43] The parent ring system of the latter, bicyclo[5.1.0]octa-3,4-dienyl anion undergoes a similar cycloreversion.[44] However, in this case deprotonation to form the cyclooctatetraenyl dianion competes with recyclization.

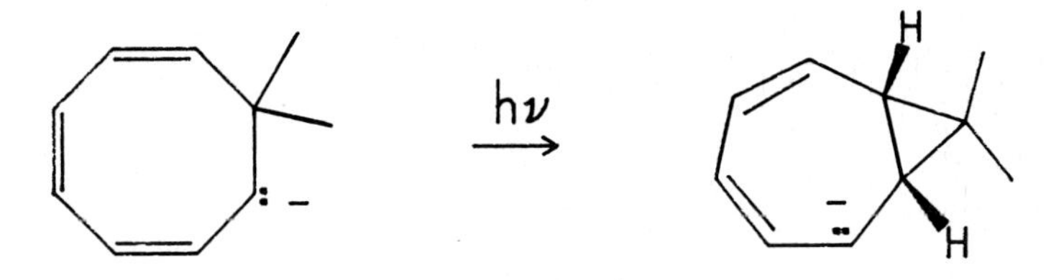

Figure 19. Photocyclization and cycloreversion of 8,8-dimethylcyclooctatrienyl anion.

Fox and coworkers have carried out extensive studies on cyclooctatrienyl anion.[45] No cyclized products were observed under their conditions; this may reflect the energetic constraints outlined above. Instead, products of electron-transfer and radical coupling were isolated. In the case of added anthracene, the product of formal nucleophilic addition was isolated, although the mechanism probably involves electron-transfer to anthracene followed by coupling of cyclooctatrienyl radical and anthracene radical anion. With bromobenzene as electron acceptor, the usual $S_{RN}1$ coupling type of product--a phenylcyclooctadiene of undetermined regiochemistry--was observed.

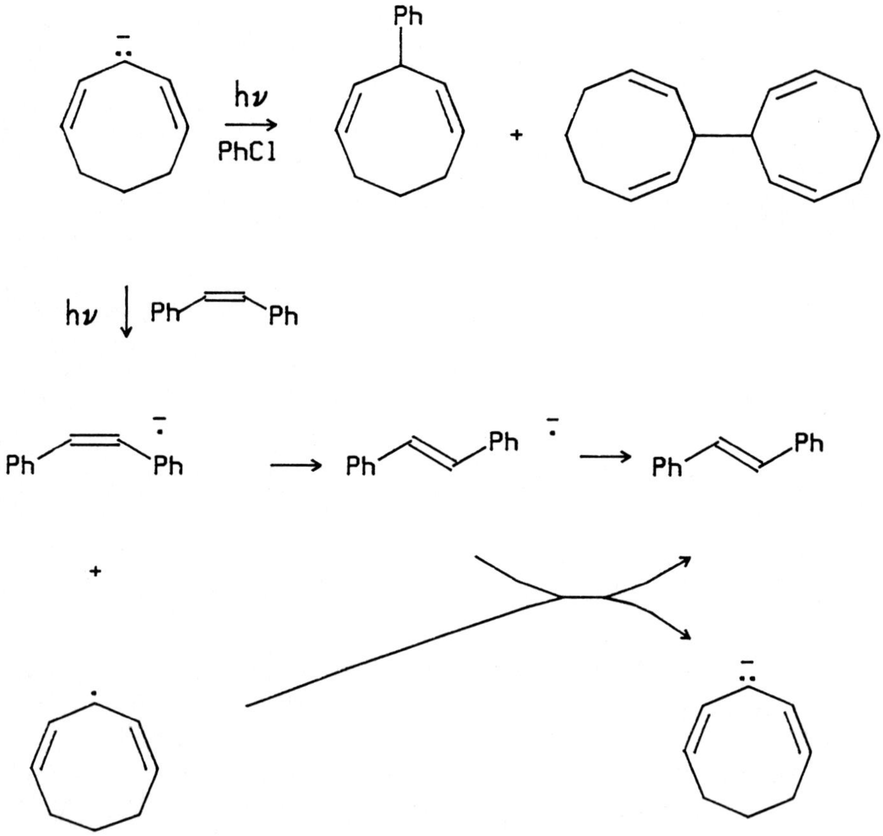

Figure 20. Photoredox chemistry of cyclooctadienyl anion.

VII. CYCLIC ANIONS

The photochemistry of aromatic anions might be anticipated to diverge from that of acyclic nonaromatic anions. In fact, the spectral shifts associated with deprotonation to yield such anions exhibit bathochromic, or even hypsochromic, behavior upon deprotonation, and such systems might be expected to have an enhanced excited state basicity. This phenomenon apparently has been observed in one case, that of cyclononatetraenyl anion.[46]

A. Cyclopentadienyl, indenyl, and fluorenyl anions.

Irradiation of cyclopentadienylsodium produces dicyclopentenyl as the major product.[47] A plausible mechanism involves hydrogen atom abstraction, proton transfer, and dimerization of the resulting radicals (see Figure 21), although other mechanisms are possible.

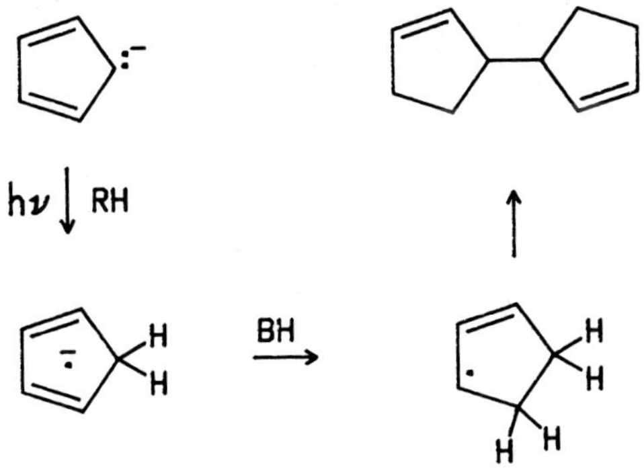

Figure 21. Photodimerization of cyclopentadienyl anion.

Indenyl anion itself apparently only exhibits photochemistry in the presence of electron acceptors. Irradiation in the presence of bromobenzene produces 1-phenylindene but not 2-phenylindene (see Fig. 22).[48] This observation raises interesting theoretical considerations. Since the reaction has all the earmarks of an $S_{RN}1$ reaction, in which the intervention of a radical anion is implicated, arylation at the 2-position might have been anticipated. Previous evidence has indicated that in the radical anion manifold, isoindenes are more stable than indenes. Thus thermodynamic control of the $S_{RN}1$ reaction should lead to the 2-phenyl isomer. This contrathermodynamic result has been interpreted by Rossi as arising from frontier orbital control involving the highest occupied molecular orbital (HOMO) of the nucleophile and a singly-occupied orbital (SOMO) on the radical.[48] Since the HOMO of the indenyl anion has a large coefficient at C-1, arylation occurs at that site.

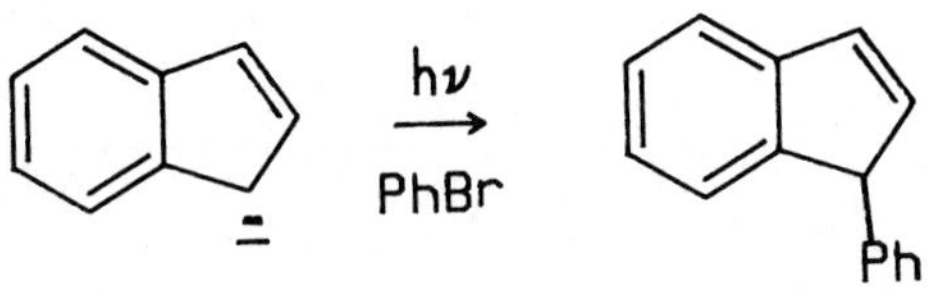

Figure 22. Photoarylation of indenyl anion.

The frontier orbital argument is complicated by the photochemistry of 1-phenylindenyl anion.[49] Here the ambident nucleophile has three sites for radical attack, with the highest Hückel coefficient at C-1, the site of phenyl substitution. Nevertheless, the major product from irradiation of this anion in the presence of bromobenzene arises from attack at C-3. An argument based upon the basicity of the various centers has been advanced, in which the bond formed fastest reflects attack at the most basic site (see Fig. 23).[15b] Of course, steric effects can also be invoked, although the basicity argument rationalizes the regiochemistry of arylation of a number of substituted indenes. More recently, Russell has uncovered examples in the alkylation of enolates in which neither reaction thermodynamics nor anion basicities seem to be controlling.[50]

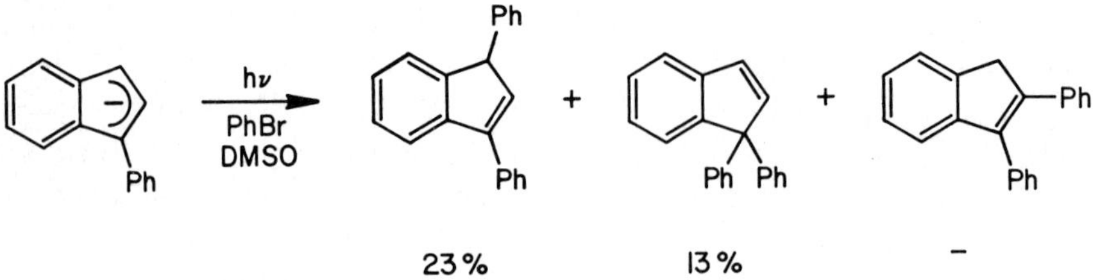

Figure 23. Photoarylation of 1-phenylindenyl anion.

The photophysics of fluorenyl anions has been studied extensively as strongly fluorescent substrates for investigating ion-pair phenomena.[51] Excited-state ion pair behavior follows the conventional pattern, with increasing cation size causing the absorption maximum to shift to the red but the emission maximum to shift to the blue . However, the magnitude of the emission shift is less than that of the absorption shift, since the perturbation of the excited-state, which is more diffuse than the ground state, will be relatively smaller.

The photochemistry of fluorenyl anions and their derivatives has received correspondingly little attention. Fluorenyl anion itself, as well as its 9-phenyl derivative, undergoes photomethylation in dimethyl sulfoxide.[15a] This electron-transfer reaction can be diverted in the presence of electron acceptors. A novel acceptor investigated by Fox is a semiconductor surface. Under those conditions, photomethylation in dimethyl sulfoxide is suppressed and oxidative dimers are produced (see Fig. 24).[52]

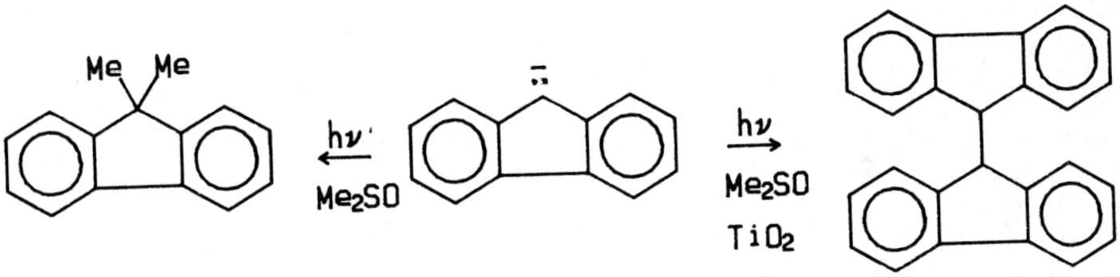

Me Me
hν
Me_2SO
hν
Me_2SO
TiO_2

Figure 24. Photooxidation of fluorenyl anion.

A case of what might be considered an intramolecular $S_{RN}1$ reaction is provided by 9-(2-chlorophenyl)fluorenyl anion.[53] This carbanion undergoes photodehalogenation to yield fluoradene in high yield (see Fig. 25). The nature of the intermediate is not yet clearly established.

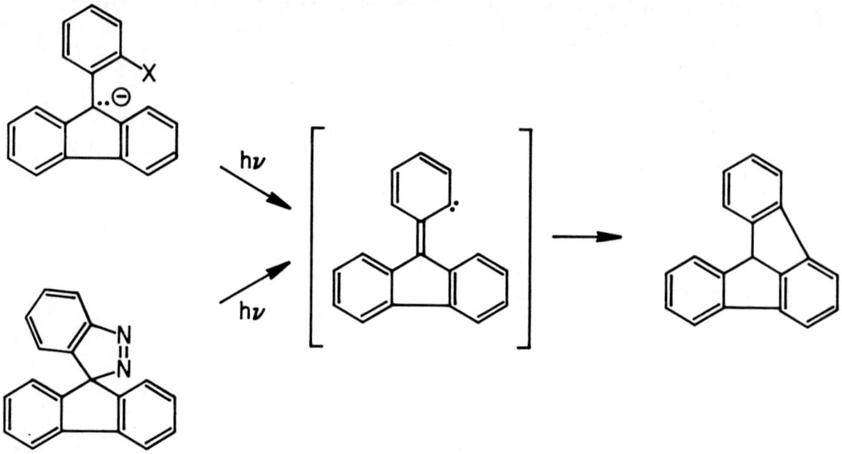

Figure 25. 9-(2-Chlorophenyl)fluorenyl anion photolysis.

B. 2-Haloindenyl anions and photochemical bond cleavage. If a 2-haloallyl anion is constrained by a ring, ground state reactions are inhibited and novel photoproducts can result. An example is provided by the 2-halo-1,3-diphenylindenyl system. Although not formally an allyl anion, the same antisymmetric/symmetric dichotomy exists for the ground and excited states. Although the anion is stable thermally, irradiation of the bromo or chloro derivative results in photodehalogenation and the formation of adducts which depend upon the identity of solvent and added substrate.[54] All products can be rationalized on the basis of the novel carbene 1,3-diphenylisoindenylidene (see Fig. 26). In the absence of added substrate, solvent adducts are produced. For dimethyl sulfoxide and tetrahydrofuran, the products from hydrogen atom abstraction and radical-radical recombination are produced, while tert-butyl alcohol produces an OH insertion product. For allylic olefins, hydrogen atom abstraction and radical-radical recombination also results. More startling are adducts produced from the electron rich olefins, 1,1-dimethoxyethylene and ethyl vinyl ether. In this case, a formal 2+7 cycloaddition with loss of alcohol from the adduct results in formation of a benzo[3,4]fluorene. Similarly, 2,3-dimethyl-1,3-butadiene produces a dihydrofluorene. In these case, the products can be rationalized on the basis of an unobserved spiroisoindene intermediate (see Fig. 27).

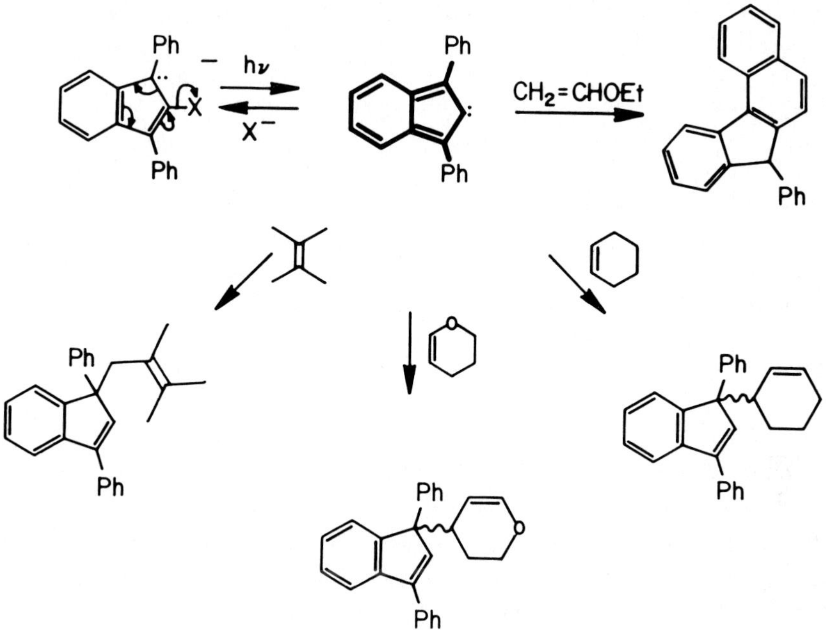

Figure 26. Photolysis of 2-halo-1,3-indenyl anion.

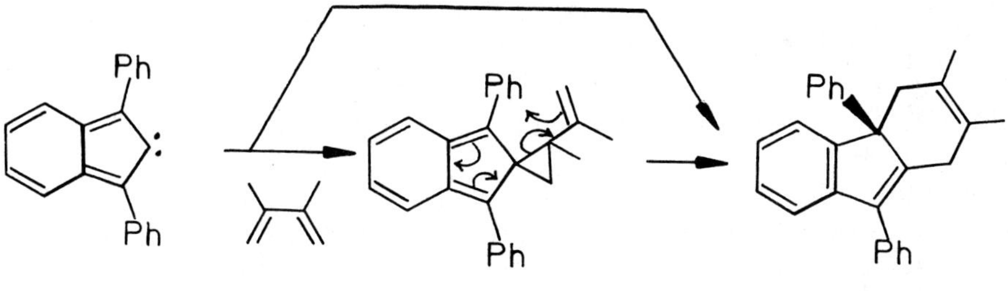

Figure 27. Cycloaddition of olefins to 1,3-diphenylisoindenylidene

C. Cyclooctatetraenyl Dianion. The cyclooctatetraenyl dianion ($COT^{=}$) has been extensively investigated by Szwarc.[55] The doubly charged nature of this aromatic anion makes electron ejection the exclusive decay pathway. In the presence of excess cyclooctatetraene, the latter is the electron-acceptor, and radical anion disproportionation leads to reformation of dianion (see Fig. 28). The reversible nature of this photochemistry has led to the proposal of the cyclooctatetraene dianion as a photosensitizer in solar energy conversion.[56]

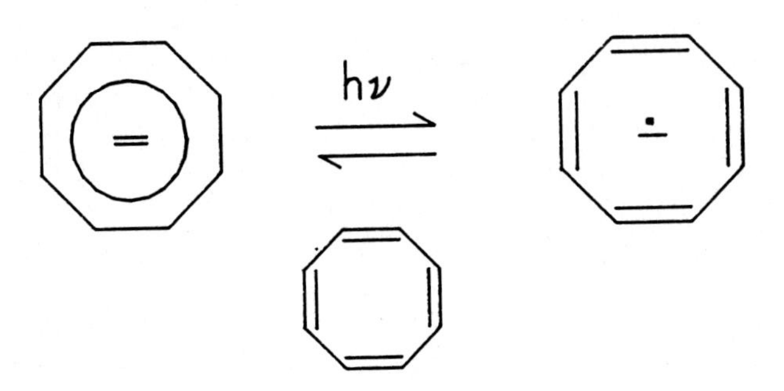

Figure 28. Photoejection from cyclooctatetraenyl anion.

D. Cyclononatetraenyl anion. Irradiation of this anion leads to the formation of dihydroindenes and other photoproducts through the presumed intervention of cis,cis,cis,cis-1,3,5,7-cyclononatetraene,[46] which is known to undergo cyclization to give the ultimate isolated product (see Fig. 29). This is apparently an example of enhanced excited state basicity leading to that all cis intermediate.

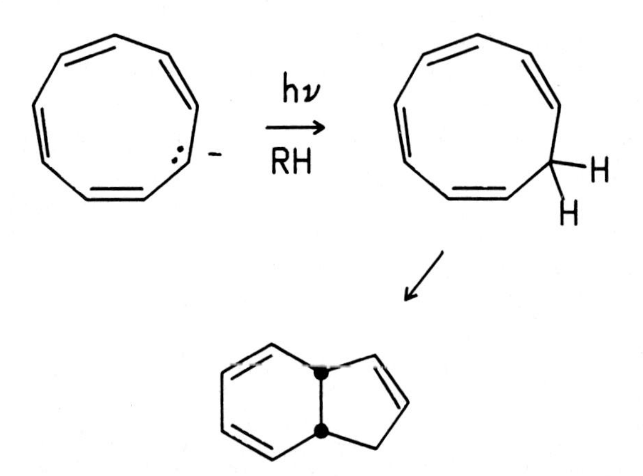

Figure 29. Photoprotonation/cyclization of cyclononatetraenyl anion

VIII. OXYGEN-CENTERED ANIONS

To the extent that enolates, phenolates, and other oxygen-centered anions are isoelectronic with odd-alternant hydrocarbon anions, the non-bonding molecular orbital has a large coefficient at oxygen. Since photoexcitation formally transfers an electron from that orbital to one with larger coefficients on carbon, photoprotonation on carbon and other results of "intramolecular charge transfer" can be expected and have been observed.

A. Enolates. Even in the absence of added electron acceptors, simple enolates undergo redox chemistry. For instance, 2-pentanone enolate undergoes formation of the products of both oxidative and reductive dimerization (see Fig. 16).[35] Although the mechanism is unknown, an attractive mechanism is one in which undeprotonated ketone acts as an electron acceptor from photoexcited enolate to yield an enolate radical and the ketone radical anion, each of which can dimerize. The cross-dimerization product, a photoaldol, would undergo the reverse reaction and so not be observed. A minor product, an epoxide, may result from excited state protonation at carbon, as predicted by Hückel theory (see Fig. 30). The intervention of photoejection in the photochemistry of enolates is not surprising, given the relatively short absorption wavelengths and resulting high excited state energies.

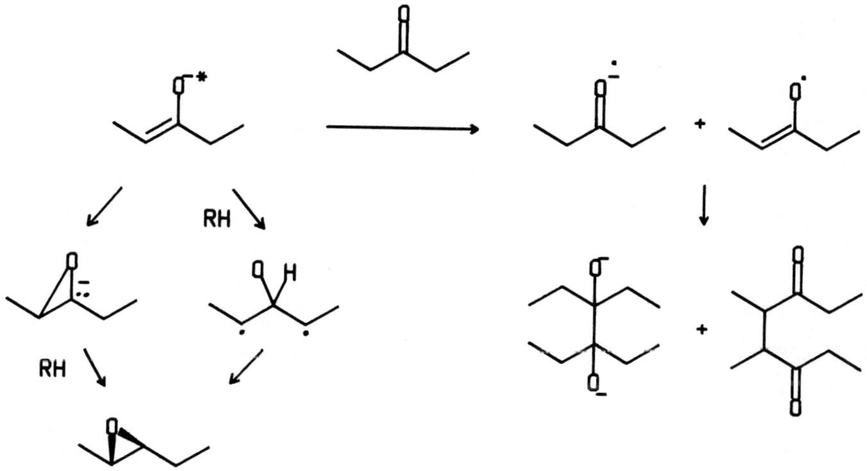

RH

RH

Figure 30. Mechanisms of epoxide formation

In the presence of reactive electron acceptors, carbon-carbon bond formation can result. With aryl halides, alpha-arylation of the enolate occurs throught the now well-known $S_{RN}1$ process (see Scheme 2),[17] and this process is one of the few

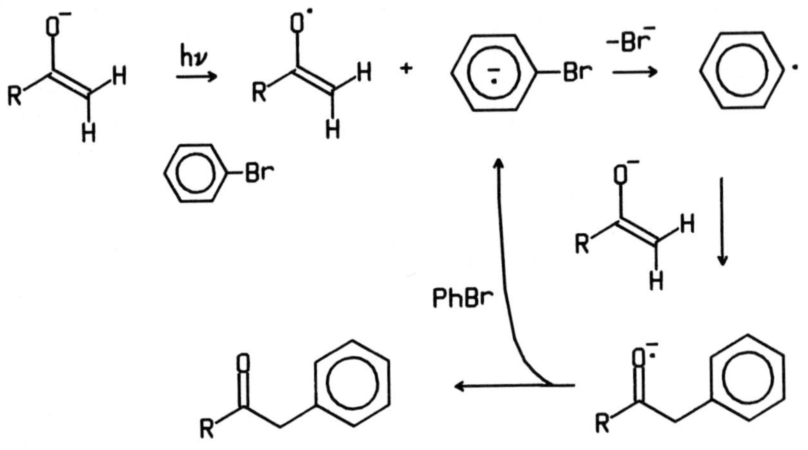

Scheme 2. $S_{RN}1$ mechanism

predictable, efficient arylation methods for enolates. The $S_{RN}1$ arylation is characterized by distinct regiochemistry. For instance, unlike traditional alkylative methods, the $S_{RN}1$ method does not produce products of O-arylation except under forcing conditions.[57] Also, for the enolate derived from phenylacetate anion, para-arylation is the exclusive product, in contrast to the corresponding ethyl phenylacetate monoanion, which undergoes attack exclusively at the α position (see Fig. 31).[58] This regiochemistry has been rationalized in various ways. For instance, the stability of the radical anion intermediate is greater for C-arylation, which produces a ketone radical anion, than for O-arylation, which would produce the radical anion of a vinyl ether. Such stability has also been invoked for the formation of products of para-arylation. Thus other electron acceptors which have been employed in $S_{RN}1$ reactions of enolates include nitroalkanes and vinyl and alkyl mercurials.[59] Particularly significant is the use of organomercurials as electron acceptors, which allows the incorporation of tert-butyl groups.[16] tert-Butylation of enolates by standard alkylative methods is, of course, impossible given the tendency of tert-butyl chloride and other alkylating agents to eliminate.

Table 3. Representative enolates and electrophores in the $S_{RN}1$ reaction.

Enolate	Electrophore	Product (% Yield)	Ref.
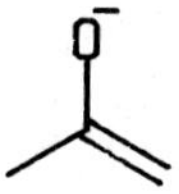	PhBr	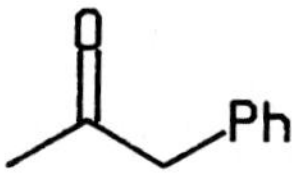 Ph (94)	71
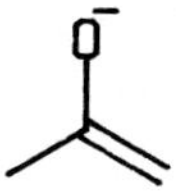	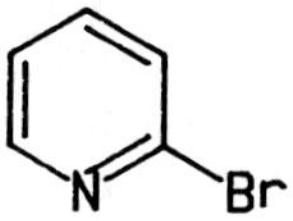 Br	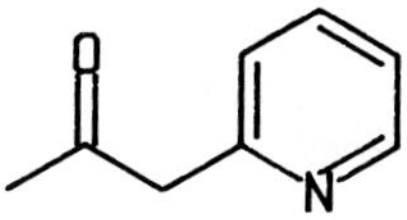 (100)	72
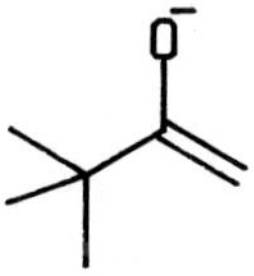	PhBr	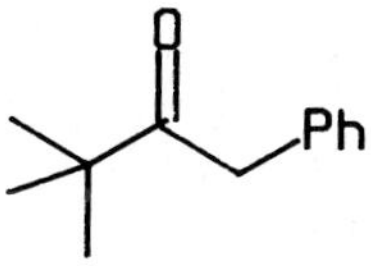 Ph (96)	73
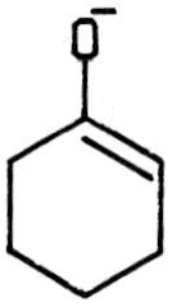	PhBr	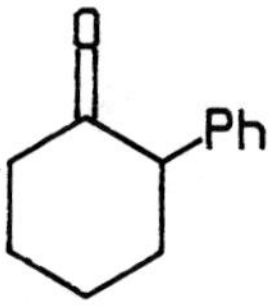 Ph (72)	73
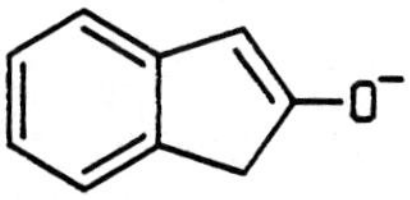	PhBr	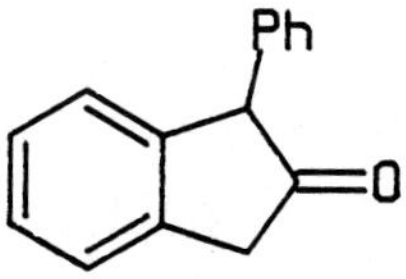 Ph (90)	73
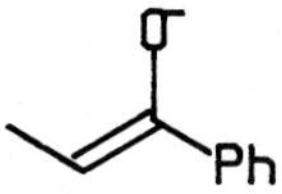 Ph	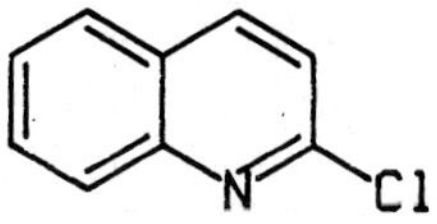 Cl	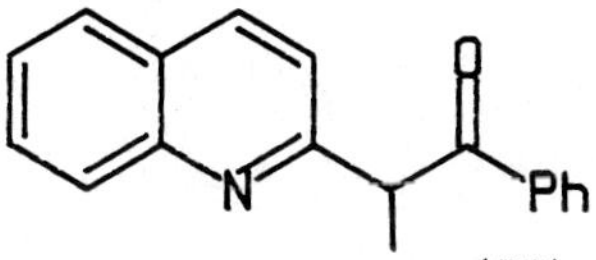 Ph (50)	74

Enolate	Electrophore	Product (% Yield)	Ref.
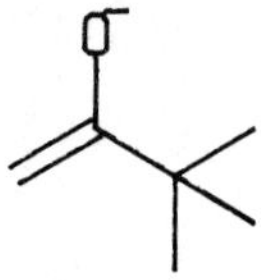 O^-	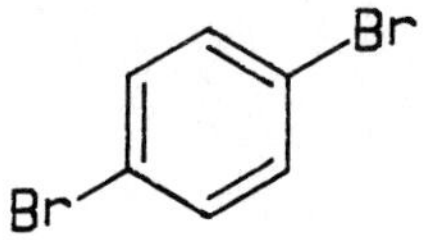 Br, Br	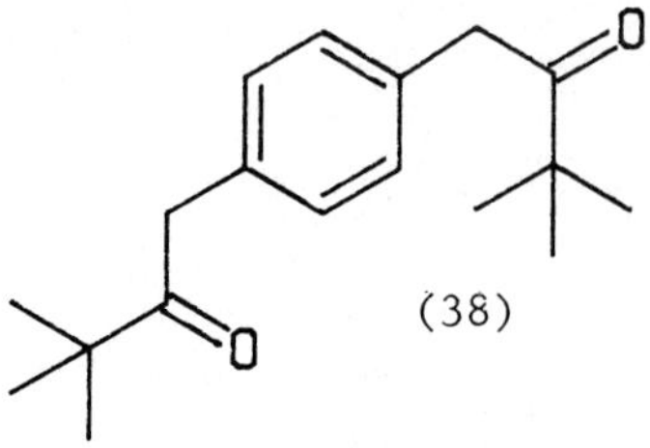 O, O (38)	75
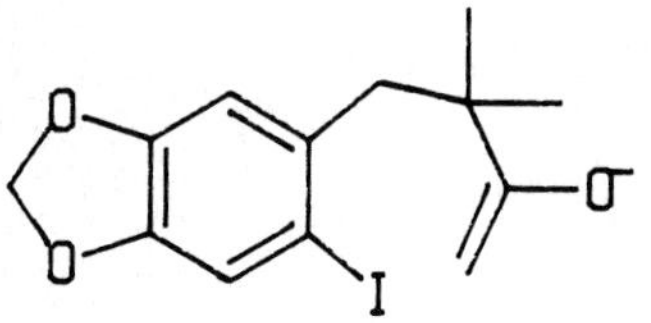 O, O, I, O^-	--	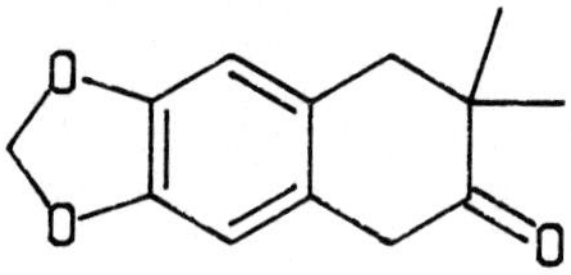 O, O, O (99)	73
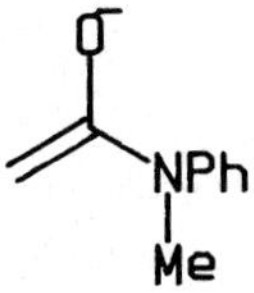 O^-, NPh, Me	PhBr	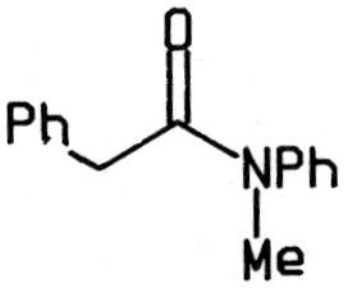 Ph, O, NPh, Me (60)	76
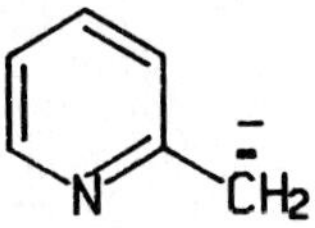 N, $\bar{C}H_2$	PhBr	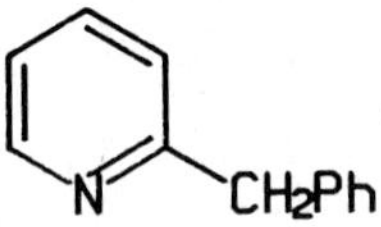 N, CH_2Ph (73)	77
$NC\bar{C}H_2$	PhBr	$PhCH_2CH_2Ph$ (18)	78
$Me_2C=NO_2^-$	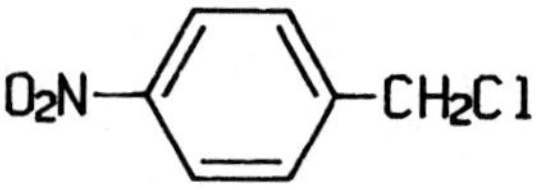 O_2N–, –CH_2Cl	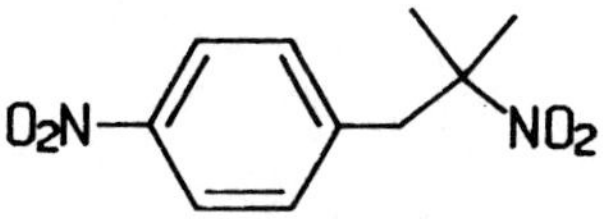 O_2N–, NO_2 (91)	79
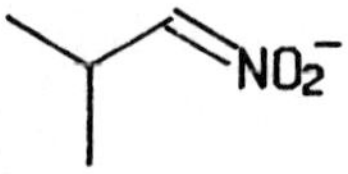 NO_2^-	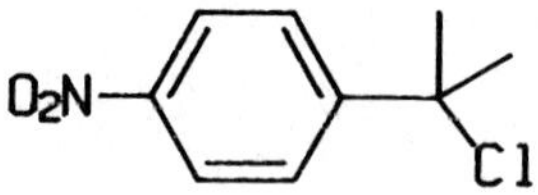 O_2N–, Cl	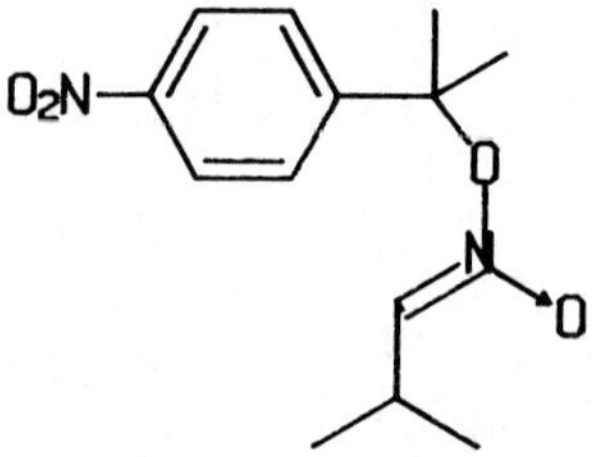 O_2N–, O, N, O (42)	80

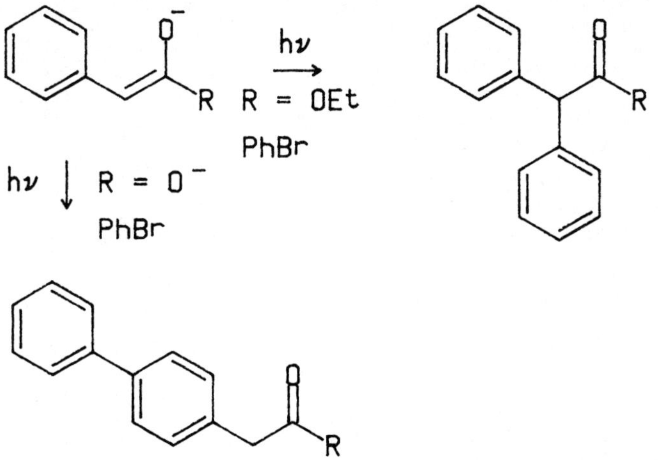

Figure 31. $S_{RN}1$ Arylation of phenylacetate anions

When the enolate and electron acceptor are incorporated in the same molecule, a cyclization can occur. This possibility has been realized in a natural product synthesis (see Fig. 32).[60] A brief listing of the various electron acceptors and reaction products is given in Table 3.

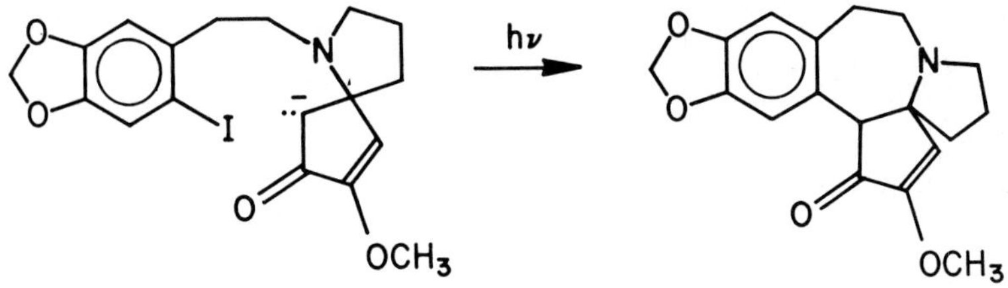

Figure 32. Intramolecular $S_{RN}1$ photocyclization.

B. Polyenolates. The chemistry of dienolates and higher polyenolates has not been well-studied, and no information exists about, say, the regiochemistry of alkylation or arylation of dienolates by a photopromoted $S_{RN}1$ reaction. An intriguing example of a photoinduced cyclization of a polyenolates is represented by White and coworkers' observation of photocyclization of a dienolate to a bicyclic ketone, although the mechanism remains obscure (see Fig. 33).[61] The conjugate acid undergoes divergent photochemistry.

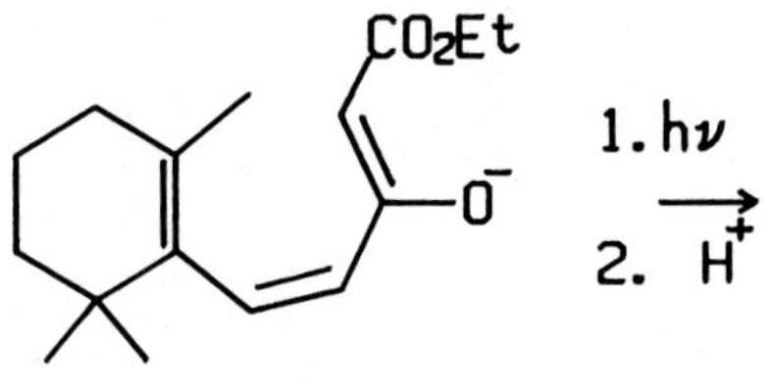

Figure 33. Photocyclization of a trienolate.

C. Phenolates. Since phenols and naphthols undergo pronounced bathochromic shifts upon deprotonation, their excited states are strong photoacids, and adiabatic deprotonation to yield the excited state of the phenolate is a facile, if indirect, way of observing the photochemistry of these species. Given this tendency, much of the photochemistry of phenols and naphthols, especially in aqueous media, undoubtedly involves intervention of the excited state of the conjugate base. For instance, irradiation of phenols in basic media produces oxidative dimers, presumably via electron ejection from the anionic excited state (see Fig. 34).[62] Formation of solvated electrons has been demonstrated spectroscopically to occur upon irradiation of, for instance, phenylphenolate anions.[63]

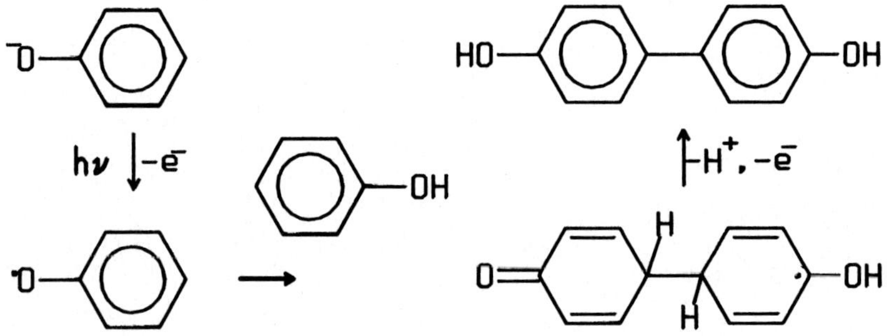

Figure 34. Oxidative photodimerization of phenolate ions.

Again, since phenolates and naphtholates are isoelectronic to odd-alternant hydrocarbons anions (i. e., benzyl and naphthylmethyl anions), photoexcitation formally removes an electron from a non-bonding molecular orbital centered on oxygen to one centered on carbon. Thus protonation and other chemical behavior reflecting that increased charge density on carbon can be anticipated. Indeed, 1-naphtholate undergoes a diffusion-controlled quenching by protons. In the presence of D_2O, deuterium incorporation at C-5 and C-8, the sites of the enhanced excited charge density predicted by Hückel theory, is observed.[64] This deuterium exchange requires the intervention of the novel tautomer 5-H-naphthalene-1-one (see Fig. 35). This observation readily accounts for the formation of the 1-naphthol dimer produced from irradiation of 1-naphtholate in methanol (see Fig. 36),[65] a result which has been attributed to enhanced excited state nucleophilicity.

In the presence of a leaving group, bond heterolysis to produce carbene or diradical intermediates can result upon phenolate irradiation. For instance, irradiation of halophenols in basic media produces substitution chemistry in the presence of nucleophiles.[65] When the nucleophile is another phenolate, bisphenols are generated (see Fig. 37).[66] Although a radical mechanism involving homolytic bond cleavage from the neutral halophenol has been suggested,[66] clearly the profound effect of

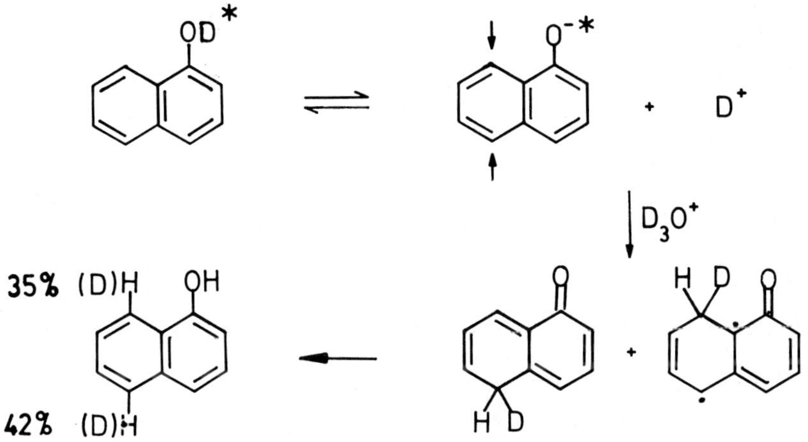

OD*
O⁻*
D⁺
D_3O^+
35% (D)H OH
42% (D)H
H D
H D

Figure 35. Deuterium exchange upon irradiation of 1-naphthol.

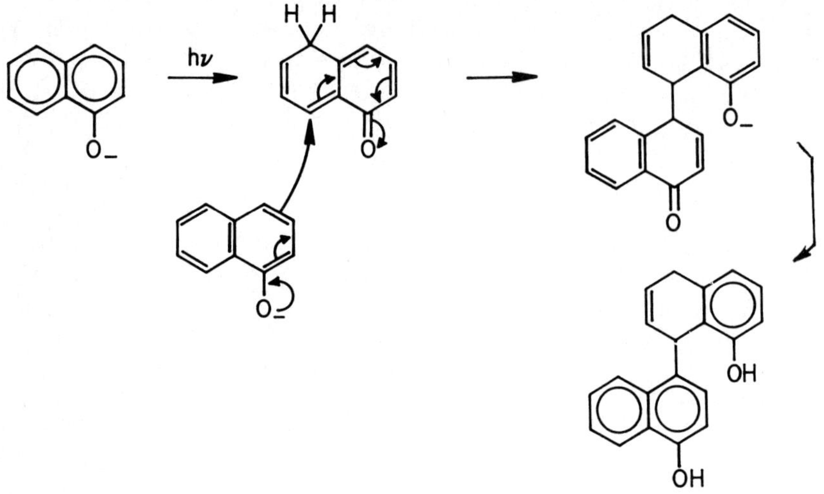

Figure 36. 1-Naphtholate photodimerization.

base concentration suggests that the phenol conjugate base is the photoactive species. Moreover, the coupling regiochemistry involves ortho/para rather than meta attack on the phenolate nucleophile, which suggests that an electrophile, e. g., a carbene, is involved (see Fig. 37). This mechanism is supported by the formation of o-aminophenols from irradiation of o-chlorophenolates in the presence of amines.[67] Irradiation of o-chlorophenolate in the absence of strong nucleophiles produces a carbene which undergoes a Wolff rearrangement (see Fig. 38).[68] The ultimate dimeric product is evidence of the power of anion photochemistry to generate complex structures from simple starting materials.

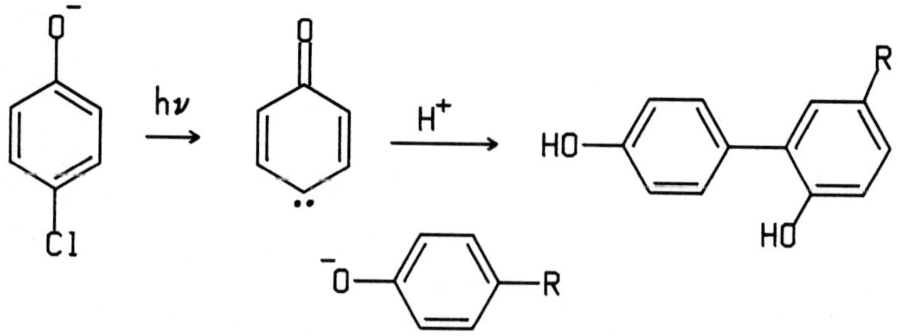

Figure 37. Photoaddition of 4-chlorophenolate to phenols.

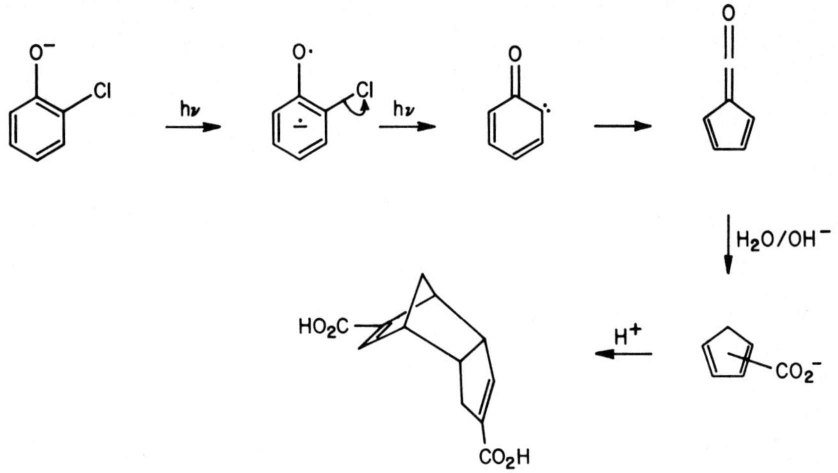

Figure 38. Photo-Wolff rearrangement from irradiation of 2-chlorophenolate.

Photodehalogenation also occurs from halomethylphenolate and naphtholate anions. In particular, trifluoromethyl groups can be photohydrolyzed (see Fig. 39).[69] Again, the experimental observations can be rationalized by an appeal to an intramolecular charge transfer.

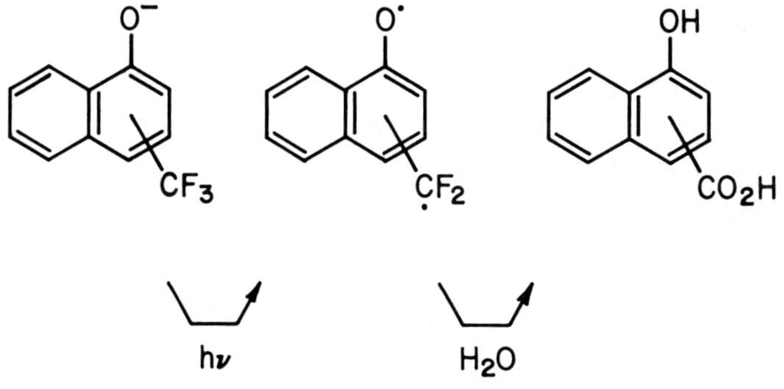

Figure 39. Photohydrolysis of trifluoromethylnaphtholates.

Similar intramolecular charge transfer behavior is exhibited by p-cyanophenolate anion. The excited state, which presumably bears an increased charge density on the nitrile carbon, undergoes hydrogen atom abstraction to yield, ultimately, a reduction/hydrolysis product (see Fig. 40).[71]

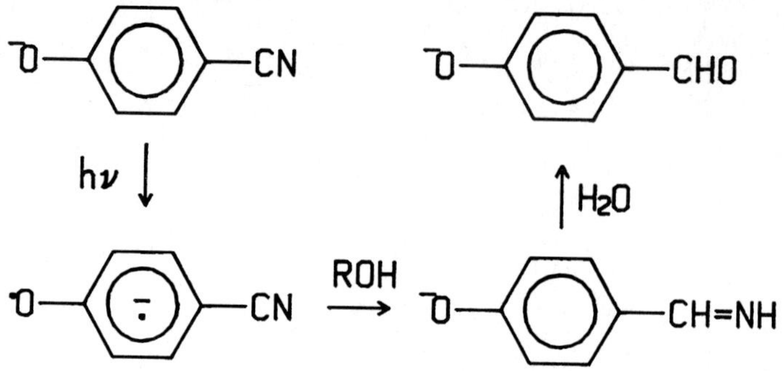

Figure 40. Photoreduction of p-cyanophenolate.

IX. CONCLUSION

With the extensive chemistry that has been developed using carbanions as reactive intermediates, the additional reactivity introduced by placing carbanions in their excited states promises to open up a new area for synthetic and mechanistic development. Currently unexplored areas are: (1) manipulation of geometric isomers of enolates photochemically for alteration of product stereochemistry; (2) use of the facile photoejection from carbanions to develop photoreducing reagents; (3) generation of strained molecules by photodehalogenation at low temperature; (4) generation of unusual tautomers as reactive intermediates through protonation of photoexcited carbanions at sites of enhanced basicity. The fact that resonance-stabilized carbanions absorb light well into the visible region makes these species attractive substrates for laser photochemistry as well. These possibilities are under investigation in several laboratories. However, few of the laboratories active in ground-state carbanion chemistry have added the simple additional procedure of turning on the lights.

ACKNOWLEDGMENT. Support of work in the laboratories of L. M. T. by the National Science Foundation, the U. S. Department of Energy, the donors of the Petroleum Research Fund, administered by the American Chemical Society, and by Research Corporation is gratefully acknowledged. Figures 6, 8, 9, 16, 25, 26, and 38 are reproduced from ref. 1(a), copyright 1986, American Chemical Society.

X. REFERENCES

1. For earlier related reviews, see (a) L. M. Tolbert, Accts. Chem. Res. 19 (1986) 268; (b) L. M. Tolbert, Org. Photochem., 6 (1983) 177; (c) M. A. Fox, Chem. Rev., 79 (1979) 253.

2. R. B. Woodward and R. Hoffmann, "The Conservation of Orbital Symmetry", Verlag Chemie, Weinheim, 1970.

3. M. J. S. Dewar and R. C. Dougherty, "The PMO Theory of Organic Chemistry", Plenum, New York, 1975.

4. A standard planar geometry with β-values set to 1.0 was used in a Huckel molecular orbital calculation.

5. (a) E. Buncel and B. Menon, J. Org. Chem., 44 (1979) 317. (b) B. Menon and E. Buncel, J. Organometal. Chem. 159 (1978) 357. (c) E. Buncel, B. C. Menon, and J. P. Colpa, Can. J. Chem., 57 (1979) 999. (d) E. Buncel and B. C. Menon, J. Am. Chem. Soc., 102 (1980) 3499. (e) E. Buncel and T. K. Venkatachalam, J. Org. Chem., 49 (1984) 413. (f) R. Asami, M. Levy, and M. Szwarc, J. Chem. Soc., (1961) 361. (g) R. Waack and M. A. Doran, J. Am. Chem. Soc., 85 (1963) 1651. (h) L. M. Dorfman, R. J. Sujdak and B. Bockrath, Acc. Chem. Res., 9 (1976) 352. (i) E. Buncel and B. C. Menon, J. Amer. Chem. Soc., 99 (1977) 4457. (j) E. Buncel and B. C. Menon, J. Organometal. Chem., 141 (1977) 1. (k) D. A. Bors, M. J. Kaufman, and A. J. Streitwieser, Jr. J. Am. Chem. Soc., 107 (1985) 6975. (l) For a general discussion, see E. Buncel and B. Menon, Comprehensive Carbanion Chemistry, E. Buncel and T. Durst, eds., Elsevier, Amsterdam, 1980, pp. 97-124.

6. (a) J. Fabian and R. Zahradnik, Wiss. Zeits. Tech. Univ. Dresden, 26 (1977), 315; (b) K. Hafner, K. U. Goliasch, Angew. Chem. 74 (1962) 118, 344.

7. M. A. Fox and T. A. Voynick, Tetrahedron Lett., (1980) 3943.

8. T. Förster, Z. Elektrochem., 54 (1950) 42.

9. (a) A. Weller, Prog. React. Kinet., 1, T. Porter and B. Stevens, eds., Pergamon, London, (1961) p. 188. (b) For a general discussion, see J. F. Ireland and P. A. H. Wyatt, Adv. Phys. Org. Chem., 12 (1976) 131.

10. R. Gygax, H. L. McPeters, and J. I. Brauman, J. Am. Chem. Soc., 101 (1979) 2567.

11. J. H. Richardson, L. M. Stephenson, and J. I. Brauman, J. Chem. Phys., 63 (1975) 74.

12. (a) D. B. Chesnut and G. J. Sloan, J. Chem. Phys., 33 (1960) 637; 35 (1961) 443. (b) P. B. Ayscough, A. P. McCann, and R. Wilson, Proc. Chem. Soc., London, (1961) 16.

13. H. O. House and P. D. Weeks, J. Am. Chem. Soc., 97 (1975) 2785.

14. E. A. Chandross, Trans. N. Y. Acad. Sci., 31 (1969) 571.

15. (a) L. M. Tolbert, J. Am. Chem. Soc., 100 (1978) 3952. (b) L. M. Tolbert, J. Am. Chem. Soc., 102 (1980) 3531. (c) L. M. Tolbert, J. Am. Chem. Soc., 102 (1980) 6808. (d) L. M. Tolbert and R. D. Merrick, J. Org. Chem., 47 (1982) 2808.

16. G. A. Russell and R. K. Khanna, Tetrahedron, 41 (1985) 4133.

17. For a general review, see R. A. Rossi and R. H. de Rossi, "Aromatic Substitution by the SRN1 Mechanism," American Chemical Society Monograph No. 178, Washington, D. C. (1983).

18. H. M. Parkes and R. N. Young, J. Chem. Soc. Perkin Trans. II (1978) 249.

19. R. J. Bushby, J. Chem. Soc. Perkin Trans. II, (1980) 1419

20. R. N. Young and M. A. Ahmad, J. Chem. Soc. Perkin Trans. II (1982) 35.

21. G. Boche, K. Buchl, D. Martens, and D. R. Schneider, Liebigs Ann. Chem. (1980) 1135.

22. G. Boche and D. R. Schneider, Angew. Chem. Int. Ed. Engl., 16 (1977) 869.

23. R. J. Bushby and Myron P. Tytko, J. Chem. Soc., Chem. Commun. (1986) 23.

24. (a) L. M. Tolbert and M. Z. Ali, J. Org. Chem., 1985, 50, 3288. (b) M. Z. Ali, Ph. D. Thesis, University of Kentucky, 1982.

25. D. R. Boate and D. H. Hunter, Org. Magn. Reson., 22 (1984) 167.

26. L. M. Tolbert and M. Z. Ali, J. Org. Chem., 47 (1982) 4793.

27. M. A. Fox, C. C. Chen, and K. A. Campbell, J. Org. Chem., 48 (1983) 321.

28. D. H. Hunter and R. A. Perry, J. Chem. Soc. Chem. Commun (1980) 877.

29. M. J. S. Dewar and D. J. Nelson, J. Org. Chem., 47 (1982) 2614.

30. D. W. Boerth and A. Streitwieser, Jr., J. Am. Chem. Soc., 103 (1981) 6443.

31 (a) A. Streitwieser, Jr., E. Juaristi, and L. L. Nebenzahl, "Comprehensive Carbanion Chemistry," E. Buncel and T. Durst, eds., (1980) 323. (b) R. A. Cox and R. Steward, J. Am. Chem. Soc., 98 (1976) 488.

32. S. Cohen and A. Yogev, J. Am. Chem. Soc., 98 (1976) 2013.

33. (a) G. W. Griffin, A. F. Marcantonio, H. Kristinsson, R. C. Petterson, and C. S. Irving, Tetrahedron Lett., (1965) 2951. (b) S. S. Hixson and P. Ruede, unpublished results. (c) P. Ruede, Ph. D. Thesis, University of Massachusetts, 1982.

34. (a) K. Yamada, T. Kanekiyo, S. Tanaka, K. Naruchi, and M. Yamamoto, J. Am. Chem. Soc., 103 (1981) 7003. (b) K. Yamada, S. Tanaka, K. Naruchi, and M. Yamamoto, J. Org. Chem., 47 (1982) 5283.

35. E. E. van Tamelen, J. Schwartz, and J. I. Brauman, J. Am. Chem. Soc., 92 (1970) 5798.

36. F. Bernardi, M. A. Robb, H. B. Schlegel, and G. Tonachini, J. Am. Chem. Soc., 106 (1984). 1198

37. (a) S. Siddiqui, Ph. D. Thesis, University of Kentucky, 1982. (b) M. Z. Ali, Ph. D. Thesis, University of Kentucky, 1982.

38. R. M. Pagni, M. N. Burnett, and H. M. Hassaneen, Tetrahedron, 38 (1982) 843.

39. M. Newcomb and W. T. Ford, J. Am. Chem. Soc., 96 (1974) 2968.

40. M. A. Fox, J. Am. Chem. Soc., 101 (1979) 4008.

41. H. M. Parkes and R. N. Young, J. Chem. Soc. Perkin Trans. II, (1980) 1137.

42. R. A. Rossi and J. F. Bunnett, J. Org. Chem., 38 (1973) 3020.

43. S. W. Staley and N. J. Pearl, J. Am. Chem. Soc., (1973) 2731.

44. H. Kloosterziel and G. M. Gorter-la Roy, Chem. Commun. (1972) 352.

45. M. A. Fox and N. J. Singletary, J. Org. Chem., 47 (1982) 3412.

46. J. Schwarz, J. Chem. Soc. Chem. Commun. (1969) 833.

47. E. E. van Tamelen, J. I. Brauman, and L. E. Ellis, J. Am. Chem. Soc., 89 (1967) 5073.

48. R. A. Rossi, R. H. deRossi, and A. F. Lopez, J. Org. Chem., 41 (1976) 3367.

49. L. M. Tolbert and S. Siddiqui, J. Org. Chem., 49 (1984) 1744.

50. G. A. Russell and R. K. Khanna, J. Am. Chem. Soc., 107 (1985) 1450.

51. (a) J. Plodinec and T. E. Hogen-Esch, J. Am. Chem. Soc., 96 (1974) 5262. (b) T. E. Hogen-Esch and M. J. Plodinec, J. Phys. Chem., 80 (1976) 1085. (c) T. E. Hogen-Esch and M. J. Plodinec, J. Phys. Chem., 80 (1976) 1090. (d) H. W. Vos, C. MacLean and N. H. Velthorst. J. Chem. Soc. Faraday Trans. 2, 72 (1976) 63.

52. M. A. Fox and R. C. Owen, J. Am. Chem. Soc., 102 (1980) 6559. (b) M. A. Fox and R. C. Owen, ACS Symposium Ser., 146 (1981) 337.

53. L. M. Tolbert and M. Z. Ali, Abstracts 184th National Meeting, American Chemical Society, Kansas City, 1982, paper ORGN-42.

54. (a) L. M. Tolbert and S. Siddiqui, J. Am. Chem. Soc., 106 (1984) 5538. (b) L. M. Tolbert and S. Siddiqui, J. Am. Chem. Soc., 104 (1982) 4273.

55. G. Levin and M. Szwarc, J. Am. Chem. Soc., 98 (1976) 4211

56. (a) M. A. Fox and N. J. Singletary, Solar Energy, 25 (1980) 225. (b) M. A. Fox and Kabir-ud-Din, J. Phys. Chem., 83 (1979) 1800. (c) J. R. Hohman and M. A. Fox, J. Am. Chem. Soc., 104 (1982) 401.

57. R. K. Norris and D. Randles, J. Org. Chem., 47 (1982) 1047.

58. J. F. Wolfe, unpublished results.

59. G. A. Russell, J. Hershberger, and K. Owens, J. Organometallic Chem., 225 (1982) 43.

60. M. F. Semmelhack, B. P. Chong, R. D. Stauffer, T. D. Rogerson, A. Chong, and L. D. Jones, J. Am. Chem. Soc., 97 (1975) 2507.

61. J. D. White and R. W. Skeean, J. Am. Chem. Soc., 100 (1978) 6296.

62. H.-I. Joschek and S. I. Miller, J. Am. Chem. Soc., 88 (1966) 3273.

63. (a) L.-M. Coulangeon, G. Perbet, P. Boule and J. Lemaire, Can. J. Chem. 58 (1980) 2230. (b) A. Matsuzaki, T. Kobayashi, and S. Nagakura, J. Phys. Chem. 82 (1978) 1201.

64. (a) S. P. Webb, S. W. Yeh, L. A. Philips, M. A. Tolbert, and J. H. Clark, J. Am. Chem. Soc., 106 (1984) 7286. (b) S. P. Webb, L. A. Philips, S. W. Yeh, L. M. Tolbert, and J. H. Clark, J. Phys. Chem., 90 (1986) 5154.

65. T. Kitamura, T. Imagawa, and M. Kawanishi, J. Chem. Soc. Chem. Commun. (1977) 81.

66. (a) K. Omura and T. Matsuura, J. Chem. Soc. Chem. Commun. (1969) 1394. (b) K. Omura and T. Matsuura, Tetrahedron, 27 (1971) 3101. (c) K. Omura and T. Matsuura, Synthesis (1971) 28.

67. L. M. Tolbert and A. Merritt, unpublished results.

68. (a) C. Guyon, P. Boule, and J. Lemaire, Tetrahedron Lett., 1581 (1982); (b) C. Guyon, P. Boule, and J. Lemaire, Nouv. J. Chim., 8 (1984) 685.

69. P. Seiler and J. Wirz, Helv. Chim. Acta, 55 (1972) 2693.

70. (a) K. Omura and T. Matsuura, J. Chem. Soc. Chem. Commun. (1969) 1516. (b) G. Boche and A. Bieberbach, Tetrahedron Lett., 1021 (1976).

71. R. A. Rossi and J. F. Bunnett, J. Org. Chem., 38, (1973) 1407.

72. A. P. Komin and J. F. Wolfe, J. Org. Chem., 42, (1977) 2481.

73. M. F. Semmelhack and T. Bargar, J. Am. Chem. Soc., 102, (1980) 7765.

74. J. V. Hay and J. F. Wolfe, J. Am. Chem. Soc. 97 (1975) 3702.

75. R. A. Alonso and R. A. Rossi, J. Org. Chem., 45 (1980) 4760.

76. R. A. Rossi and R. A. Alonso, J. Org. Chem., 45 (1980) 1239.

77. J. F. Bunnett and B. F. Gloor, J. Org. Chem., 39 (1974) 382.

78. J. F. Bunnett and B. F. Gloor, J. Org. Chem, 38 (1973) 4156.

79. G. A. Russell and W. C. Danen, J. Am. Chem. Soc., 88 (1966) 5663.

80. R. K. Norris and D. Randles, J. Org. Chem., 47 (1982) 1047.

Chapter 5

FLUORO-CARBANIONS

R.D. CHAMBERS and M.R. BRYCE
Department of Chemistry, University of Durham, Durham, DH1 3LE, U.K.

CONTENTS

I. EFFECT OF FLUORINE ON CARBANION STABILITIES 272
A. Electronic Effects Associated with Fluorine Substitution 276
B. Olefinic Systems 282
C. Aromatic Systems 283
D. Stereochemistry of Fluorocarbanions 285
E. Delocalised Anions 286

II. NUCLEOPHILIC ATTACK ON UNSATURATED FLUOROCARBONS 287
A. Fluorinated Alkenes 287
B. Fluorinated Arenes 289

III. FLUORIDE ION INDUCED PROCESSES 292
A. Oligomerisation of Fluorinated Alkenes 293
B. Equilibration by Fluoride Ion 295
C. Formation of Observable Anions 297
D. Observable σ-Complexes 300
E. Nucleophilic Friedel Crafts Reactions 301
F. Skeletal Rearrangements 303
G. Reactions Induced by Other Ions 307

IV. ORGANOMETALLIC COMPOUNDS 307
V. YLIDS 312
VI. REFERENCES 315

I. EFFECT OF FLUORINE ON CARBANION STABILITIES

The relationship between structure and acid strength in fluorocarbon systems has been extensively studied.[1] Rates of base-catalysed H/D exchange for a series of haloforms clearly indicate that halogen attached to a carbanion centre facilitates carbanion formation in the order I ~ Br > Cl > F (Table 1).[2]

TABLE 1

Haloform	Rate of exchange (10^5k) ℓ mol^{-1} sec^{-1}
CHF_3	Too slow to measure
$CHCl_3$	820
$CHBr_3$	101,000
CHI_3	105,000
$CHCl_2F$	16
$CHBr_2F$	3,600
CHI_2F	8,800

In early work with simple saturated fluorocarbon systems (1) - (4) Andreades studied base-catalysed H/D exchange in NaOMe/MeOH solutions.[3] Rates of exchange were measured for

CF_3H	$CF_3(CF_2)_5CF_2H$	$(CF_3)_2CFH$	$(CF_3)_3CH$
(1)	(2)	(3)	(4)

both the forward and reverse directions starting with the appropriate isotopically-labelled substrates and using ^{19}F n.m.r. and mass spectroscopic techniques. Relative reactivities for exchange are given in Table 2. An activation energy difference of 7.8 kcal mol^{-1} is calculated on changing from

TABLE 2

Compound	(1)	(2)	(3)	(4)
Derived Anion	CF_3^-	$CF_3(CF_2)_5CF_2^-$	$(CF_3)_2CF^-$	$(CF_3)_3C^-$
Relative Reactivity	1	6	2×10^5	10^9
Approx. pK_a	31	30	20	11

the primary (2) to the secondary system (3). These observed

reactivity differences reflect decreasing stability of the carbanions, viz. tertiary > secondary > primary, and clearly indicate that β-fluorine ($\bar{C}$-CF) is far more stabilising than α-fluorine ($\bar{C}$F). Based on second order rate constants for (1) - (4) pK_a values were estimated (Table 2).

Substituted nitromethanes (5) have been studied and pK_a values determined spectrophotometrically (Table 3).[4] For this series it can be seen that α-Cl leads to the expected increase in CH acidity relative to X=H, while the effect of α-F is in the opposite direction. Lowering of acid strength by α-F increases in the order Y = CO_2Et, $CONH_2$, Cl, NO_2.

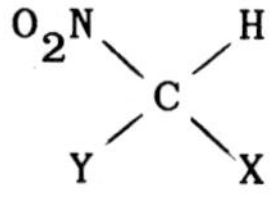

$$\begin{array}{ccc} O_2N & & H \\ & C & \\ Y & & X \end{array}$$

(5)

TABLE 3

pK_a values for substituted nitromethanes (5)

	Y = CO_2Et	Y = $CONH_2$	Y = Cl	Y = NO_2
X = Cl	4.16	3.50	5.99	3.80
= H	5.75	5.18	7.20	3.57
= F	6.28	5.89	10.14	7.70

In 1972 Burton[5,6] carried out kinetic acidity studies on a series of halogenated carbon acids (6) - (9) and established the effect of various substituents on the stabilities of fluoro- and halo-carbanions. Relative k_D and k_T values showed that loss of hydrogen from the substrate to form a carbanion was the rate determining step in the exchange. The effects of substituents on the carbanions derived from (6), (7) and (8) are very significant since the substituents are bonded directly to the carbanion centre. This is reflected in the kinetic acidity data given in Table 4.[6]

CF_3CHXCF_3 (6)

(a) X = F
(b) Cl
(c) Br
(d) I
(e) OMe
(f) CF_3

CF_3CX_2H (7)

(a) X = F
(b) Cl
(c) Br
(d) I

CHX_3 (8)

(a) X = F
(b) Cl
(c) Br
(d) I

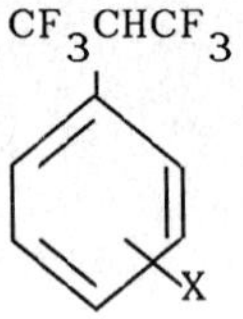

CF_3CHCF_3 attached to a benzene ring bearing X (9)

(a) X = H
(b) = m-F
(c) = p-F
(d) = m-Cl
(e) = p-Cl
(f) = m-Br
(g) = p-Br
(h) = m-I
(i) = p-I
(j) = m-CF_3
(k) = p-CF_3
(l) = m-$CF(CF_3)_2$
(m) = p-$CF(CF_3)_2$

TABLE 4

Compound	pK_a
(9a)	17.9
(6a)	18.0
(6b)	12.6
(6c)	11.5
(6d)	13.7
(6e)	>22
(6f)	11.0
(7a)	27
(7b)	17.2
(7c)	16.9
(7d)	17.1
(8a)	28
(8b)	15.5
(8c)	13.7
(8d)	13.7

These pK_a data indicate that stabilities of the carbanions derived from (6) are in the order X = CF_3 > Br > Cl > I > Ph ~ F > OMe. Table 5 gives the relative isotopic exchange rates for compounds (9).[5]

TABLE 5

Compound	Relative rate of exchange
(9a)	1.00
(9b)	85.5
(9c)	7.28
(9d)	129
(9e)	27.6
(9f)	218
(9g)	74.9
(9h)	155
(9i)	88.5
(9j)	42.4
(9k)	102
(9l)	39.2
(9m)	92.3

Streitwieser et al[7] have determined kinetic and equilibrium acidities of fluorinated bicycloalkanes (10) - (12). These compounds provide an unusual opportunity to study F-substituent effects on carbanions as elimination of F^- from the carbanions derived from (10) - (12) is inhibited by the very unfavourable formation of a bridgehead olefin. Results are given in Table 6.

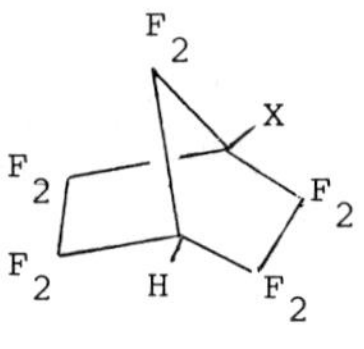

F_2 X F_2 F_2 F_2 H F_2 F_2 F_2 F F_2 F_2 F_2 H F_2 F_2 F F F_2 F H F_2

(10) (a) X = F
(b) X = H
(c) X = CH_3
(d) X = CF_3

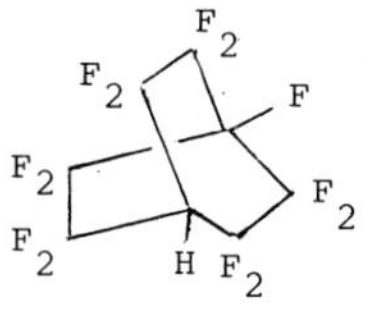

(11)

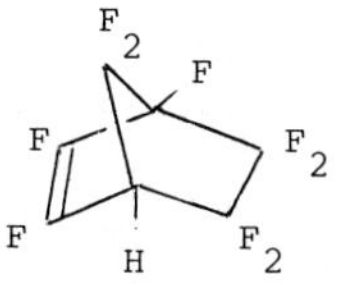

(12)

TABLE 6

Compound	pK_a
(10a)	20.5 ± 0.3
(10b)	22.2 ± 0.2
(10c)	23.4 ± 0.1
(10d)	20.7 ± 0.2
(11)	18.3 ± 0.3
(12)	21.6 ± 0.7

A. Electronic Effects Associated with Fluorine Substitution

Based on the acidity data presented in Tables (1) - (6) it is clear that fluorine facilitates carbanion formation relative to hydrogen but it is not necessarily the electronegativity of the halogen that has the dominant influence on carbanion stability. In principle, a variety of factors could have an influence on the anions derived from compounds (6) - (9) i.e. inductive effects (I_σ), hybridisation effects, influence of 'd' orbitals, polarisation, steric factors and 'negative hyperconjugation' (see later).

For a fluorine substituent we have two opposing inductive effects, I_σ and electron-pair repulsion I_π; the resultant of these effects will be dependent upon the geometry of the system. Consideration of the geometries illustrated in Figure 1 leads to the conclusion that

$$I_\sigma \quad \bar{C}\longrightarrow F \text{ (stabilising)} \qquad I_\pi \quad \bar{C}\text{—}\ddot{F} \text{ (destabilising)}$$

I_π repulsion is greater for a planar sp^2 hybridised carbon than if the carbanion is tetrahedral. The lower acidifying effect of F than Cl can therefore be explained by the enhanced I_π effect of F, while the stronger acidifying

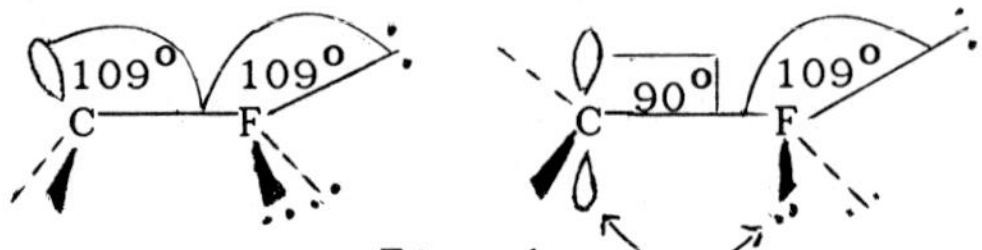

Fig. 1

effect of Br and I could result from increased availability of 'd' orbitals and increased polarisability of these larger halogens. Steric effects will be at a minimum with F, but may be significant or even dominant with the larger halogens. It is well known that carbanions favour pyramidal sp^3 geometries except when distorted by conjugation or steric factors. For the nitrocarbanions (Table 3) conjugation of the nitro group with the carbanionic centre will lead to considerable sp^2 character. Consequently, it is understandable that in this system I_π destabilisation by F dominates over I_σ stabilisation. The nature of group Y (Table 3) will also affect the degree of conjugation between the carbanion centre and the nitro group.

Table 2 dramatically demonstrates that a fluorocarbon group is much more effective at carbanion stabilisation than is a fluorine atom. Part of this stabilisation has been attributed to "C–F negative hyperconjugation" which was first proposed by Roberts in 1950[8] and has been a subject of lively debate for both theoreticians and experimentalists. Negative hyperconjugation is depicted by structures (9) and involves "no-bond resonance". Andreades[3] considered the dependence of exchange rate on the number of β-fluorines (Table 2) to be too high to be accountable

$$\underset{(9a)}{F\text{-}CR_2\text{-}\bar{C}R_2} \longleftrightarrow \underset{(9b)}{F^- \; CR_2{=}CR_2}$$

solely by inductive effects and invoked negative hyperconjugation through perfluoroalkyl groups to explain his results.

$$CF_3\text{-}\bar{C}\langle \longleftrightarrow F^-CF_2{=}C\langle \qquad \ddot{F} \longleftarrow \bar{C}\langle$$

Nevertheless, the difference in stability of a carbanion with fluorine and trifluoromethyl attached does not <u>require</u> such an effect. We have already discussed and explained the reasons for the limited stabilising, and sometimes the overall destabilising influence of fluorine attached to a carbanion centre, arising from I_π repulsion.[6,9]

Burton[5,6] has given detailed consideration to the factors influencing the stability of carbanions (6) - (9) (Table 4). The importance of inductive effects is shown by comparison of kinetic acidities of compounds (6) and (9) (Table 7); for the latter series the substituent X is separated from the carbanionic site by a phenyl group and so the acidities of (9) are lower than (6) for the same substituent. Thus decreasing the inductive stabilisation of the carbanion is not compensated for by conjugation with the phenyl ring. Polarisation does not seem to be a dominant factor in stabilising carbanions from (6) as this would predict stabilities in the order I > Br > Cl > F rather than the observed order, viz. Br > Cl > I > F. It has also been argued that 'd' orbital effects are of little importance[6] and that two mechanisms are operative in the stabilisation of these carbanions, i.e. inductive effects and I_π destabilisation. Indeed the observed order for the series (6) (Table 4) is that predicted for I_π destabilisation except for substituent X = I (6d). Additional evidence for a strong I_π effect is found in comparison of the acidities of compounds (9a), (6a), (9b) and (9c) (Table 7). The inductive effect of the αF of (6a) should

TABLE 7

Compound	Relative rate
$PhCH(CF_3)_2$ (9a)	1.00
$FCH(CF_3)_2$ (6a)	0.86
m-F-$C_6H_4CH(CF_3)_2$ (9b)	18.4
p-F-$C_6H_4CH(CF_3)_2$ (9c)	1.57

outweigh that of the meta and para aromatic fluorines of (9b) and (9c), yet the relative exchange rates of (9a), (6a) and (9c) are all very similar, while compound (9b) is significantly more acidic. These effects are, however, understandable in the context of the delocalised ion (10) where a fluorine substituent in the ring located at the <u>para</u>-position is only slightly stabilising. Fluorine adjacent to the centres of high charge

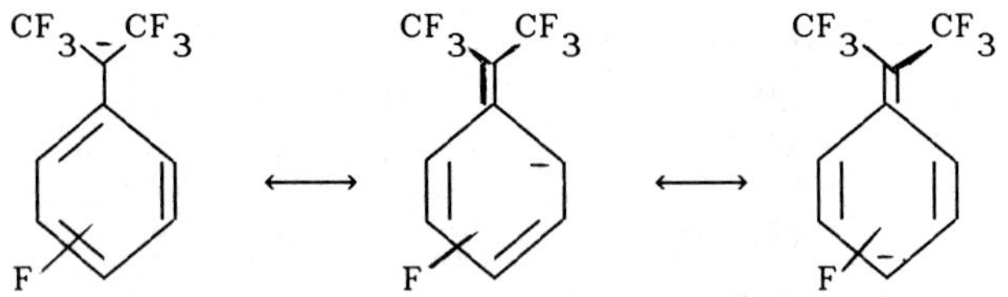

(10a) ⟷ (10b) ⟷ (10c)

density, i.e. at the <u>meta</u>-positions is, predictably, very stabilising. Remarkably, however, fluorine at the <u>ortho</u>-position is also significantly stabilising.[9] The interesting aspect of this effect is that it parallels exactly the observations for nucleophilic aromatic substitution (see later).

To account for the discrepancy for (6d) a steric argument has been advanced.[6] Steric strain, due to the large iodine atom, probably forces the anion derived from (6d) out of the normal pyramidal configuration which will be that adopted for the series involving the smaller halogens.

A review by Holtz[10] published in 1971 critically appraised quantitative data relating to negative hyperconjugation and he concluded that σ- and π-inductive effects alone can explain all the data. Similarly, Streitwieser[7] considered that the acidities of the bicycloalkanes (10) - (12) can be explained completely by a classical field effect. He re-estimated the pK_a value for $(CF_3)_3CH$ (4) to be ca. 21, very similar to those of the bridgehead compounds (10) - (12), (cf. Andreades' value of 11 for (4))[3] and asserted that there was no need to invoke any special effects such as negative hyperconjugation to explain the stability of the anion $(CF_3)_3C^-$. However, these conclusions have recently been challenged by Tatlow[11] on the basis of D uptake from neutral media $[(CD_3)_2CO/D_2O]$ into mixtures of (4) and (10a) and (4) and (11). At room temperature over 19 days D uptake into (4) was significantly faster than into (11) and was not detectable for (10). If this neutral reaction is mechanistically similar to base-catalysed exchange, it was argued that the higher acidity of (4) over (10) and (11) may, after all, result from negative hyperconjugation.

No manifestation of negative hyperconjugation in enhanced rate-constants was noted, in the measured rate constants for nucleophilic attack by ammonia on a series of perfluoro-alkylbenzenes $C_6F_5R_F$ [R_F = CF_3; CF_2CF_3; $CF(CF_3)_2$; $C(CF_3)_3$] (13). This series might be expected to show enhanced reactivity for perfluorotoluene, due to negative hyperconjugation, but the rate constants for the series are remarkably similar.[12]

NH_3 + R_F-C$_6$F$_5$ ⟶ [$\overset{+}{N}H_3$, F / F, F / F, F / R_F anion] ⟶ NH_2-C$_6$F$_4$-R_F + HF

(13)

It is only recently that <u>ab initio</u> molecular orbital calculations have been used to probe the concept of negative hyperconjugation.[13-15] Apeloig[13] has presented calculations at the 4-31G basis set level that strongly support F-hyperconjugation; he suggested that both β-F and β-CF_3 substituents stabilise an adjacent carbanionic centre by this effect and to a similar extent. The higher inductive effect of CF_3 relative to F may fully compensate for the smaller anionic hyperconjugative stabilisation by the former. Therefore it is argued, somewhat tenuously, that the small changes on rate constants, observed in comparisons between β-F and β-CF_3 may not be indicative of the absence of hyperconjugation, as is sometimes claimed.[5,7,12] Energies were compared for syn, anti-periplanar and perpendicular conformations of $FCH_2CH_2^-$ and $CF_3CH_2CH_2^-$. Large stabilisation energies are found for both F and CF_3 substituents in the anti geometry: both C-F and C-CF_3 bonds lengthen while C-C bonds shorten. Independent calculations by Taylor[15] support these conclusions. However, Streitwieser[16] considers that his theoretical treatment of $FCH_2CH_2^-$ using STO-2G and 4-31G basis sets does not support hyperconjugation. Instead, energy differences are attributed to polarisation.

In an extensive theoretical study of energy stabilisation, charge transfer, and bond elongation in $FCH_2CH_2^-$ and $CF_3CH_2^-$ Schleyer[17] concludes that negative hyperconjugation is important. This study uses <u>ab initio</u> split valence sets 3-21+G and 4-31+G, augmented by a set of diffuse s and p functions on non-hydrogen atoms. During geometry optimisation complete charge transfer to F leads to cleavage of the C-F bond and formation of a hydrogen bonded F^- ethylene complex (14a). Alternative structures (14b) and (14c) are disfavoured. These calculations suggest that fluorine hyperconjugation is also important in $CF_3CH_2^-$; the C-C bond is shortened by 0.10Å and the C-F bond antiparallel to the carbanion lone pair is lengthened by 0.13Å compared to CF_3CH_3. F-Negative hyperconjugation is at a maximum when the dihedral angle between the carbon lone pair and the CF bond is at 180° and this represents an energy minimum configuration (15a). Negative

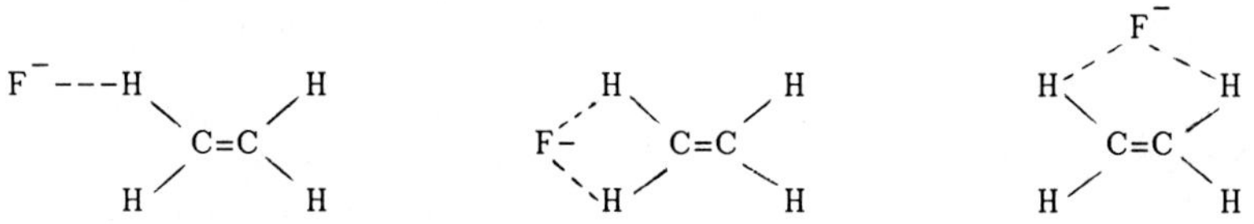

(14a)

(14b)

(14c)

hyperconjugation is at a minimum at ca. 30° and 90° (15b) (Table

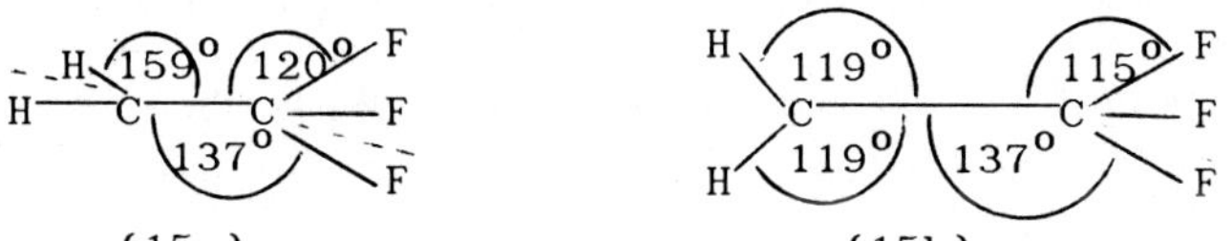

(15a)

(15b)

8). Atomic charges were obtained from Mulliken population analyses and increased charge transfer is observed to those fluorines which can interact hyperconjugatively with the carbanion.

TABLE 8

Data for (15a) and (15b)

Dihedral angle of lone pair-CF bond	C-F Bond length (Å)	Charge on F
180°	1.499	-0.369
ca. 30	1.454	-0.341
ca. 60	1.407	-0.330
90	1.399	-0.344
CH_3CF_3	1.370	-0.267

The contribution of the hyperconjugating F atom in (15a) (180°) to overall anion stabilisation was calculated by using

$CF_3CH_2^-$ as a model with a 180° lone pair-H dihedral angle. The third F of (15a), anticoplanar to the carbanion lone pair was found to contribute 20 kcal mol^{-1} (one half) of the total stabilisation energy of the CF_3 group in (11a).

Further theoretical work concerning $CF_3CH_2^-$ supports the work of Apeloig and Schleyer, and concludes that negative hyperconjugation, not induction, is dominant in the anti configuration for which molecular orbital plots clearly show in plane HOMO-LUMO mixing of a C-F $\sigma*$ orbital with the carbon lone pair.[18]

It seems clear that there is a sound theoretical basis for accepting the concept of negative hyperconjugation but, so far, there is no unambiguous evidence to indicate that such an effect leads to substantial rate enhancements. Perhaps one of the most convincing physical manifestation of the effect arises from bond lengths measured for the salt $[(CH_3)_2N]_3S^+CF_3O^-$ by X-ray analysis.[19] The C-F bond lengths are exceptionally long and the C-O bond lengths short compared with gas-phase experimental values for CF_3OR (R = F, Cl, CF_3) derivatives. The role of F-hyperconjugation was estimated by comparing the electronic charge on F in CF_3O^- with that for model CF_3 compounds. The excess charge on each F in CF_3O^- implies a significant contribution from hyperconjugation resonance structures. Also, C-13 and F-19 chemical shift data show substantial upfield shifts for positions adjacent to charge (see IIIC), i.e. $^-$C-CF, and the data are now interpreted as indicating increased charge density at these positions.

B. Olefinic Systems

The influence of halogen on the acidity of hydrogen on a vinylic carbon has been studied by Viehe, Daloze and co-workers. Thus for cis and trans 1-chloro-2-fluoroethene the kinetic acidity of the hydrogen α to chlorine is greater than that of the hydrogen α to fluorine.[20] Quantitative data have been obtained[21] for a range of halogenated ethenes in $CD_3OD/^-OMe$ and pK_a values determined[7] (Table 9). It can be seen that α-halogen substituents facilitate vinyl carbanion formation in the same order as that of haloforms (Table 1) viz. Br > Cl > F.

Schleyer[17] concludes that negative hyperconjugation is also operative in olefinic systems as two local minima corresponding to trans and cis β-fluorovinyl anions (16a and 16b) were located on the $C_2H_2F^-$ potential energy surface. However, both (16a) and

TABLE 9

Carbon acid	pK_a
CCl_2 = CHBr	24.6
CCl_2 = CHCl	25.0
CF_2 = CHCl	25.3
CCl_2 = CHF	26.3
CF_2 = CHF	27.2

(16b) are thermodynamically unstable relative to fully dissociated acetylene and F^- by 10.3 and 13.7 kcal mol^{-1}, respectively. The calculated structures of (16a) and (16b) also support negative hyperconjugation; extreme elongation of the CF bond and a short C-C bond is observed, especially in (16a) where

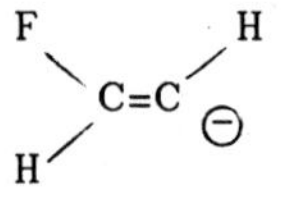

(16a) 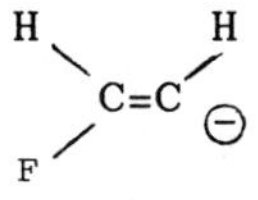 (16b)

negative hyperconjugation is maximised.

C. <u>Aromatic Systems</u>

Early data showed that the acidifying influence of fluorine in substituted benzenes decreases in the order ortho > meta > para (Table 10).[22] Ortho-F exerts a greater acidifying effect

TABLE 10

Relative Rates of Base-Catalysed Deuterium Exchange

Compound	Rate, relative to Benzene	Toluene
Fluorobenzene - 2d	6.3×10^5	-
Fluorobenzene - 3d	107	-
Fluorobenzene - 4d	11.2	-
Benzotrifluoride - 3d	580	-
2,5-Difluorotoluene -αd	-	350

than o-Cl as indicated by the relative acidities of 1,3-dihalobenzenes.[23] Polyfluorobenzenes are very acidic (cf. C_6F_5H pK_a 26 and benzene pK_a 43).[24] Streitwieser has determined the rates of T exchange and of nucleophilic substitution by sodium methoxide for a range of fluorinated benzenes (Table 11).[25] It has been shown that pentafluoro- and 1,2,4,5-tetrafluorobenzene are metallated with BuLi instead of undergoing nucleophilic substitution.[26]

TABLE 11

Polyfluorobenzene	$10^4 k$ ℓ $mol^{-1} sec^{-1}$ Exchange, 40°	Displacement, 50°
C_6HF_5	1360	1.05
1,2,3,4-Tetrafluoro	0.0053	0.018
1,2,4,5-Tetrafluoro	58	>10
1,2,3,5-Tetrafluoro	5.6	0.049
1,3-Difluoro	0.0061	

As mentioned previously F-substitution in an aromatic ring enhances the acidity of hydrogen at a benzylic position (Table 5).[5,9] Indeed a comparison of equilibrium acidities of di- and tri-substituted pentafluorophenyl methanes with their hydrocarbon analogues shows that substitution of each phenyl group by a pentafluorophenyl group results in an enhancement of the acidity (cf. Ph_3CH pK_a 31.5, and $(C_6F_5)_3CH$ pK_a 15.8); this can be readily explained by inductive effects of the C_6F_5 group.[27]

D. Stereochemistry of Fluorocarbanions

As described earlier there is considerable experimental evidence that fluorine substitution stabilises pyramidal carbanions but destabilises planar carbanions and, using *ab initio* calculations, Burdon has shown that CH_2F^-, like CH_3^- is more stable in the pyramidal form than in the planar form. The difference in energy between these two forms is greater for CH_2F^- (13.2 Kcal mol^{-1}) than for CH_3^- (1.1 Kcal mol^{-1}). Mulliken population analyses show that the fall in the net charge on carbon in passing from planar to pyramidal conformation is much greater for CH_2F^- than for CH_3^- (Table 12).[28] This can readily be explained by the increased ability of fluorine to accept negative

TABLE 12

Mulliken Population Analyses

Net charge on	CH_3^-		CH_2F^-	
	Planar	Pyramidal	Planar	Pyramidal
C	-1.3646	-1.2345	-0.9926	-0.6318
H	+0.1215	+0.0782	+0.1465	+0.0304
F			-0.3004	-0.4290

charge in the pyramidal form where there will be less repulsion between fluorine lone pairs and the filled $2p_z$ carbon orbital than there is in the planar form.

A recent *ab initio* study on cyclopropyl, α-fluorocyclopropyl and α-chlorocyclopropyl anions has determined the relative energies of planar and bent $C_3H_4X^-$ anions, and hence barriers to inversion have been calculated.[29] Results of geometry optimisation for all three anions indicate that atom X in $C_3H_4X^-$ is significantly out of the plane of the cyclopropyl ring; that is, the bent conformation is of minimum energy. Barriers to inversion of X are substantial, especially for the fluoro- and chloro-derivatives, indicating a likelihood of retention of configuration in reactions of these anions (Table 13). Further calculations suggest that substitution of H by Cl in this system slightly destabilises the cyclopropyl anion, whereas substitution by F significantly destabilises the anion.[29]

TABLE 13

Data for Cyclopropyl Anions $C_3H_4X^-$

	Angle of C–X bond with plane of ring	Barrier to inversion of X (Kcal mol^{-1})
X = H	61	17.4
F	71	41.8
Cl	65	39.7

E. Delocalised Anions

Anions derived from cyclopentadiene derivatives (17)[30] and (18)[31] and from cyclooctatetraene (COT) derivatives (19)[32] and (20)[33] have recently been studied. The ability of CF_3 groups to enhance the acidity of weak carbon acids is evident in compound (17), $pK_a \leq -2$. This compound, although less acidic than pentacyanocyclopentadiene, is eighteen orders of magnitude more acidic than cyclopentadiene.[30] However, perfluorocyclopentadiene (18) is only slightly more acidic than cyclopentadiene.[31] This is a dramatic illustration of the different effects of F and perfluoroalkyl substituents on carbanion stabilities, as described in Section IA, especially when the system is planar.

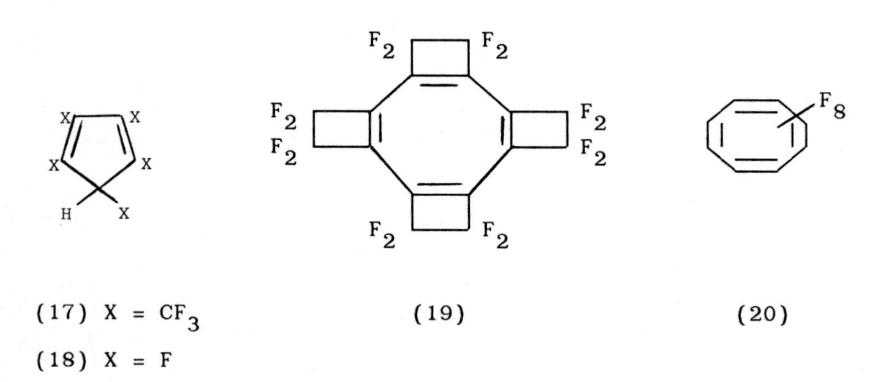

(17) X = CF_3 (19) (20)

(18) X = F

Compound (19) is, remarkably, a planar COT derivative and the electrochemistry of (19) shows two reversible reductions. The increased electron transfer rate for (19) compared with COT is probably due to the planarity of neutral (19) and the electronic influence of fluorine is manifest in the unusually high value for the reduction potential of (19) which is 2.3V more positive than COT.[32] The radical anion of perfluoro-COT (20) has been generated by γ-irradiation of (20) at 77K in a matrix, and e.s.r. parameters suggest that the radical anion is a planar aromatic species with most of the spin density on the carbon atoms.[33]

II. NUCLEOPHILIC ATTACK ON UNSATURATED FLUOROCARBONS[1,34,35,36]

Carbanionic species are intermediates in reactions that involve nucleophilic attack on fluorocarbon-alkenes and -arenes. Consequently, an understanding of the factors that determine reactivity and orientation of attack in these systems is important to the discussion of some of the other carbanionic processes, described later. A fascinating aspect of the chemistry of unsaturated fluorocarbons is the 'mirror-image' relationship with the chemistry of unsaturated hydrocarbons, and more will be said on this later. However, just as electrophilic attack on alkenes, proceeding via carbocationic species, is the dominant process in the chemistry of alkenes, so the converse is true for unsaturated fluorocarbons, whose chemistry is dominated by nucleophilic attack.

e.g. $Nuc^- + CF_2{=}CFCF_3 \longrightarrow [Nuc\text{-}CF_2\bar{C}FCF_3] \longrightarrow$ product

e.g. Nuc^- + [F-ring, CF_3] $\longrightarrow$ [Nuc, F; F, F, F, F; CF_3; anion] $\longrightarrow$ product

A. Fluorinated Alkenes

Some of the factors that influence reactivity and orientation of attack are:

(a) There is a significant <u>polar contribution</u> to the activating influence of fluorine and this needs to be noted, to account for the greater reactivity of alkenes bearing fluorine vs chlorine, at comparable sites.

i.e. $\overset{\delta+}{=C} \longrightarrow \overset{\delta-}{F} \quad >> \quad \overset{\delta+}{=C} \longrightarrow \overset{\delta-}{Cl}$

(b) We have established earlier, that fluorine attached to carbon, which is itself adjacent to a carbanionic centre, is strongly stabilising $\bar{C}\text{-}C \longrightarrow F$.

(c) When fluorine is directly attached to the carbanionic centre, then electron withdrawal is offset by electron-pair repulsion. The resultant may be stabilising or slightly de-stabilising, depending on the geometry of the carbanion.

F-propene is attacked exclusively at the difluoromethylene group and is more reactive than F-ethene.[35] This regiospecificity we attribute to the polar contribution from fluorine, making the difluoromethylene group especially susceptible to nucleophilic attack, coupled with the fact that trifluoromethyl will stabilise adjacent charge. This also accounts for the relative reactivities of F-propene and F-ethene, since the intermediate (21a) will be more stable than (21b). Nevertheless, this type of argument is insufficient to account for the much greater reactivity of F-propene than F-2-butene (22) because the corresponding intermediate (23) could have only

$$[Nuc\text{-}CF_2\text{-}\bar{C}(CF_3)F] \qquad [Nuc\text{-}CF_2\text{-}\bar{C}F_2]$$

(21a) (21b)

$$Nuc^- + CF_3CF{=}CFCF_3 \longrightarrow [Nuc\text{-}CF(CF_3)\text{-}\bar{C}FCF_3] \longrightarrow \text{Products}$$

(22) (23)

marginally different stability from that of (21a). Consequently, a Frontier-Orbital approach has also been used to account for reactivity and orientation of attack.[37] This approach recognises that HOMO-LUMO interaction between nucleophile and F-alkene respectively, will be important and that replacing fluorine in a F-alkene by trifluoromethyl reduces the LUMO energy. This increases reactivity, providing that the trifluoromethyl groups are on the same side of the double bond i.e. the reactivity order is $CF_2{=}CF_2 < CF_2{=}CFCF_3 < CF_2{=}C(CF_3)_2$. However, coefficients are also, apparently, important and introduction of trifluoromethyl increases the coefficient in the LUMO at the carbon opposite i.e. (24). This will be enhanced for two trifluoromethyl groups on the same side of the double bond (24) but for trifluoromethyl groups on opposite sides i.e. (25), their effect on coefficients,

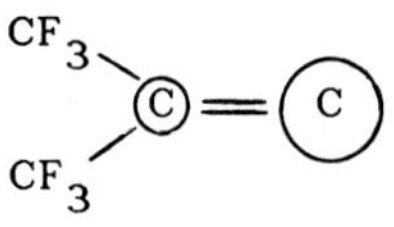

$(CF_3)_2$Ⓒ=Ⓒ CF_3Ⓒ=ⒸCF_3

(24) 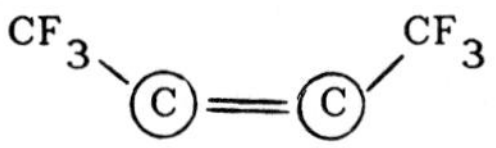 (25)

and hence reactivity, is opposing. We therefore observe the reactivity order $CF_2{=}CF_2 < CF_2{=}CFCF_3 > CF_3CF{=}CFCF_3$.

B. Fluorinated Arenes

The influence of fluorine as a substituent on reactivity and orientation in aromatic systems is much more difficult to define and, indeed, to determine. Nevertheless, by comparing a range of kinetic data for substitution in various fluoroaromatic compounds[38,39] it is possible to disentangle quite different activating effects of fluorine atoms, located at _ortho_, _meta_- and _para_-positions (26), with respect to the position of nucleophilic attack, as outlined below. From the results, it has been concluded that _ortho_- and _meta_-fluorine atoms are activating (a range of ca. 30 - 100x, with respect to H at the equivalent position, depending on the system) while _para_-fluorine is slightly deactivating (ca. 0.2 - 0.5x, with respect to H). Some difficulty arises, however, in interpreting these findings. For charge delocalised into the ring in the transition-state, it is easy to appreciate that, for _para_-F,

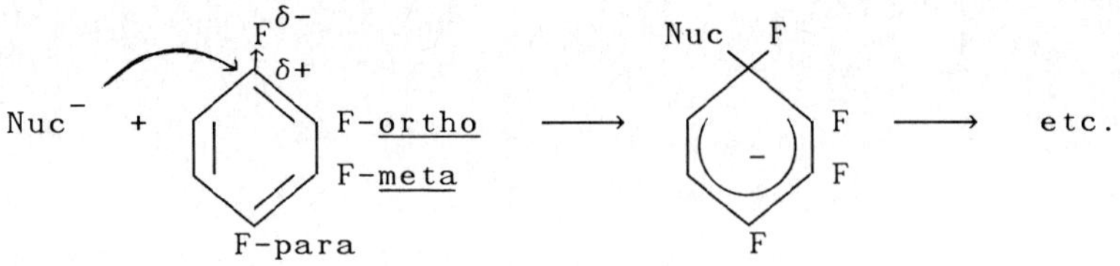

$F^{\delta-}$, $\delta+$, Nuc$^-$ + F-ortho, F-meta, F-para ⟶ Nuc F, F, F, F ⟶ etc.

(26)

the balance between electron-pair repulsion and inductive

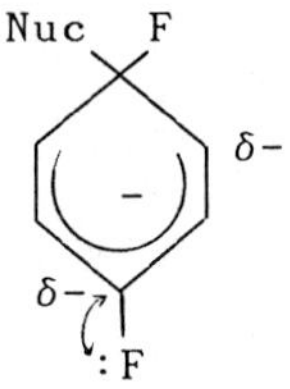

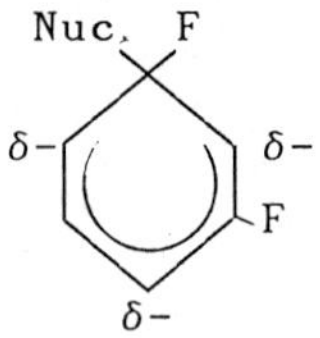

Nuc F, $\delta-$, $\delta-$, :F   Nuc F, $\delta-$, $\delta-$, F, $\delta-$

(27)   (28)

para-F < H (small)   meta-F >> H

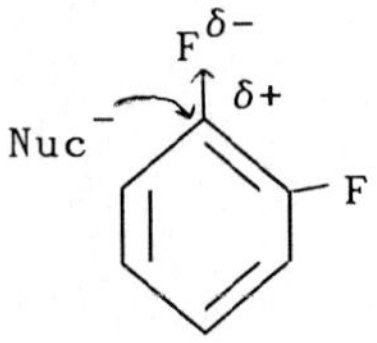

$F^{\delta-}$, $\delta+$, Nuc$^-$, F   ortho-F polar contribution strongly activating

(29)

electron-withdrawal (27) could result in fluorine being slightly destabilising. Also, a fluorine atom meta- to the position of attack (28), and hence adjacent to the centres of highest charge density, would be strongly stabilising. However, the difficulty arises in interpreting the activating influence of ortho-fluorine (29) because, for charge delocalised into the ring in the transition state, we might anticipate that ortho- and para-fluorine atoms would have similar effects. This is, in fact, not the case because equilibrium constants have been determined[40-42] for the formation of σ-complexes (30) and (31) and the effect of ortho- and para-fluorine may be compared. Quite clearly, fluorine has a strong stabilising influence, with respect to hydrogen, at the ortho-position (30), whereas there is little difference between X = H and F, in (31). This is a surprising result but it is also borne out by calculation on simple σ-complexes.[43,44] It would be simple to argue that charge is much greater at the para- than ortho-positions in σ-complexes (30) and (31), but then this would

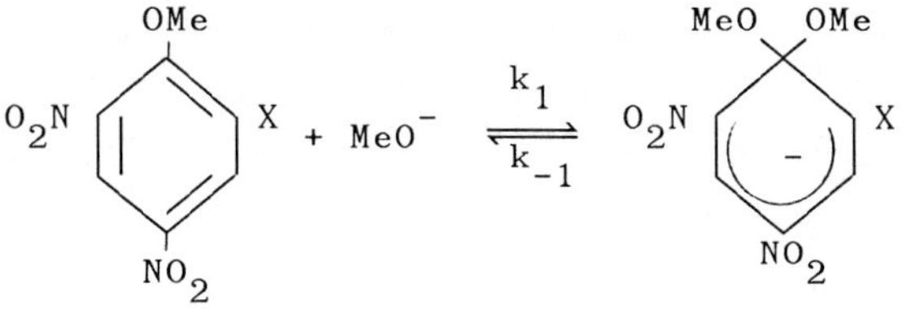

$$\text{2,4-}(O_2N)_2\text{-6-X-}C_6H_2OMe + MeO^- \underset{k_{-1}}{\overset{k_1}{\rightleftharpoons}} (30) \quad (25^oC,\ MeO^-/MeOH)$$

(30)

X =	k_1/mol^{-1}s^{-1}	k_{-1}/s^{-1}	K_1/ℓ mol^{-1}
H	2×10^{-3}	42	5×10^{-5}
Cl	0.18	0.06	3
F	0.1	0.4	0.3

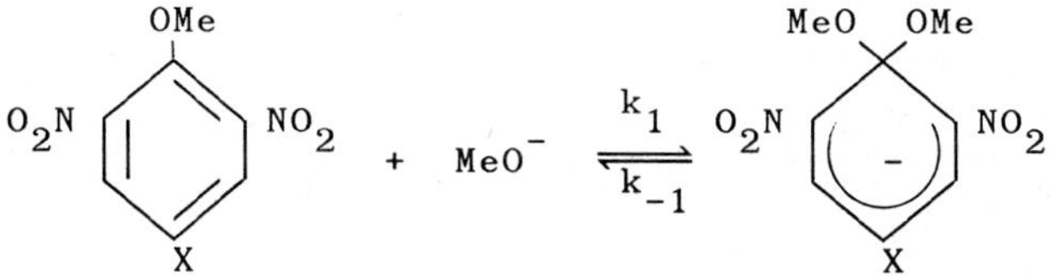

$$\text{2,6-}(O_2N)_2\text{-4-X-}C_6H_2OMe + MeO^- \underset{k_{-1}}{\overset{k_1}{\rightleftharpoons}} (31)$$

(31)

X =	k_1/mol^{-1}s^{-1}	k_{-1}/s^{-1}	K_1/ℓ mol^{-1}
H	1.5×10^{-3}	20	7.5×10^{-5}
Cl	1.2×10^{-2}	5	2.5×10^{-3}
F	2.5×10^{-3}	30	8.5×10^{-5}

require similar charge density at ortho- and meta-positions in order to have similar effects of fluorine at these sites. In fact calculations[45] reveal a slight preference for charge density at the para-position (32) but a major difference in charge density at ortho- in comparison with meta-positions.

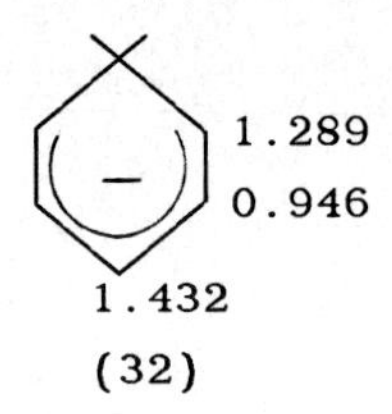

(32)

We feel therefore, that there is not a qualitatively satisfying explanation for these observations that has been advanced so far, although Schleyer has[46] pointed out that most electron density lies near the ipso-position in σ-complexes and argued that this could account for ortho-fluorine having the greatest stabilising influence.

It has also been suggested that a fluorine atom located ortho- to the reaction centre has a large polar influence that is strongly activating and, therefore, the dominant influence. There is evidence to support this view because the influence of ortho-F relative to meta-F increases with reactivity of the system.[47] A further approach would be to consider the system subject to control by Frontier Orbitals. If on fluorine substitution (33) is now the LUMO, rather than (33) and (34) being degenerate, and Epiotis[48] has argued that this is so, then comparable activating effects of fluorine atoms ortho- and meta- to the site of nucleophilic attack would follow. Although somewhat extreme positions have been taken up[48,49] on the value Frontier-orbitals

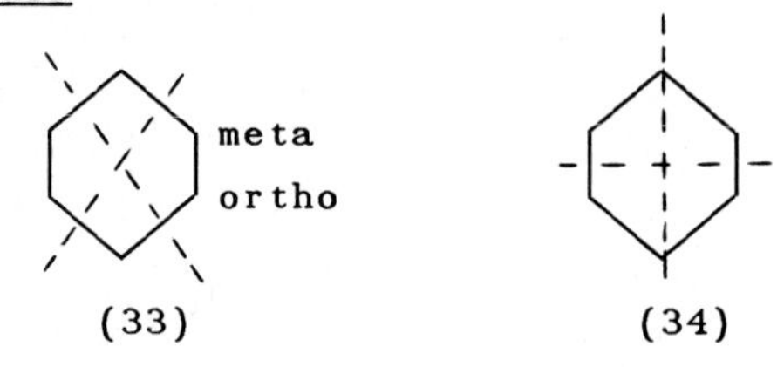

(33) (34)

of the Frontier Orbital approach to this problem, further consideration and calculations will help to resolve their relative merits.

III. FLUORIDE ION INDUCED PROCESSES

An analogy may be advanced between the role of fluoride ion, in its reactions with unsaturated fluorocarbons, and the role of the proton in its reactions with unsaturated hydrocarbons. Miller and his co-workers in pioneering studies showed that carbanions may be generated by reactions of fluoride-ion with fluorinated alkenes[50] and trapping with a variety of species has been achieved (Scheme 1).[51,52] At first sight, it might be expected that this analogy would be neither helpful, nor could it

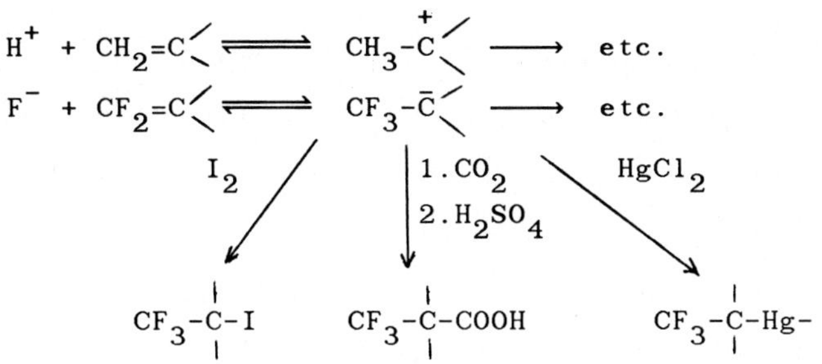

$$H^+ + CH_2{=}C\langle \rightleftharpoons CH_3{-}\overset{+}{C}\langle \longrightarrow \text{etc.}$$

$$F^- + CF_2{=}C\langle \rightleftharpoons CF_3{-}\bar{C}\langle \longrightarrow \text{etc.}$$

$\xrightarrow{I_2}$ $CF_3{-}\overset{|}{\underset{|}{C}}{-}I$; $\xrightarrow{1.\,CO_2;\ 2.\,H_2SO_4}$ $CF_3{-}\overset{|}{\underset{|}{C}}{-}COOH$; $\xrightarrow{HgCl_2}$ $CF_3{-}\overset{|}{\underset{|}{C}}{-}Hg{-}$

SCHEME 1.

be pursued very far because reactions of carbanions are being compared with those of carbocations. However, the analogy has been helpful and is surprisingly extensive; furthermore, its pursuit has led to a range of interesting chemistry.

A. Oligomerisation of Fluorinated Alkenes[51,53]

The oligomerisation of F-ethene[54-56] is a useful example to take, since it illustrates how processes like this are useful for building up synthetically more sophisticated systems from readily available small molecules, by providing a range of oligomers (Scheme 2). The tetramer (42) may be formed by dimerisation

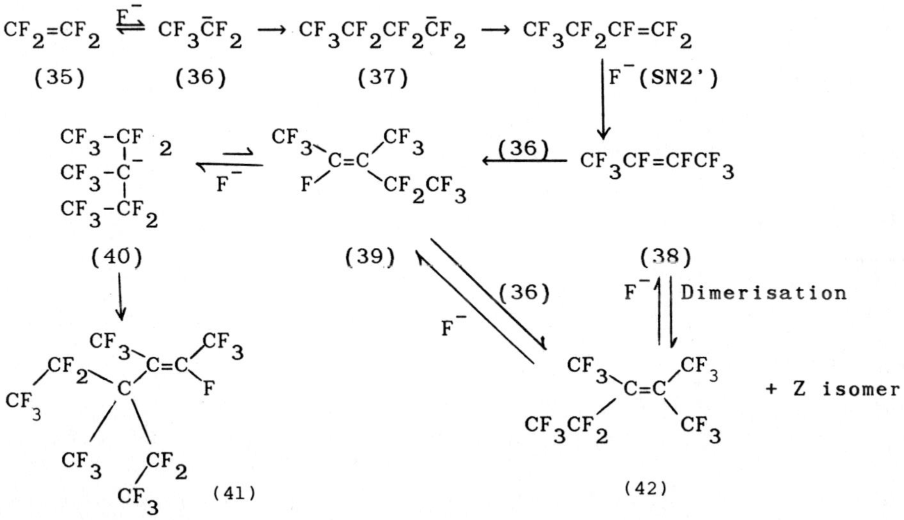

$$\underset{(35)}{CF_2{=}CF_2} \overset{F^-}{\rightleftharpoons} \underset{(36)}{CF_3\bar{C}F_2} \longrightarrow \underset{(37)}{CF_3CF_2CF_2\bar{C}F_2} \longrightarrow CF_3CF_2CF{=}CF_2$$

$$CF_3CF_2CF{=}CF_2 \xrightarrow{F^-(S_N2')} \underset{(38)}{CF_3CF{=}CFCF_3}$$

$$\underset{(38)}{CF_3CF{=}CFCF_3} \xrightarrow{(36)} \underset{(39)}{(CF_3)(F)C{=}C(CF_3)(CF_2CF_3)} \underset{F^-}{\rightleftharpoons} \underset{(40)}{(CF_3CF_2)(CF_3)\bar{C}{-}CF(CF_3)_2}$$

$$(40) \longrightarrow \underset{(41)}{(CF_3)(F)C{=}C(CF_3){-}C(CF_3)(CF_2CF_3)_2}$$

$$(39) \underset{F^-}{\overset{(36)}{\rightleftharpoons}} \underset{(42)}{(CF_3)(CF_3CF_2)C{=}C(CF_3)_2}\ \text{+ Z isomer}$$

$$(38) \underset{F^-}{\overset{\text{Dimerisation}}{\rightleftharpoons}} (42)$$

SCHEME 2

of (38), a process which has been established,[56] or via the trimer (39) by further reaction with anion (36). Furthermore, the trimer (39) forms a very stable tertiary anion[57] (40) which will react with F-butene (38) to form a pentamer (41). Similar fluoride-ion induced oligomerisation reactions of F-propene,[53,58] and chlorotrifluoroethene[59] have been described; in all cases highly branched systems are produced, due to the tendency of the extending carbanion to lose fluoride ion, e.g. from (37), giving an alkene which reacts further by adding fluoride ion or carbanion to the terminal site. In contrast, a fluoride- elimination process does not occur for the growing carbanion (45), produced in the reaction of fluoride-ion with F-butyne (43) and, so far, (46) is the only polymer that has been produced by a fluoride-ion induced process.[60] F-Cyclopentene and

$$\underset{(43)}{CF_3C{\equiv}CCF_3} \overset{F^-}{\rightleftharpoons} \underset{(44)}{CF_3CF{=}\bar{C}CF_3} \xrightarrow{(43)} \underset{(45)}{CF_3CF{=}C(CF_3)C(CF_3){=}\bar{C}CF_3}$$

$$\downarrow \text{(43) etc. etc.}$$

$$\underset{(46)}{[C(CF_3){=}C(CF_3)]_n}$$

F-cyclohexene give dimers only[61] while F-cyclobutene (47) reacts very readily to give mainly a trimer (51) with a smaller proportion of an equimolar mixture of the dimers (48) and (49)

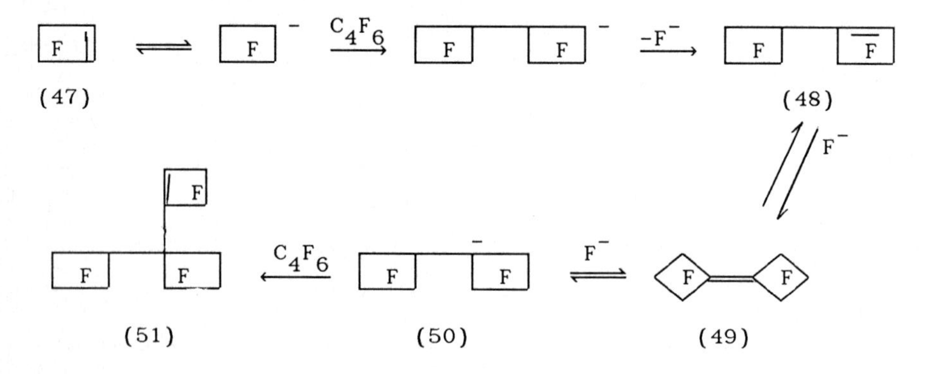

SCHEME 3

(Scheme 3).[62,63] The tertiary carbanion (50) has been observed directly.[64] Pyridine will also induce oligomerisation of (47)[65] but the trimer formed in this case (53) is different from (51) (Scheme 4). The work of Burton and his co-workers[66] on the observation of stable ylides from (47) using trialkylamines allows us to understand the formation of (53) via the intermediacy of an ylide (52).[66] The difference between this process and the fluoride-ion induced reaction lies principally in the fact that the trimer is now produced by nucleophilic substitution in the dimer (48), and hence involves displacement of vinylic fluorine, whereas in the process leading to (51), the trimer is obtained via a carbanion (50), which is itself derived from (48) by nucleophilic addition of fluoride ion.

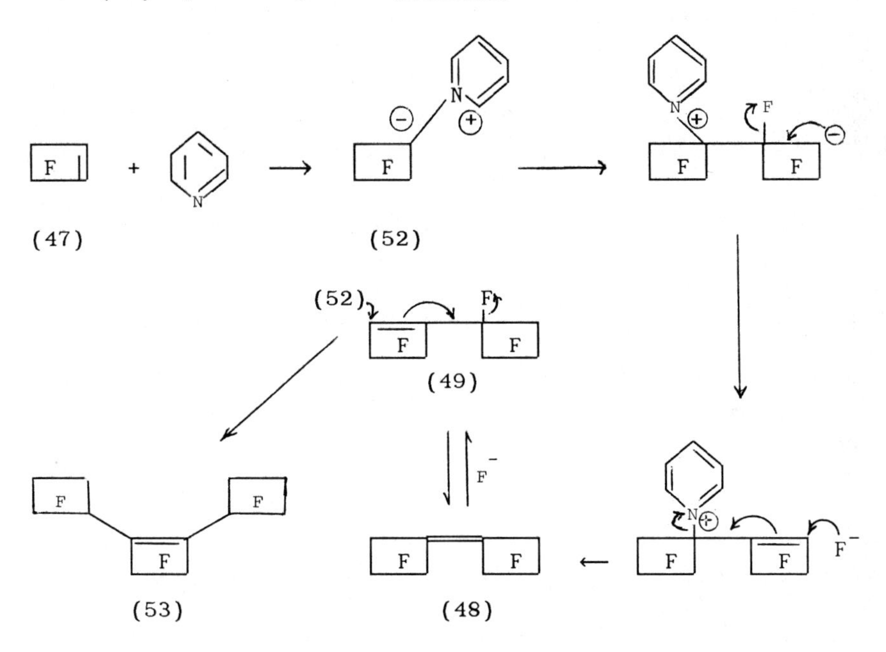

SCHEME 4

B. Equilibration by Fluoride Ion

In the absence of other factors, structures without vinylic fluorine are generally the most stable; this may be attributed to the fact that vinylic fluorine (54) does not significantly affect π-orbital energies, relative to vinylic hydrogen (55) but a perfluoroalkyl group (56) is strongly stabilising.[37]

Consequently, alkene (57) is the predominant isomer under equilibrium conditions i.e. in the presence of fluoride ion,

$$FC{=}C \sim HC{=}C \gg CF_3C{=}C \quad (\pi\text{-orbital energies})$$

(54) (55) (56)

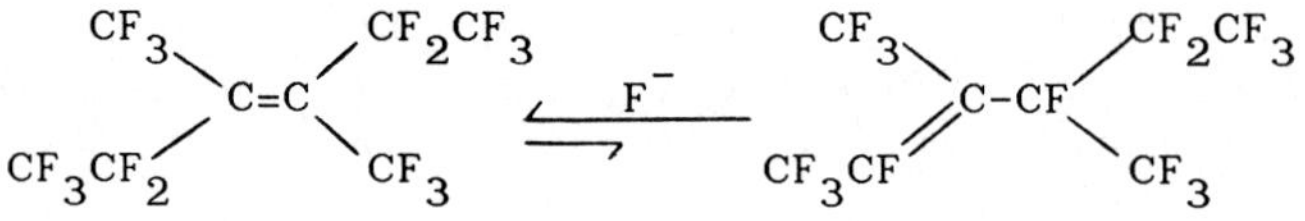

$$(CF_3)(CF_3CF_2)C{=}C(CF_2CF_3)(CF_3) \underset{}{\overset{F^-}{\rightleftharpoons}} CF_3CF{=}C(CF_3){-}CF(CF_2CF_3)(CF_3)$$

(57) (58)

although isomer (58) is the more reactive[67] and products derived exclusively from this isomer have been observed under appropriate conditions.[68]

The balance between exo- and endo-structures is indicated below, for F-cycloalkene dimers.[63] Clearly, the role of vinyl fluorine is not dominant in determining the position of equilibrium in these systems. In spite of a vinylic fluorine atom, isomer (59), with a saturated six-membered ring, is the more stable, while the converse is true for the five-membered ring system (60). Here, eclipsing interactions are minimised in the isomer with the exocyclic double-bond. In principle, we would expect the same situation to obtain in the four-membered

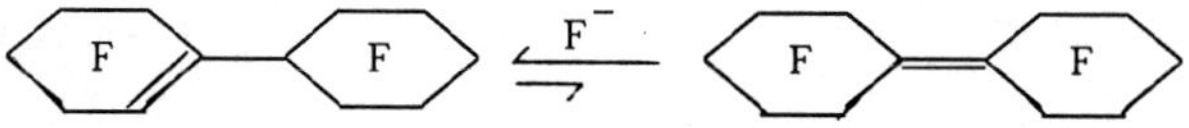

(59) $\overset{F^-}{\rightleftharpoons}$ (Not observed)

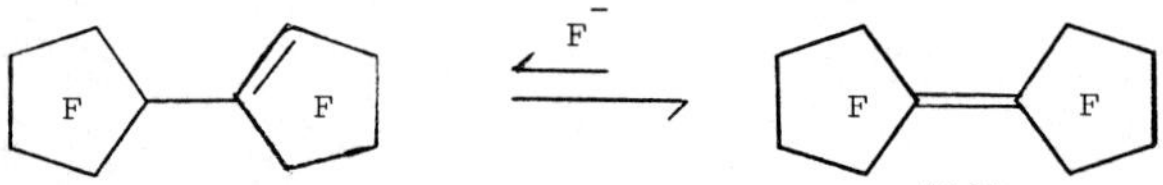

$\overset{F^-}{\rightleftharpoons}$ (60) (Formed exclusively)

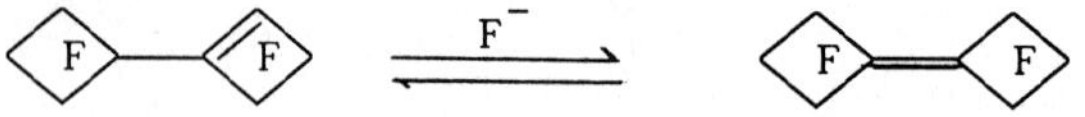

(48) $\overset{F^-}{\rightleftharpoons}$ (49) (Equimolar mixture)

ring system (48). However, angle strain appears to be the compensating factor because in (48) this is minimised, rather than in (49) where both rings are subjected to additional angle strain by having a planar carbon atom. These factors, just outlined, enable the formation of observable carbanions (see above) from (59), (48) and (49) but not (60);[64] this is easy to understand since addition to (59) forms another strain-free six-membered ring, and addition to (48) and (49) relieves angle strain.

C. Formation of Observable Anions

The spectacular developments in superacid chemistry, amongst which are the generation of observable cations, naturally excited the prospect of generating observable carbanions via fluoride-ion reactions. Clearly, one difficulty here lies in the propensity of F-alkenes to undergo oligomerisation with fluoride ion. There are, however, a number of cases now,[57,64] where observable anions have been generated, and some of these are shown below (Table 14).

e.g. $CsF + CF_2{=}C(CF_3)_2 \xrightarrow{\text{Tetraglyme}} (CF_3)_3C^-$.

The remarkable feature about the n.m.r. spectra of the anions that have been described[57,64] is the fact that the C-13 chemical shifts associated with the carbon bearing charge are, understandably shifted upfield, with respect to appropriate precursors or models, but the carbon atoms adjacent to the charged centres i.e. $CF\text{-}\bar{C}$ are associated with a downfield shift, which is also reflected in the fluorine spectra, for fluorine atoms attached to those carbon atoms. However Adcock and co-workers[72a] have emphasised the fact that basic theory of substituent chemical shifts is in some disarray since it is now established that increased electron density may be associated with shifts to highfield for positions remote from the substituent. On this basis it is clear, therefore, that low field shifts arising from both C-13 and F-19 data for positions adjacent to a carbanion centre $^-C\text{-}CF$ are indicative of increased electron-density at these positions. Consequently, these data provide some of the physical manifestation of 'negative hyperconjugation', that has been so lacking (see section IA) in rate-constant data.

A contrast exists between the C-13 chemical shifts recorded for ions shown in Table 14, where the carbanionic carbon atoms are associated with significant upfield shifts,[57] and the downfield shifts recorded by Seebach and co-workers for a wide

TABLE 14

Precursor	Counter-Ion	Anion	Ref.
(48,49)	A,B,	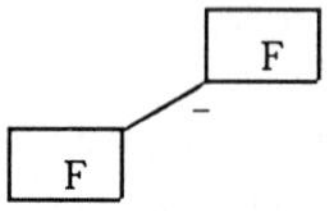 [structure: F-ring – F-ring anion]	64,70a
(59)	A	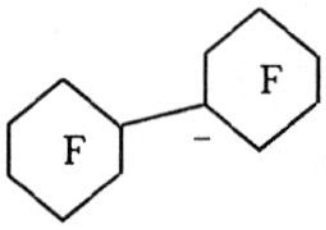 [structure: F-ring – F-ring anion]	64
(63)	A	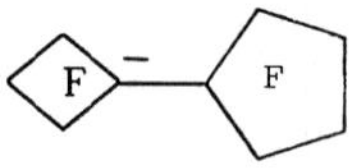 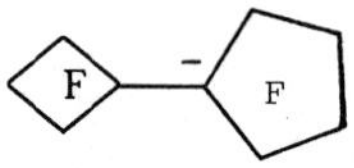 [structure: F-ring – F-ring anion] + [structure: F-ring – F-ring anion]	64
$(CF_3)_2C{=}CF_2$	A,B,C	$(CF_3)_3C^-$	57,70a,71
$C_2F_5(CF_3)C{=}CFCF_3$	A	$C_2F_5(CF_3)\bar{C}C_2F_5$	57
$(CF_3)_2C{=}CFC_2F_5$	A,C	$(CF_3)_2\bar{C}CF_2C_2F_5$	57,71
$(CF_3)_2C{=}C{=}C(CF_3)_2$	A,B	$(CF_3)_2C{=}CF{-}\bar{C}(CF_3)_2$	69,70b
$(CF_3)_2C{=}C(C_2F_5)_2$	B	$(CF_3)_2CFC^-(C_2F_5)_2$	70
$CF_3(C_2F_5)C{=}C(C_2F_5)CF_3$	B	$CF_3(C_2F_5)CFC^-(C_2F_5)CF_3$	70
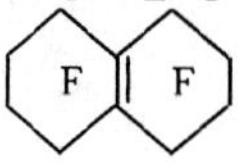 [structure: F-ring = F-ring]	B	 [structure: F-ring F-ring anion]	70

A, Cs^+. B, $(Me_2N)_3S^+$. C, $[(Me_2N)_2CH]^+$

range of organo-lithium derivatives,[72b] i.e. on replacing H or halogen by lithium. It has been established by these studies that organo-lithium compounds are highly associated and this is undoubtedly contrasting with e.g. the caesium salts shown in Table 14. The very small downfield for an α-fluoro anion (61a) in comparison with the precursor (61) is explained by association of the lithium atom with oxygen (61b).[72c]

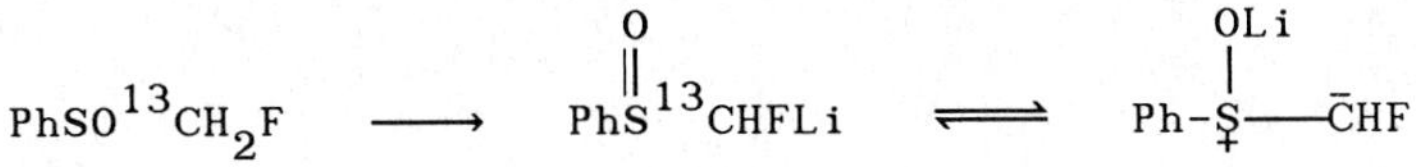

$$PhSO^{13}CH_2F \longrightarrow PhS(O)^{13}CHFLi \rightleftharpoons Ph\text{-}\overset{+}{S}(OLi)\text{-}\bar{C}HF$$

(61) (61a) (61b)

A choice of carbanions is available in some cases where the orientation of addition of fluoride ion is less obvious.[64] From the unsymmetrical system (62), anion (62a) is formed exclusively and this is probably a reflection of the electronegativity of carbon contained in a cyclobutane ring. More surprisingly, both ions (64a) and (64b) were generated from (63), in spite of the fact that no observable anion could be generated from (60), due to unfavourable eclipsing effects (Scheme 5). Both anions (64a)

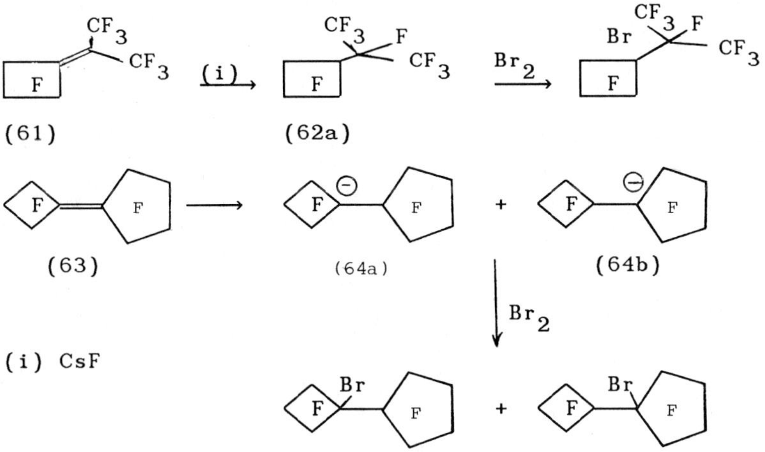

(61) (62a)

(63) (64a) (64b)

(i) CsF

SCHEME 5.

and (64b) have been trapped with bromine. Obviously, relief of angle strain allows formation of (64a) and (64b) but the destabilising effects of eclipsing interactions become apparent in the fact that line broadening occurs in the F-19 n.m.r. spectrum, as the temperature is raised, and a single sharp resonance was observed at 80°C.[64] This indicates very rapid equilibration of all sites, with fluoride ion in solution, and strongly resembles the very familiar exchange of protons with acid (Scheme 6).

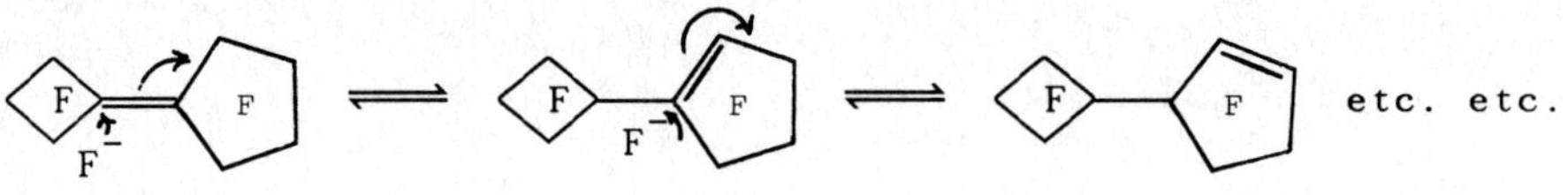

Rapid at 80°C

SCHEME 6.

The effect of the counter-ion has not been fully explored but recent results from the du Pont Company[70a,b] suggest that $(Me_3N)_3\overset{+}{N}Me_3SiF_2^-$ is an excellent source of soluble fluoride ion[70c,d] and that a range of observable anions may be generated using this source. Early interesting studies, using $[(Me_2N)_2CH]^+HF_2^-$ as a soluble fluoride ion source are complicated by the problems of exchange reactions involving proton-transfer.[71]

D. Observable σ-Complexes

Addition of fluoride ion to activated aromatic systems may lead to observable σ-complexes.[73-75] Caesium fluoride dissolves readily in a solution of F-sym-triazine (65) in sulpholan, giving a σ-complex (66). At 0°C, exchange of fluoride is slow, whereas, as the temperature is raised, exchange becomes apparent and leads to a single broad resonance.[73] Similar complexes, e.g. (68), have been observed for the corresponding perfluoroalkyl derivatives e.g. (67) but, so far, there is no report of observable complexes from other, less activated perfluorinated aromatic systems.

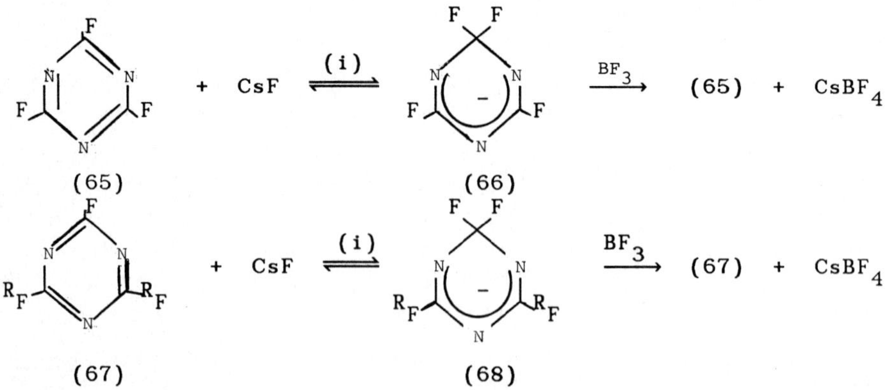

$R_F = CF(CF_3)_2$

(i) Sulpholane

E. 'Nucleophilic Friedel Crafts Reactions'[76]

Carbanions, generated by reaction of fluoride-ion with unsaturated fluorocarbons, may be trapped by reaction with activated polyfluoroaromatic compounds resulting in the introduction of perfluoroalkyl groups. This is, of course, reminiscent of familiar cationic processes.

$$H^+ + CH_2=C< \rightleftharpoons CH_3-\overset{+}{C}< \xrightarrow{Ar-H} Ar-\underset{|}{\overset{|}{C}}-CH_3 + H^+$$

$$F^- + CF_2=C< \rightleftharpoons CF_3-\bar{C}< \xrightarrow{Ar-F} Ar-\underset{|}{\overset{|}{C}}-CF_3 + F^-$$

The carbanions are generated from the corresponding F-alkene and processes involving, e.g. $CF_3CF_2^-$,[77-79] $(CF_3)_2CF^-$,[80,81] $(CF_3)_3C^-$,[80,82] $CF_3(C_2F_5)CF^-$[83] and CF_3CFCl^-[84] have been described. Mono- and di-substituted derivatives (70) and (71) of F-pyridine (69) and F-pyridazine (72), viz. (73) and (74) may be obtained directly and examples are shown. Orientation of substitution, first at the 4-position in F-pyridine and at the 4,5-positions in F-pyridazine, is consistent with the model for orientation and reactivity for nucleophilic aromatic substitution, discussed earlier.

Observed reactivity is the product of an equilibrium constant K, for the generation of a carbanion by the addition of fluoride ion, and a rate constant k which depends on the reactivity of the carbanion involved. F-alkylation is, of course, in competition with oligomerisation of the F-alkene itself. Relative stabilities of the F-alkylanions are in the order shown (and reactivities in the reverse order), with F-t-butyl being the most stable. Nevertheless, F-alkylation by

$$\text{(69)} + (CF_3)_2CF^- \longrightarrow \text{(70) [4-}CF(CF_3)_2\text{]} \xrightarrow{(CF_3)_2\bar{C}F} \text{(71) [3,4-di-}CF(CF_3)_2\text{]}$$

(69) (70) (71)

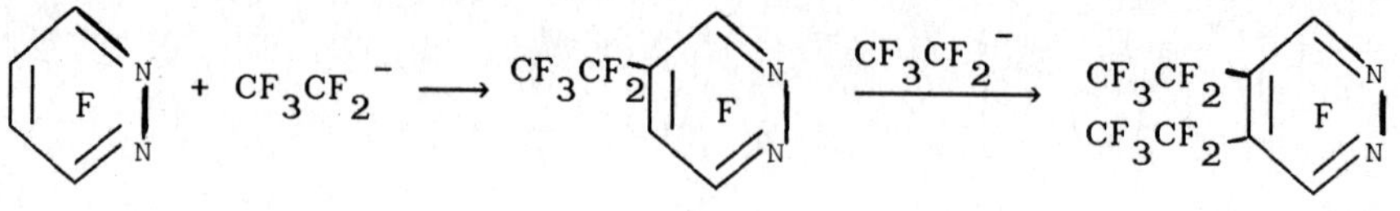

$$\text{(72)} + CF_3CF_2^- \longrightarrow CF_3CF_2\text{-(73)} \xrightarrow{CF_3CF_2^-} (CF_3CF_2)_2\text{-(74)}$$

(72) (73) (74)

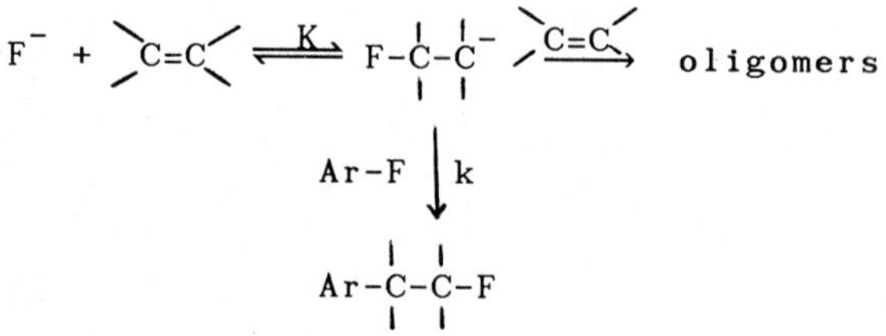

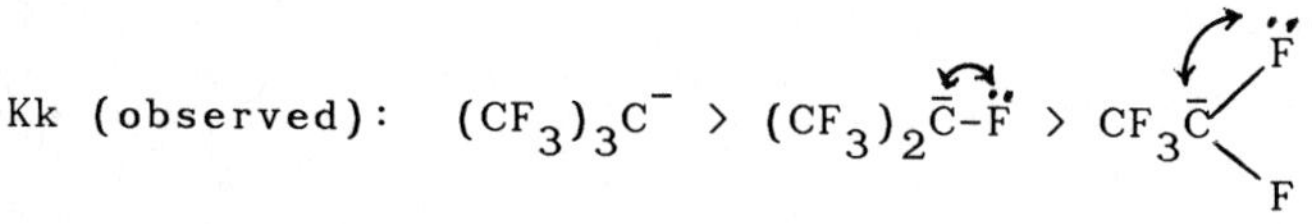

$$F^- + \rangle C{=}C\langle \underset{}{\overset{K}{\rightleftharpoons}} F{-}\overset{|}{\underset{|}{C}}{-}\overset{|}{\underset{|}{C}}^- \xrightarrow{\rangle C=C\langle} \text{oligomers}$$

$$\downarrow k \ (Ar{-}F)$$

$$Ar{-}\overset{|}{\underset{|}{C}}{-}\overset{|}{\underset{|}{C}}{-}F$$

Kk (observed): $(CF_3)_3C^- > (CF_3)_2\bar{C}{-}F > CF_3\bar{C}F_2$

(← order of anion stability →)

this anion is also the most efficient of the series[80] and we can conclude, therefore, that a favourable equilibrium constant, K, is the dominating influence affecting efficiency.

Polysubstitution is more complicated; F-tri-isopropyl-sym-triazine (76) may be obtained from (75)[85] but crowding limits the number of F-isopropyl groups that can be usefully introduced into the pyridine ring (78, 79).[86] In contrast, however, the less sterically demanding F-ethyl group may be introduced in greater numbers and modest yields of F-pentaethylpyridine (80) may be obtained.[77] A highly substituent-labelled pyridine system (81) has been extremely valuable in photochemical studies[87] and, indeed, the polyfluoroalkylation process has been the key to a range of interesting systems for photolysis.[87,88]

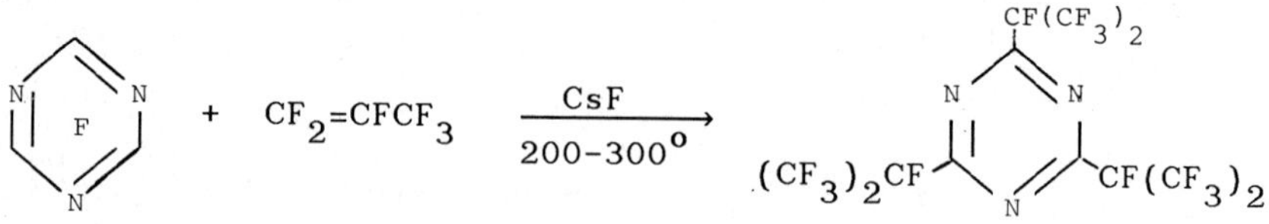

$$\text{(75)} + CF_2{=}CFCF_3 \xrightarrow[200-300^o]{CsF} \text{(76) } [CF(CF_3)_2]_3\text{-triazine}$$

(75) (76)

$$\text{(77)} + CF_2{=}CFCF_3 \xrightarrow[\text{sulpholan}]{CsF,\ 70^o} \text{(78) } [CF(CF_3)_2]_3\text{-pyridine} +$$

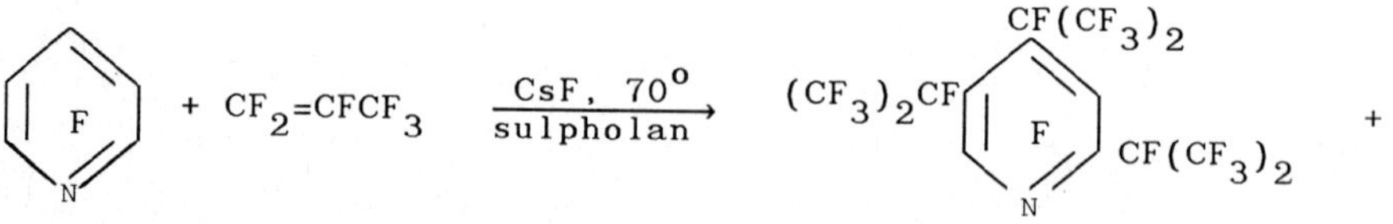

(77) (78)

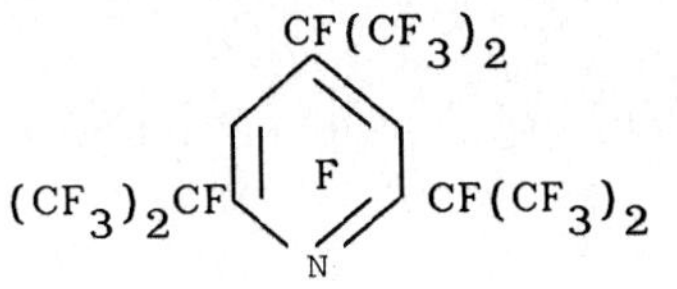

$CF(CF_3)_2$

$(CF_3)_2CF$ F $CF(CF_3)_2$ etc.

N

(79)

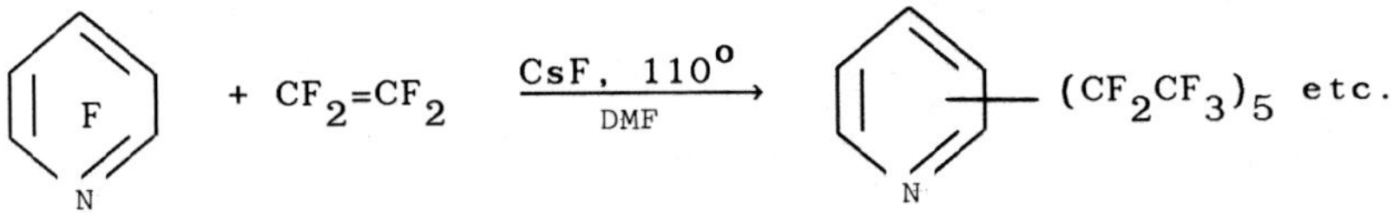

F + $CF_2=CF_2$ $\xrightarrow[\text{DMF}]{\text{CsF, 110}^{\circ}}$ $(CF_2CF_3)_5$ etc.

N N

(77) (80)

F

CF_3 F CF_3 $\xrightarrow{(i)}$ CF_3 CF_3 $(CF_3)_2CF$ $CF(CF_3)_2$

N N

(ii)

CF_2CF_3

CF_3 CF_3

$(CF_3)_2CF$ $CF(CF_3)_2$

N

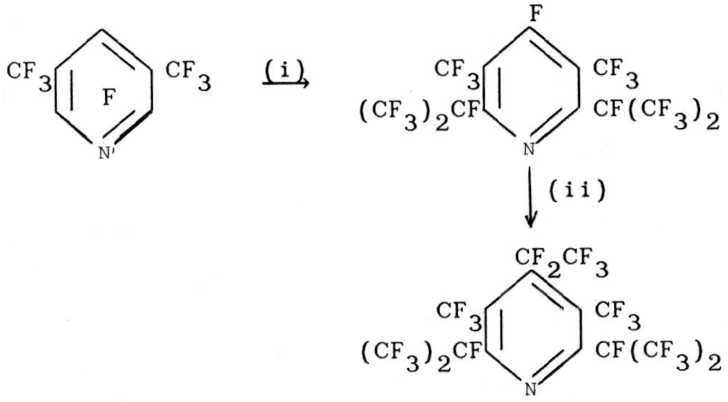

(81)

(i) $CF_2=CFCF_3$, Sulpholan (ii) $CF_2=CF_2$, Sulpholan

F. Skeletal Rearrangements

Just as polyalkylbenzenes e.g. (82) may be rearranged in the presence of acid, so analogous rearrangements[86] may also be induced by heating F-polyalkylaromatic compounds in the presence of fluoride ion (i.e. using more severe conditions than those employed for the original preparation). F-2,4,5-tri-isopropylpyridine (83) may be converted to the corresponding 2,4,6-isomer (84). These skeletal rearrangements are, however, intermolecular processes because the migrating group may be trapped by excess of some other reagent. For example, in the presence of excess F-pyridazine, F-4-isopropylpyridazine (86) instead of (87) is obtained from (85) (Scheme 8). A complex balance of effects operates and the products depend on (i) the stability of the F-alkyl-anion being generated, (ii) steric requirements of the F-alkyl group introduced, and (iii) the conditions employed. The subtleties of these points have been dealt with elsewhere.[80]

Nevertheless, the systems do provide for organic chemistry in general, some very nice examples of competition between kinetic and thermodynamic control of reaction products.

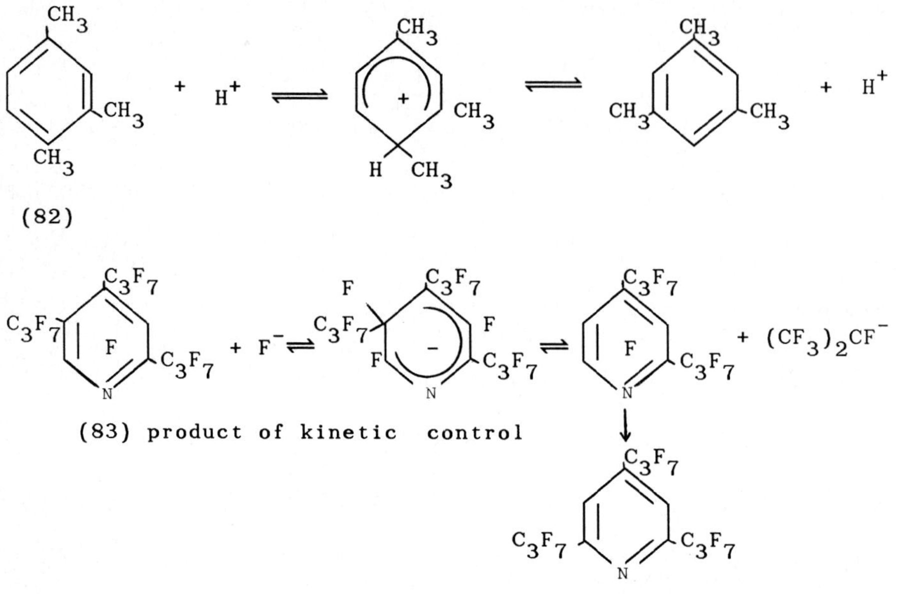

CH_3 CH_3 CH_3 + H^+ ⇌ CH_3 + CH_3 H CH_3 ⇌ CH_3 CH_3 CH_3 + H^+

(82)

C_3F_7 C_3F_7 F C_3F_7 N + F^- ⇌ F C_3F_7 C_3F_7 F − F C_3F_7 N ⇌ C_3F_7 F C_3F_7 N + $(CF_3)_2CF^-$

(83) product of kinetic control

C_3F_7 C_3F_7 C_3F_7 N

(84) product of thermodynamic control

SCHEME 7.

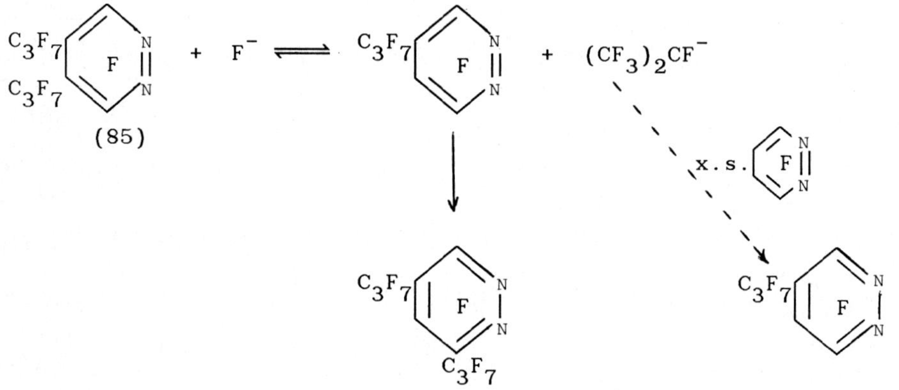

C_3F_7 C_3F_7 F N N + F^- ⇌ C_3F_7 F N N + $(CF_3)_2CF^-$

(85)

x.s. F N N

C_3F_7 F N N C_3F_7

(87)

C_3F_7 F N N

(86)

SCHEME 8.

A more mechanistically impressive rearrangement induced by fluoride ion involves the, at first sight absurd, conversion of (48) to (88) (Scheme 9).[89] However, a process can be described, based on sound precedent, involving combinations of allylic displacements by fluoride ion and the reversible cyclobutene-butadiene process.

An even more dramatic skeletal rearrangement occurs, however, in the reaction of trimer (53) with caesium fluoride at room temperature.[90] This provides a direct synthetic route to a seven-membered ring system (93) in a remarkable process, which involves a well-known anionic ring-opening of cyclobutylcarbinyl systems. More controversially, however, a mechanism was suggested that involves an internal nucleophilic displacement of fluorine from a saturated site. It has been shown previously, that under favourable circumstances i.e. where the internal nucleophile is generated close to the C-F site, then such a displacement can occur.[91] However, two recent pieces of evidence point to an

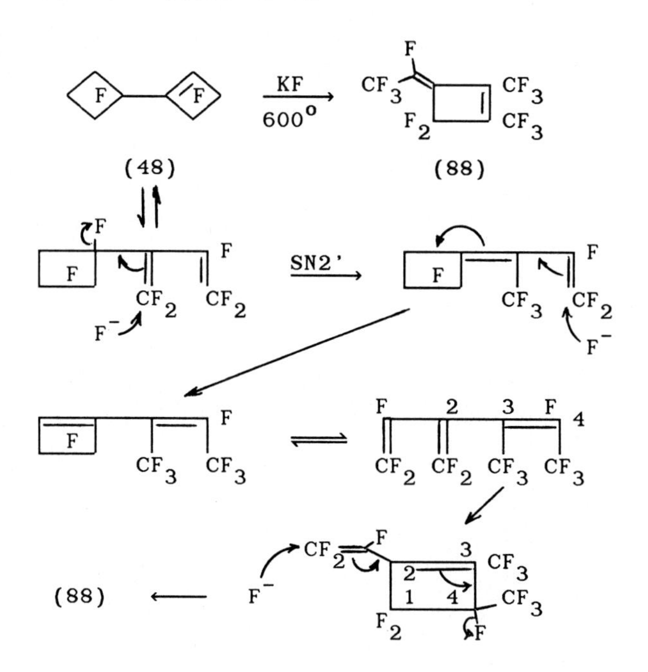

SCHEME 9.

alternative and perhaps more acceptable mechanism. First, it has been shown[92] that ready conversion of the diene (89) to the cyclic system (91) occurs with great ease and demonstrates that intramolecular attack of a crowded anion at a crowded site (90) can occur (Scheme 10). Secondly, there has been a report[93] of

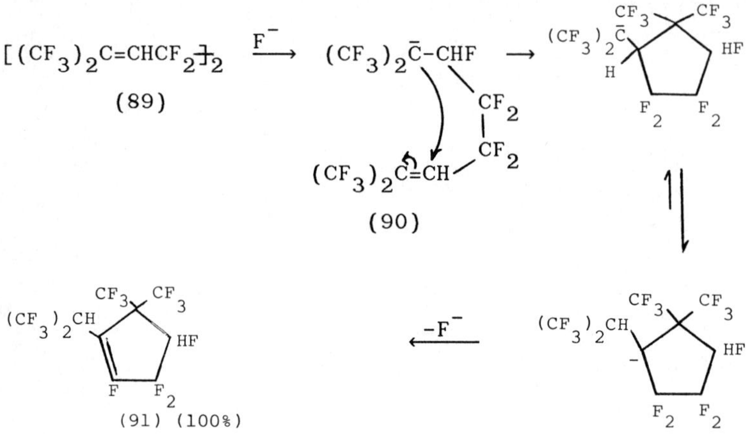

$[(CF_3)_2C{=}CHCF_2]_2$ (89) $\xrightarrow{F^-}$ $(CF_3)_2\bar{C}{-}CHF{-}CF_2{-}CF_2{-}CH{=}C(CF_3)_2$ (90) $\rightarrow$ ... $\xrightarrow{-F^-}$ (91) (100%)

SCHEME 10.

direct observation of the di-anion (92). Consequently, another mechanism (Scheme 11), additional to that suggested previously[90] is shown below, which avoids the suggestion of nucleophilic displacement from saturated carbon.

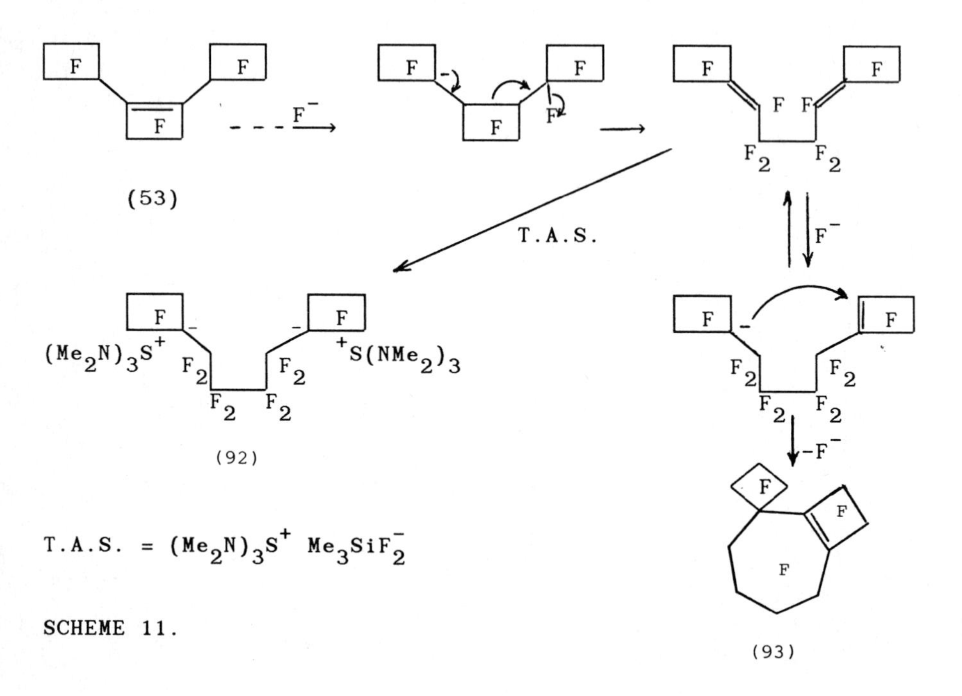

SCHEME 11.

G. Reactions Induced by Other Ions

In typically innovative work, Krespan and co-workers[94a,d] have generated carbanions by reaction of a nucleophile with tetrafluoroethylene and then trapped these intermediate carbanions with carbon dioxide, esters etc. This constitutes an elegant route to a range of difunctional fluorocarbon derivatives

$$B^- + CF_2{=}CFX \longrightarrow [B{-}CF_2CFX^-] \xrightarrow{CO_2} BCF_2CFXCO_2^-$$

$$[B{-}CF_2CFX^-] \xrightarrow{R_FCO_2R} BCF_2CFX\overset{O^-}{\underset{OR}{C}}R_F \xrightarrow{H^+} BCF_2CFXCOR_F$$

(e.g. B=CN, N_3, OC_6H_5, OCH_3, OCH_2CF_3)

IV. ORGANOMETALLIC COMPOUNDS[95a,95b]

The stability of fluorinated organometallic compounds follows the order of stability, $R_F^- < CF_2{=}CF^- \ll Ar_F^-$, reflecting the ease of elimination of metal fluoride. Early attempts to generate CF_3Li lead to formation of tetrafluoroethene probably via a carbene mechanism,[96] but higher alkyl derivatives, e.g. perfluoroisopropyl- and perfluoro-n-heptyllithium are more stable.[97] However, trifluorovinyl-lithium can be obtained via halogen exchange procedures or by direct metallation of a hydro derivative,[98] and a Grignard reagent and a lithium compound can be readily obtained from trifluoropropyne.[99] More recently, Wakselman has described several reactions of fluorinated Grignard reagents that are of synthetic importance. For example, reaction of R_FMgX (R_F = C_4F_9, C_6F_{13}, C_8F_{17} and C_6F_5) with phenyl cyanate produces perfluorinated nitriles.[100] Similarly, reaction of R_FMgX with α-chloroalkylthiocyanates (94) yields sulphides (95) which can be oxidised and then transformed into alkenylsulphones (96) by dehydrochlorination.[101] Reactions of perfluorooctenyl-magnesium bromide with alkyl tin halides have been shown to yield unsaturated perfluoroalkyl tin

$$PhCHCl\text{-}CH_2SCN \xrightarrow{C_4F_9MgBr} PhCHCl\text{-}CH_2SC_4F_9 \xrightarrow[2)-HCl]{1)mcpba} PhCH{=}CH\text{-}SO_2C_4F_9$$

(94) (95) (96)

derivatives (97) where the fluorinated double bond is α to the metal atom.[102]

$$C_6F_{13}CF{=}CFMgBr + (C_4H_9)_3SnCl \longrightarrow C_6F_{13}CF{=}CFSn(C_4H_9)_3$$

(97)

Perfluoroaryl-lithium derivatives are best obtained by metal-halogen exchange and they are well known precursors for other perfluoroaryl-metal compounds.[103] Conversion of C_6F_5Li to a range of monosubstituted pentafluorobenzene derivatives, e.g. perfluoro-phenol, -thiophenol and -benzaldehyde, has been described.[104] Tetrafluorobenzene is liberated from C_6F_5Li at $-10^{o}C$ in ether[105] and from C_6F_5MgBr at ca. $100^{o}C$ in polar solvents;[106] in both cases the aryne has been trapped by furan. Fluorine substitution can also stabilise two negative charges on a single aromatic ring as demonstrated by the interception at $-65^{o}C$ of dilithio-species (98) and (99) by carboxylation.[107]

Reactions of the dilithio-derivative of perfluorobiphenyl (100) deserve attention: biphenylene derivative (101) is formed via intramolecular trapping of the aryne followed by protonation,[14,108] and the dibenzothiophen (102)[109] and fluorenone systems (103)[110] can be formed as shown (Scheme 12).

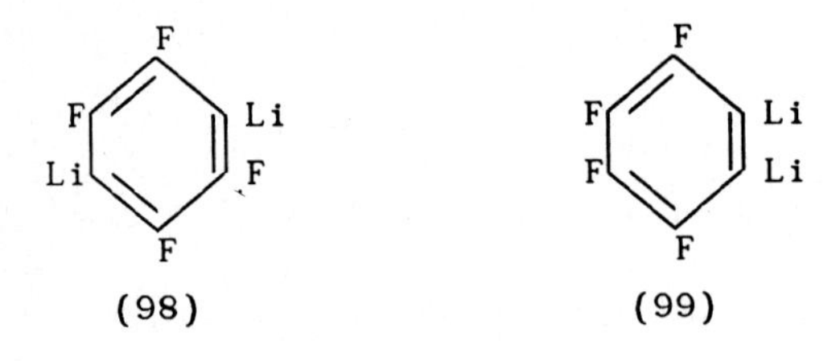

(98) (99)

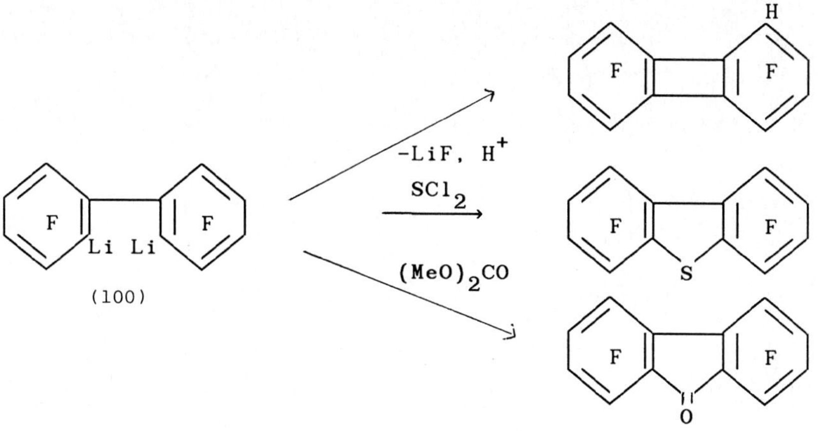

(100) (101) (102) (103)

SCHEME 12.

Reaction of C_6F_5Li with dimethyloxalate gave decafluorbenzil (104) which rearranged to the benzilic ester (105). This rearrangement (104) → (105) occurs relatively rapidly even at $-78^{o}C$ in the presence of methoxide, whereas the corresponding rearrangement in the hydrocarbon requires heating. To explain this massive rate enhancement for aryl group migration in decafluorobenzil relative to benzil a transition state involving a tight ion pair (106a) or even charge delocalisation onto the pentafluorophenyl ring (106b) has been suggested, rather than a transition state which is close to the initial state.[111] This difference in mobility between phenyl and pentafluorophenyl groups has been underlined by further reactions of alkyl- and aryl-lithiums with decafluorobenzil, where the migratory aptitude of the aryl group is closely related to the stability of that group as an anion.[112]

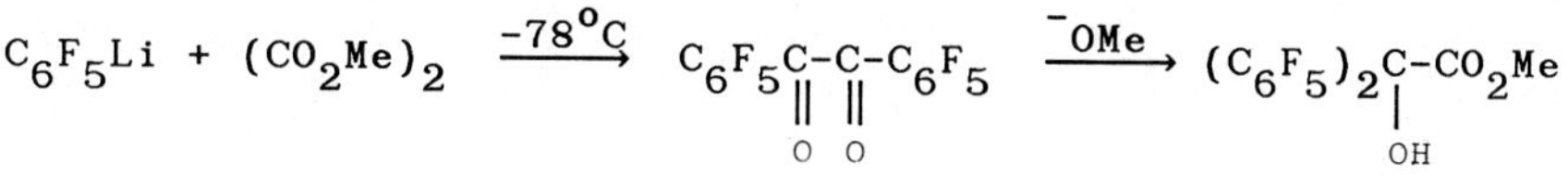

(104) (105)

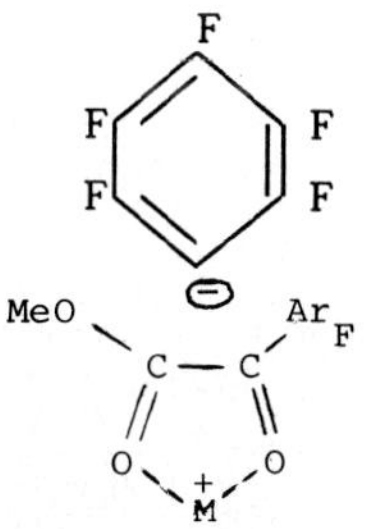

(106a)

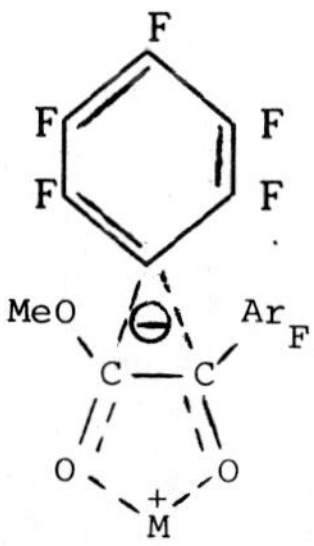

(106b)

In contrast to lithium and magnesium derivatives several other metals form stable perfluoroalkyl organometallic species.[1] Reactions of CF_3I with metal atoms, or reactions of CF_3 radicals with metallic halides or metal atoms have resulted in the preparation of per(trifluoromethyl) derivatives of a number of main group elements, e.g. $(CF_3)_4Ge$, $(CF_3)_4Sn$, $(CF_3)_3Bi$ and $(CF_3)_2Te$,[113] and the formation of partially substituted transition metal compounds, e.g. CF_3PdI.[114] Perfluoroalkyl-Hg, Zn, Cd and Cu species are currently attracting attention because of their stability and potential applications in organic synthesis. $(CF_3)_2Hg$ has been shown to exchange ligands with the halides of several main-group elements,[115] and complexes with Lewis bases of $(CF_3)_2Cd$ can be readily isolated from the interaction of $(CF_3)_2Hg$ with $(CH_3)_2Cd$ in suitable solvents.[116] The $(R_F)_2Cd$ species are much more reactive than $(R_F)_2Hg$ in ligand exchange reactions and in reactions with acyl halides. Thus $(CF_3)_2Cd$-glyme reacts with acyl halides at below room temperature to generate the corresponding acyl fluorides in excellent yield along with CF_2 which was trapped with alkenes.[116] Burton has recently described new routes to $(CF_3)_2M$ (M = Cd or Zn) that avoid ligand exchange processes, by reaction of difluorodihalo-methanes with Cd or Zn powder. This reaction involves a remarkable difluoromethylene to trifluoromethyl group transformation.[117a] The proposed mechanism is shown in Scheme 13. In the initial steps electron transfer from the metal produces anion (107) which loses halide to give difluorocarbene which is

$$M + CF_2X_2 \longrightarrow M^+[CF_2X_2]^{\cdot-} \longrightarrow \underset{(107)}{[CF_2X]^-} + X^- + M^{2+}$$

$$[CF_2X]^- \longrightarrow [:CF_2] + X^-$$

$$(CF_3)_2M + CF_3MX + X^- \xleftarrow{MX} [CF_3]^- \overset{F^-}{\rightleftharpoons} [:CF_2] + X^-$$

$$[:CF_2] \xrightarrow{Me_2NCHO} Me_2NCF_2H\ (108) + CO$$

$$Me_2NCF_2H \rightleftharpoons [Me_2N{=}CFH]^+F^-$$

SCHEME 13.

trapped by solvent DMF to produce amine (108). It has been demonstrated that amine (108) is an excellent source of fluoride ion and thus CF_3^- is formed in the reaction mixture, and can react with metal halide to form product $(CF_3)_2M$.

A very useful metathesis process has been described[117b] in which CuBr was added to a solution of $[CF_3CdX]$ in DMF.

$$2M + 2CF_2X_2 \xrightarrow[\text{Room Temp}]{\text{DMF}} [CF_3MX + (CF_3)_2M] \xrightarrow[-80^oC \text{ to R.T.}]{\text{CuY}} [CF_3Cu]$$

80-90% 90-100%

M = Cd, Zn Y = I, Br, Cl, CN

X = Br, Cl

Coupling with aryl iodides occurs in high yields and unequivocal evidence for $[CF_3Cu]$ is presented. A remarkable conversion process is also observed, if the system is not stabilised by addition of HMPA.

$$2[CF_3Cu] \longrightarrow [CF_3CF_2Cu] + CuF$$

Perfluoroalkyl-copper reagents can be used to replace hydrogen or halogen (usually I) in an aromatic ring with R_F. Thus a one-step synthesis of perfluorooctylbenzene from perfluorooctyl iodide, benzene and copper in DMSO proceeds in moderate yield.[118] Replacement of hydrogen probably proceeds via a free radical mechanism.[119] The ease of halogen replacement by R_FCu reagents follows the sequence I > Br >> Cl and is thought to proceed via the mechanism shown in Scheme 14, without the involvement of free carbanions.[120] Trifluoromethylation of purine and pyrimidine bases of nucleosides[121] and thiophenes[122] using this method has also been described.

$$R_FI \xrightarrow[\text{solvent (L)}]{\text{Cu}} R_FCuL_3 + ICuL_3$$

$$R_FCuL_3 \xrightarrow{\text{ArI}} [Ar\langle^{I}_{R_F}\rangle CuL_3] \longrightarrow ArR_F + ICuL_3$$

SCHEME 14.

Perfluoroalkylzinc iodides have also found application in perfluoroalkylation of organic molecules especially under conditions of ultrasonic promotion. For example, with the aid of this technique ketones and aldehydes can be converted to perfluoroalkylcarbinols in an efficient, one-step reaction (Scheme 15).[123] The range of perfluoroalkylations that have been accomplished by this technique is shown in Scheme 16.[124]

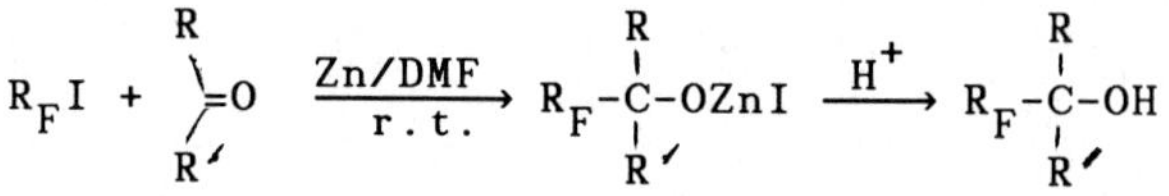

$$R_FI + RR'C{=}O \xrightarrow[r.t.]{Zn/DMF} R_F{-}C(R)(R'){-}OZnI \xrightarrow{H^+} R_F{-}C(R)(R'){-}OH$$

R_F = CF_3, n-C_3F_7, i-C_3F_7, n-C_4F_9, n-C_6F_{13} and n-C_8F_{17}

SCHEME 15.

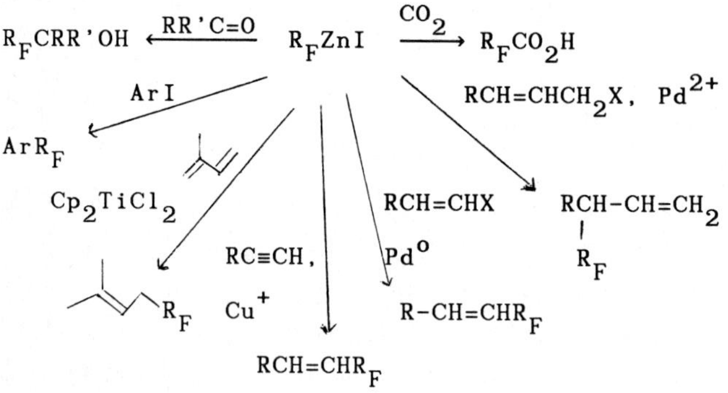

$R_FCRR'OH$ ← ($RR'C{=}O$) — R_FZnI — (CO_2) → R_FCO_2H

R_FZnI + ArI → ArR_F

R_FZnI + isoprene, Cp_2TiCl_2 → $(CH_3)_2C{=}CHCH_2R_F$

R_FZnI + RC≡CH, Cu^+ → $RCH{=}CHR_F$

R_FZnI + RCH=CHX, Pd^o → $R{-}CH{=}CHR_F$

R_FZnI + $RCH{=}CHCH_2X$, Pd^{2+} → $RCH(R_F){-}CH{=}CH_2$

SCHEME 16.

V. YLIDS

The first reported stable polyhalogenated phosphonium ylid was hexafluoro(triphenylphosphoranylidene)cyclobutene (109) obtained from the reaction of hexafluorocyclobutene (47) and triphenylphosphine. Ylid (109) provides a rare example of an isolable F-carbanion and the remarkable stability of (109) was attributed to the overlap of phosphorus 'd' orbitals with the ylid carbon 'p' orbitals, combined with the strong inductive effect of the adjacent CF_2 groups.[125] Burton and co-workers have reported several stable ylids derived from cyclic and acyclic F-alkenes. In particular phosphonium, ammonium and arsonium ylids have been studied.[66,126] Phosphonium ylids are readily cleaved by halogens to give F-dihalocycloalkanes, e.g.

(109) → (110), whereas arsonium ylid (111) reacts with bromine and iodine to give cyclobutene derivatives (112). Hydrolysis reactions occur readily, e.g. (111) → (113) → (114) (Scheme 17).[127] The role of α-fluorine in the mechanistic course of the reaction of F-1-alkenes with tertiary phosphines has been emphasised by the formation of terminal vinylphosphoranes e.g. (115) instead of ylids. It thus appears that when only β-fluorines are present, e.g. cyclic F-alkenes, the ylid is formed, whereas when both α- and β-fluorines are present the destabilising effect of fluorine

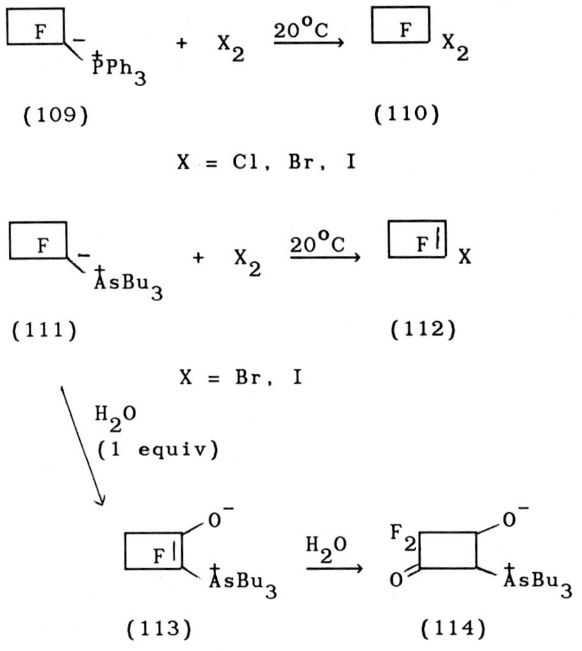

(109) + X_2 $\xrightarrow{20^oC}$ (110)

X = Cl, Br, I

(111) + X_2 $\xrightarrow{20^oC}$ (112)

X = Br, I

(111) $\xrightarrow{H_2O \text{ (1 equiv)}}$ (113) $\xrightarrow{H_2O}$ (114)

SCHEME 17.

attached to a carbanionic site is sufficient to disfavour formation of the ylid.[127]

$$CF_3(CF_2)_2CF{=}CF_2 + Bu_3P \longrightarrow CF_3(CF_2)_2CF{=}CFP(F)Bu_3$$

(115) (100% Z-isomer)

However, fluorine-containing phosphoranium salts (116) can be isolated from reaction of fluorotrihalomethanes and phosphines. When R = butyl, salt (116) readily undergoes a

Wittig reaction with F-acylfluorides to yield F-vinylphosphonium salts (117). Subsequent hydrolysis of salt (117) provides a stereo-specific route to chain extended E-1-hydro-F-olefins (118) (Scheme 18).[128]

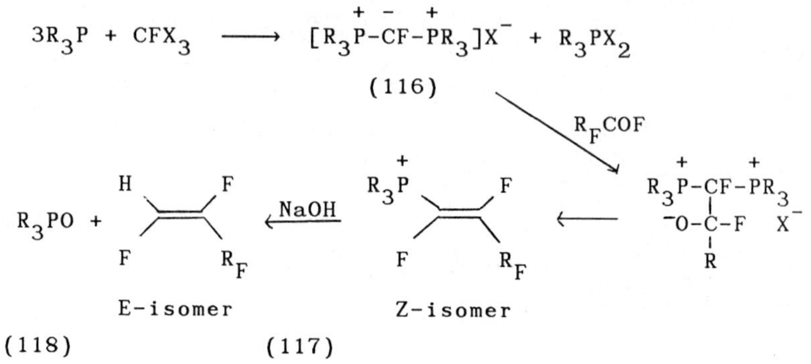

$$3R_3P + CFX_3 \longrightarrow [R_3\overset{+}{P}-\overset{-}{C}F-\overset{+}{P}R_3]X^- + R_3PX_2$$

(116) $\xrightarrow{R_FCOF}$ $R_3\overset{+}{P}-CF(-\overset{+}{P}R_3)-C(-O^-)(F)(R)\ X^-$ $\longrightarrow$ (117) Z-isomer: $R_3\overset{+}{P}(F)C{=}C(F)(R_F)$ $\xrightarrow{NaOH}$ R_3PO + (118) E-isomer: $H(F)C{=}C(F)(R_F)$

SCHEME 18.

Conclusion

For a period during the early development of the chemistry of organo-fluorine compounds, these systems were curiosities and sowewhat apart from organic chemistry. However, we hope that this review on carbanions has demonstrated that fluorine-containing systems are now integrated with organic chemistry and provide a significant challenge to current models of structure and reactivity. Furthermore, the complementary nature of the chemistry of unsaturated fluorocarbons, compared with that of hydrocarbons, helps to extend the mechanistic basis of organic chemistry.

REFERENCES

1 R.D. Chambers, Fluorine in Organic Chemistry, Wiley-Interscience, New York, 1973.
2 J. Hine, N.W. Burske, M. Hine, P.B. Langford, J. Amer. Chem. Soc., 79 (1957) 1406; J. Hine and N.W. Burske, J. Amer. Chem. Soc., 78 (1956) 3337.
3 S. Andreades, J. Amer. Chem. Soc., 86 (1964) 2003.
4 H.G. Adolph and M.J. Kamlet, J. Amer. Chem. Soc., 88 (1966) 4761.
5 K.J. Klabunde and D.J. Burton, J. Amer. Chem. Soc., 94 (1972) 820.
6 K.J. Klabunde and D.J. Burton, J. Amer. Chem. Soc., 94 (1972) 5985.
7 A. Streitwieser, D. Holtz, G.R. Ziegler, J.O. Stoffer, M.L. Brokaw and F. Guibe, J. Amer. Chem. Soc., 98 (1976) 5229.
8 J.D. Roberts, R.L. Webbe and E.A. McElhill, J. Amer. Chem. Soc.,72 (1950) 408.
9 A. Streitwieser and H.F. Koch, J. Amer. Chem. Soc., 86 (1964) 404.
10 D. Holtz, Prog. Phys. Org. Chem., 8 (1971) 1.
11 J.H. Sleigh, R. Stephens and J.C. Tatlow, J. Fluorine Chem., 15 (1980) 411.
12 R.D. Chambers, J.S. Waterhouse and D.L.H. Williams, Tet. Lett. (1974) 743.
13 Y. Apeloig, J. Chem. Soc. Chem. Comm. (1981) 396.
14 R.C. Bingham, J. Amer. Chem. Soc., 97 (1975) 6743.
15 J.G. Stamper and R. Taylor, J. Chem. Res. (5), 128(M) (1980) 1930.
16 A. Streitwieser, C.M. Berke, G.W. Schriver, D. Grier and J.B. Collins, Tetrahedron, supp. 1, 37 (1981) 345.
17 P. von R. Schleyer and A.J. Kos, Tetrahedron, 39 (1983) 1141.
18 D.S. Friedman, M.M. Frand and L.C. Allen, Tetrahedron, 41 (1985) 449.
19 W.B. Farnham, B.E. Smart, W.J. Middleton, J.C. Calabrese and D.A. Dixon, J. Amer. Chem. Soc., 107 (1985) 4565.
20 H.G. Viehe, Angew. Chem. Int. Ed., 6 (1967) 767.
21 D. Daloze, H.G. Viehe and G. Chiurdoglu, Tetrahedron Lett. (1969) 3925.
22 A. Streitwieser and F. Mares, J. Amer. Chem. Soc., 90 (1968) 644.
23 J. Hine and P.B. Langford, J. Org. Chem., 27 (1962) 4149.
24 A. Streitwieser, P.J. Scannon and H.M. Niemeyer, J. Amer. Chem. Soc., 94 (1972) 7936.
25 A. Streitwieser, J.A. Hudson and F. Mares, J. Amer. Chem. Soc., 90 (1968) 648. See also A. Streitwieser and C.C.C. Shen, Tetrahedron Lett. (1979) 327.
26 R.J. Harper, E.J. Soloski and T. Tamborski, J. Org. Chem., 29 (1964) 2385.
27 R. Filler and C-S. Wang, Chem. Comm. (1968) 287.
28 J. Burdon D.W. Davies, and G. del Cande, J. Chem. Soc. Perkin Trans. 2 (1979) 1205.
29 J. Tyrrell, V.M. Kolb and C.Y. Meyers, J. Amer. Chem. Soc., 101 (1979) 3497.
30 E.D. Langanis and D.M. Lemal, J. Amer. Chem. Soc., 102 (1980) 6634.

31 G. Paprott and K. Seppelt, J. Amer. Chem. Soc., 106 (1984) 4060.
32 W.E. Britton, J.P. Ferraris and R.L. Soulen, J. Amer. Chem. Soc., 104 (1982) 5322.
33 B.W. Walther, F. Wilhains and D.M. Lemal, J. Amer. Chem. Soc., 106 (1984) 548.
34a R.D. Chambers and S.R. James, Chapter 3, in Comprehensive Organic Chemistry, Vol. 1, Ed. Sir Derek Barton and W.D. Ollis, Pergamon, Oxford, 1978.
34b R.D. Chambers and R.H. Mobbs, Advances in Fluorine Chemistry, Butterworths, London, 1965, Ch. 4, p.50.
34c R.D. Chambers and C.R. Sargent, Advances in Heterocyclic Chemistry, 28 (1981) 1.
35 B.E. Smart, The Chemistry of Functional Groups Supplement D: The Chemistry of Halides, Pseudo-Halides and Azides; S. Patai and Z. Rapport Eds., Wiley, Chichester 1983, Part I, Ch. 14.
36 B.L. Dyatkin, N.I. Delyagina, and S.R. Sterlin, Russ. Chem. Rev. (Engl. Transl.), 45 (1976) 607.
37 M.R. Bryce, R.D. Chambers and G. Taylor, J. Chem. Soc. Perkin Trans. 1, (1984) 509.
38 R.D. Chambers, D. Close and D.L.H. Williams, J. Chem. Soc. Perkin Trans. 2, (1980) 778.
39 R.D. Chambers, Dyes and Pigments, 3 (1982) 183 and references contained.
40 M.R. Crampton and H.A. Khan, J. Chem. Soc. Perkin Trans. 2, (1972) 1173.
41 M.R. Crampton, M.E. El-Ghariani and H.A. Khan, J. Chem. Soc. Perkin Trans. 2, (1972) 1178.
42 F. Millot and F. Terrier, C.R. Acad. Sci. Ser. C, 284 (1977) 979; F. Terrier, Chem. Rev., 82 (1982) 77.
43 A.J. Birch, A.L. Hinde and L. Radom, J. Amer. Chem. Soc., 102 (1980) 6430.
44 J. Burdon, I.W. Parson and E.J. Avramides, J. Chem. Soc. Perkin Trans. 2, (1979) 1201.
45 J. Burdon, I.W. Parsons and E.J. Auramides, J. Chem. Soc. Perkin Trans. 1, (1979) 1268.
46 P. von R. Schleyer, discussions at International Symposium on Carbanions, Durham, 1984.
47 R.D. Chambers and M.J. Seabury, forthcoming publication.
48 N.D. Epiotis, R.L. Yates, F. Bernardi and S. Wolfe, J. Amer. Chem. Soc., 98 (1976) 5435.
49 J. Burdon and I.W. Parsons, J. Amer. Chem. Soc., 99 (1979) 7445.
50 W.T. Miller, J.H. Fried and H. Goldwhite, J. Amer. Chem. Soc., 82 (1960) 3091.
51 Reference 1, p.164 and references contained.
52 I.L. Gervits, K.N. Makarov, Yu. A. Cherburkov and I.L. Knunyants, J. Fluorine Chem., 9 (1977) 45.
53 J.A. Young, Fluorine Chem. Revs., 1 (1967) 359 and references contained.
54 H.C. Fielding and A.J. Rudge, Brit. Pat., (1967) 1,082,127.
55 D.P. Graham, J. Org. Chem., 31 (1966) 955.
56 R.D. Chambers, J.A. Jackson, S. Partington, P.D. Philpot and A.C. Young, J. Fluorine Chem., 6 (1975) 5.
57 A.E. Bayliff, M.R. Bryce, R.D. Chambers and R.S. Matthews, J. Chem. Soc. Chem. Comm., (1985) 1018.
58 Reference 1, p.168 and references contained.
59 R.D. Chambers, A.A. Lindley, P.D. Philpot, H.C. Fielding and J. Hutchinson, Israel J. Chem., 17 (1978) 150.

60 R.D. Chambers, S. Partington and D.B. Speight, J. Chem. Soc. Perkin I, (1974) 2673.
61 R.D. Chambers, M.Y. Gribble and E. Marper, J. Chem. Soc. Perkin I, (1973) 1710.
62 R.D. Chambers, G. Taylor and R.L. Powell, J. Chem. Soc. Perkin I, (1980) 426.
63 R.D. Chambers, G. Taylor and R.L. Powell, J. Chem. Soc. Perkin I (1980) 429.
64 R.D. Chambers, R.S. Matthews, G. Taylor and R.L. Powell, J. Chem. Soc. Perkin I, (1980) 435.
65 R.L. Pruett, C.T. Bahner and H.A. Smith, J. Amer. Chem. Soc., 74 (1952) 1638.
66 D.J. Burton, R.D. Howells and P.D. Vander Valk, J. Amer. Chem. Soc., 99 (1977) 4830.
67 R.D. Chambers, A.A. Lindley and H.C. Fielding, J. Fluorine Chem., 12 (1978) 85.
68 R.D. Chambers, A.A. Lindley, P.D. Philpot, H.C. Fielding, J. Hutchinson and G. Whittaker, J. Chem. Soc. Perkin I, (1979) 214.
69 L.L. Gervits, L.A. Rozov, N.S. Mirzabekyants, K.N. Makarov, Yu. V. Zeifman, Yu. A. Cheburkov and I.L. Knunyants, Isv. Akad. Nauk SSSR, Ser. Khim., (1976) 1676; Engl. Transl. p.1582.
70a B.E. Smart, W.J. Middleton, and W.B. Farnham, J. Amer. Chem. Soc., 108 (1986) 4905.
70b W.B. Farnham, W.J. Middleton, W.C. Fultz, and B.E. Smart, J. Amer. Chem. Soc., 108 (1986) 3125.
70c W.A. Middleton, U.S. Pat., (1976) 3,940,402.
70d M. Fujita and T. Hiyama, J. Amer. Chem. Soc., 107 (1985) 4085.
71 N.I. Delagina, S.M. Igumnov, V.F. Snegirev, and I.L. Knunyants, Bull. Acad. Sc. U.S.S.R. Div. Chem. Soc. (Engl. Transl.), 30 (1981) 1836.
72a See W. Adcock and V.S. Iyer, J. Org. Chem., 50 (1985) 1535, and references contained.
72b D. Seebach, R. Hassig, and J. Gabriel, Helv. Chim. Acta, 66 (1983) 308.
72c C. Najera, M. Yus, R. Hassig, and D. Seebach, Helv. Chim. Acta, 67 (1984) 1100.
73 R.D. Chambers, P.D. Philpot and P.L. Russell, J. Chem. Soc. Perkin. I, (1977) 1605.
74 A.R. Butler, J. Chem. Soc. Perkin I, (1975) 1557.
75 F. Terrier, G. Ah-Kow, M.J. Pouet and M.P. Simmonnin, Tetrahedron Lett., (1976) 227.
76 R.D. Chambers, R.A. Storey and W.K.R. Musgrave, Chem. Comms., (1966), 384; R.D. Chambers, J.A. Jackson, W.K.R. Musgrave and R.A. Storey, J. Chem. Soc. (C), (1968) 2221.
77 H.C. Fielding, British Pat., (1968) 1,133,492; British Pat., (1967) 1,076,357; (1969) 1,148,676.
78 R.D. Chambers and M.Y. Gribble, J. Chem. Soc. Perkin I, (1973) 1404.
79 C.J. Drayton, W.T. Flowers and R.N. Haszeldine, J. Chem. Soc. Perkin I, (1971) 2750.
80 See e.g. S.L. Bell, R.D. Chambers, M.Y. Gribble and J.R. Maslakiewicz, J. Chem. Soc. Perkin I, (1973) 1716, and earlier parts of a series.
81 C.J. Drayton, W.T. Flowers and R.N. Haszeldine, J. Chem. Soc. Perkin I, (1975) 1035 and earlier parts of a series.

82 N.I. Delyagina, E. Ya. Perova, B.L. Dyatkin and I.L. Knunyants, Zhur. Org. Khim. USSR, 8 (1972) 851; Engl. Transl. p.859.
83 R.D. Chambers, J.A. Jackson, S. Partington, P.D. Philpot and A.C. Young, J. Fluorine Chem., 6 (1975) 5.
84 R.D. Chambers and M.Y. Gribble, J. Chem. Soc. Perkin I, (1973) 1411.
85 R.L. Dressler and J.A. Young, J. Org. Chem., 32 (1967) 2004.
86 R.D. Chambers, R.P. Corbally and W.K.R. Musgrave, J. Chem. Soc. Perkin I, (1972) 1281.
87 R.D. Chambers and R. Middleton, J. Chem. Soc. Perkin I, (1977) 1500.
88 R.N. Barnes, R.D. Chambers, R.D. Hercliffe and R. Middleton, J. Chem. Soc. Perkin I, (1981) 3289, and references contained.
89 R.D. Chambers, C.G.P. Jones, G. Taylor and R.L. Powell, J. Chem. Soc. Chem. Comm., (1979) 964.
90 R.D. Chambers, J.R. Kirk, G. Taylor and R.L. Powell, J. Chem. Soc. Perkin I, (1982) 673.
91 R.D. Chambers, A.A. Lindley, P.D. Philpot, H.C. Fielding, J. Hutchinson and G. Whittaker, J. Chem. Soc. Perkin I, (1979) 214.
92 R.D. Chambers and M.J. Salisbury, Unpublished Observations.
93 K. Scherer, Abstracts of Papers, 191st Meeting of the Amer. Chem. Soc., New York, April, 1986.
94a C.G. Krespan, F.A. Van-Catledge, and B.E. Smart, J. Amer. Chem. Soc., 106 (1984) 5544.
94b C.G. Krespan and B.E. Smart, J. Org. Chem., 51 (1986) 320.
94c C.G. Krespan, J. Org. Chem., 51 (1986) 326.
94d C.G. Krespan, J. Org. Chem., 51 (1986) 332.
95a Reference 1, Chapter 10.
95b S.C. Cohen and A.G. Massey, Adv. Fluorine Chem., 6 (1970) 185.
96 O.R. Pierce, E.T. McBee and G.F. Judd, J. Amer. Chem. Soc., 76 (1954) 474.
97 R.D. Chambers, W.K.R. Musgrave and J. Savoy, J. Chem. Soc., (1962) 1993; S. Andreades, J. Amer. Chem. Soc., 86 (1964) 2003.
98 F.G. Drakesmith, R.D. Richardson, O.J. Stewart and P. Tarrant, J. Org. Chem., 33 (1968) 286.
99 F.G. Drakesmith, O.J. Stewart and P. Tarrant, J. Org. Chem., 33 (1967) 280.
100 N. Thoai, M. Rubinstein and C. Wakselman, J. Fluorine Chem., 20 (1981) 271.
101 N. Thoai, M. Rubinstein and C. Wakselman, J. Fluorine Chem., 21 (1982) 437.
102 N. Redwane, P. Moreau and A. Commeyras, J. Fluorine Chem., 20 (1982) 699.
103 R.D. Chambers and T. Chivers, Organomet. Chem. Rev., 1 (1966) 279.
104 G.M. Brooke and B.S. Furniss, J. Chem. Soc. (C), (1967) 869 and references therein.
105 P.L. Coe, R. Stephens and J.C. Tatlow, J. Chem. Soc. (C), (1962) 3227.
106 N.N. Vorozhtsov, V.A. Barkhasch, N.G. Ivanova and A.K. Petrov, Tetrahedron Lett., (1964) 3575.
107 R.J. Harper, E.J. Soloski and C. Tamborski, J. Org. Chem., 29 (1964) 2385.
108 S.C. Cohen, M.L.N. Reddy, D.M. Roe, A.J. Tomlinson and A.G. Massey, J. Organomet. Chem., 14 (1968) 241.

109 R.D. Chambers, J.A. Cunningham and D.J. Spring, Tetrahedron, 24 (1968) 3997.

110 R.D. Chambers and D.J. Spring, J. Chem. Soc. (C), (1968) 2394.

111 R.D. Chambers, M. Clark and D.J. Spring, J. Chem. Soc. Perkin Trans. 1, (1972) 2464.

112 R.D. Chambers and M. Clark, J. Chem. Soc. Perkin Trans. 1, (1972) 2469.

113 R.J. Lagow and J.A. Morrison, Adv. Inorg. Chem. Radiochem., 23 (1980) 177.

114 K.J. Klabunde, B.B. Anderson and K. Neuenschwander, Inorg. Chem., 19 (1980) 3719.

115 R.J. Lagow, R. Eujen, L.L. Gerchman and J.A. Morrison, J. Amer. Chem. Soc., 100 (1978) 1722.

116 L.J. Kranse and J.A. Morrison, J. Amer. Chem. Soc., 103 (1981) 2995.

117a D.J. Burton and D.M. Wiemers, J. Amer. Chem. Soc., 107 (1985) 5014.

117b D.M. Wiemers and D.J. Burton, J. Amer. Chem. Soc., 108 (1986) 832.

118 T. Fuchikami and I. Ojima, J. Fluorine Chem., 22 (1983) 541.

119 P.L. Coe and N.E. Miller, J. Fluorine Chem., 2 (1973) 167.

120 V.C.R. McLoughlin and J. Thrower, Tetrahedron, 25 (1969) 5921.

121 Y. Kobayashi, K. Yamamoto, T. Asai, M. Nakano and I. Kumadaki, J. Chem. Soc. Perkin Trans. 1, (1980) 2755.

122 J. Leroy, M. Rubinstein and C. Wakselman, J. Fluorine Chem., 27 (1985) 291.

123 T. Kitazume and N. Ishikawa, Chem. Lett., (1981) 1679.

124 N. Ishikawa, Proceedings of the Third Regular Meeting of Soviet-Japanese Fluorine Chemistry, Tokyo, 1983, p.241; T. Kitazume and N. Ishikawa, Chem. Lett., (1982) 1453.

125 M.A. Howells, R.D. Howells, N.C. Baenziger and D.J. Burton, J. Amer. Chem. Soc., 95 (1973) 5366.

126 D.J. Burton and P.D. van der Valk, J. Fluorine Chem., 18 (1981) 413.

127 D.J. Burton, S. Shinya and R.D. Howells, J. Amer. Chem. Soc., (1979) 3689.

128 D.J. Burton and D.G. Cox, J. Amer. Chem. Soc., 105 (1983) 650.

CHAPTER 6

REACTIONS OF HYDROGEN-BONDED CARBANION INTERMEDIATES

HEINZ F. KOCH
DEPARTMENT OF CHEMISTRY, ITHACA COLLEGE, ITHACA, NY 14850

CONTENTS

I. INTRODUCTION 322
II. INTERNAL RETURN: EFFECTS ON PKIE AND KINETIC ACIDITIES 323
III. REACTIONS OF CARBANIONS GENERATED IN ALCOHOLS 326
A. GEM-DIFLUOROALKENES WITH ALKOXIDES 326
B. DEHYDROHALOGENATIONS 330
C. HYDROGEN EXCHANGE REACTIONS 337
D. TRIMETHYL, ALKYLSILANES WITH METHOXIDE 339
E. PI-DELOCALIZED VS. LOCALIZED CARBANIONS 342
F. VINYL HALIDES WITH ALKOXIDES 346
IV. CARBON ACIDS 352
V. CONCLUSIONS 354
VI. REFERENCES 356

I. INTRODUCTION

Proton transfer from one atom to another is one of the basic reactions in chemistry. Since many reactions are initiated by transfer of proton from carbon to a base, the study of carbon acids is of particular importance to organic and biochemistry. It would be convenient to have pK_a values for a large variety of organic compounds. Bordwell and co-workers have undertaken a comprehensive study of carbon acids. Their list of pK_a values measured in dimethyl sulfoxide (DMSO) now exceeds 17 pages of computer printout. In an early paper,[1] they discuss their choice of DMSO and analyze the problems with assigning pK_a values from measurements in solvents that have a low dielectric constant. The measurement of pK_a values for many carbon acids is not possible by an equilibrium method, and the determination of "kinetic acidities", often determined by base-catalyzed exchange of hydron with solvent molecules, has attempted to fill this void.[2] Over the last twenty years, Streitwieser and co-workers have studied both equilibrium and kinetic acidities using cyclohexylamine,[3] methanol,[3] and tetrahydrofuran[4] as solvents.

The relationship between pK_a and rates of proton exchange can be a complicated one, and using experimental rate constants as a measure of the proton-transfer step may not be reliable. We have been interested in proton transfer between carbon and oxygen, and used the following five step process as a working model for the mechanism of alkoxide-catalyzed exchange reactions:[5]

$$\text{R-H} + {}^{\ominus}\text{OR} \rightleftharpoons \text{R-H}\cdots{}^{\ominus}\text{OR} \quad (1)$$

$$\text{R-H}\cdots{}^{\ominus}\text{OR} \rightleftharpoons \text{R}^{\ominus}\cdots\text{H-OR} \quad (2)$$

$$\text{D-OR} + \text{R}^{\ominus}\cdots\text{H-OR} \rightleftharpoons \text{R}^{\ominus}\cdots\text{D-OR} + \text{H-OR} \quad (3)$$

$$\text{R}^{\ominus}\cdots\text{D-OR} \rightleftharpoons \text{R-D}\cdots{}^{\ominus}\text{OR} \quad (4)$$

$$\text{R-D}\cdots{}^{\ominus}\text{OR} \rightleftharpoons \text{R-D} + {}^{\ominus}\text{OR} \quad (5)$$

Eqs. (1) and (5) represent the formation and destruction of encounter complexes that are necessary to align the donor and acceptor sites. After the necessary desolvation of alkoxide ion, the proton-transfer reaction, eq. (2), can take place. Exchange of protium-conjugate acid, H-OR, and deuteroacid, D-OR, occurs in eq. (3). Little is known about the detailed mechanism of this step. Does R⊖•••H-OR dissociate to give an alcohol, RO-H, and a carbanion, R⊖, which can add RO-D to form R⊖•••D-OR, or is the exchange step a synchronous process, **(3)**, with ROD entering as ROH departs?

$$R^{\ominus}\cdots H\text{-}OR \rightleftharpoons R^{\ominus} + H\text{-}OR \quad (3a)$$

$$R^{\ominus} + D\text{-}OR \rightleftharpoons R^{\ominus}\cdots D\text{-}OR \quad (3b)$$

Which step is rate limiting: Hydron transfer or the exchange process? Primary kinetic isotope effects (PKIE) associated with alkoxide-promoted dehydrohalogenations and exchange reactions as well as nucleophilic reactions of alkenes with alkoxide ions were investigated to answer some of these questions.

II. INTERNAL RETURN: EFFECTS ON PKIE AND KINETIC ACIDITIES

Reactions **(1)** to **(3)** or **(3a)** can be written in a more general form, Scheme 1. For simplicity, encounter complex formation and desolvation of alkoxide are incorporated in the first step, which represents all processes prior to the transition structure for hydron-transfer:

SCHEME 1

$$\gt C\text{-}H + {}^{\ominus}OR \underset{k_{-1}}{\overset{k_1}{\rightleftharpoons}} \gt C^{\ominus}\cdots H\text{-}OR \xrightarrow{k_2} \text{products}$$

Step 2 can be any forward reaction, and could include expulsion of a leaving group, exchange of alcohol with solvent molecules, or formation of a second intermediate. Scheme 1 represents an internal-return mechanism[2a] where proton transfer back to carbon, k_{-1}, competes with a forward step, k_2. Rate constants measured for reactions proceeding by Scheme 1 are really a combination of the individual rate constants:

$$k_{exp} = (k_1 k_2) / (k_{-1} + k_2) \quad (6)$$

Equation **(6)** can be simplified. If k_2 is 100 times greater than k_{-1}, the denominator, $k_{-1} + k_2$, will be only 1% larger than k_2, and equation **6** would be experimentally indistinguishable from:

$$k_{exp} = k_1 \qquad (7)$$

When k_{-1} is 100 time greater than k_2, equation **6** becomes:

$$k_{exp} = (k_1 / k_{-1})\, k_2 \qquad (8)$$

When eq. **(8)** holds, experimental results do not measure rates of hydron transfer, k_1, unless a relationship between k_{-1} to k_2 is known from other data. When $k_{-1} = k_2 \times 10^2$ then $k_{exp} = k_1 \times 10^{-2}$, and if, $k_{-1} = k_2 \times 10^5$ then $k_{exp} = k_1 \times 10^{-5}$. Therefore reactions with the same k_1 could result in k_{exp} values that differ by 10^3 due to the large difference in internal return. The measurement of near unity values for k^H/k^D is the best experimental evidence for substantial internal return; however to determine the actual amount of return is quite difficult in such cases.

When k_{-1} and k_2 are within a factor of 50, eq. **(6)** holds and moderate isotope effects would be measured. Experimental k^H/k^D's of such magnitude are normally attributed to the hydron transfer occurring with an asymmetric transition structure.[6] Correlations of the magnitude of PKIE and the amount of proton transfer in a transition structure have been made.[7] Streitwieser et al.[8] made use of all three isotopes of hydrogen to differentiate between these two alternative explanations. Reactions with internal return give experimental isotope effects that deviate from the Swain-Schaad relationship:[9]

$$k^H/k^D = (k^D/k^T)^y \qquad (9)$$

where y can be 2.26 to 2.344 depending on the assumptions made in the derivation.[10] Rate constants can be corrected for internal return, and the true isotope effect associated with k_1 can be calculated. When this treatment was applied to rates obtained for the exchange of a some hydrocarbons,[11] covering a range of 13 pK_a units, only small changes in the magnitude of the corrected PKIE associated with k_1 were found. Bordwell and Boyle[12] also report small changes in PKIE for hydron exchange of a wide range of nitroalkanes. These results caused both groups to question that variation of C-H pK_a results in systematic changes of the transition structure for proton-transfer reactions, which would then be reflected in the magnitude of experimental PKIE.

The Streitwieser method does not predict Arrhenius behavior of PKIE for reactions that feature internal return. Our studies of alkoxide-promoted dehydrohalogenation reactions resulted in PKIE that were virtually temperature independent.[13] Calculation of temperature dependence using models where neither the first nor second step in Scheme 1 is clearly rate limiting suggest that this anomalous Arrhenius behavior of isotope effects could be due to such a mechanism.[14]

When eq. (7) holds, normal PKIE should result. The necessity of a tunnelling correction can arise and this has been adequately discussed by Bell[15] and Saunders.[16] Proton transfer occurring with small amounts of internal return can have similar Arrhenius behavior as that resulting from tunnelling.[14] When equating the experimental hydrogen isotope effects to the extent of proton transfer in a transition structure, caution must be taken unless the amount of internal return or tunnelling is known.

When base-catalyzed hydrogen exchange rates are used to study kinetic acidities, the occurrence of small amounts of internal return will not be a major problem. If $k_2 = 10\ k^H{}_{-1}$, then k_{exp} would be within 10% of the true value of $k^H{}_1$. Since there is a negligible isotope effect associated with k_2, internal return for deuterium and tritium is very small, and k_{exp} is a good measure for k_1. Comparing compounds with large differences in internal return can present problems. For example tritium exchange for pentafluorobenzene (<u>PFB</u>) in methanolic methoxide[17] at 25°C is 15 times faster than for 9-phenylfluorene (<u>9-PhFl</u>),[10] even though <u>9-PhFl</u> has a pK_a of 18.5 much lower than for <u>PFB</u>, 25.8.[18] It should be noted that k^D/k^T associated with the exchange reaction for <u>PFB</u> is unity, while the PKIE measured for <u>9-PhFl</u> is quite normal. This point will be discussed later in the text.

III. REACTIONS OF CARBANIONS GENERATED IN ALCOHOLS

A. GEM-DIFLUOROALKENES WITH ALKOXIDES

To investigate the chemistry of carbanions in protic solvents, one must have an appropriate reaction to generate them *in situ*. If there is substantial internal return, base-catalyzed isotopic exchange with acidic hydrogens of solvent molecules will not tell when a carbanion is formed. To avoid the complications caused by internal return, alternate reactions that form carbanions must be used. Since alkenes that feature a *gem*-difluoromethylene react rapidly with alkoxide, their study was ideal. Miller et al.[19] reported that the reaction of $CF_2{=}CCl_2$ with ethanolic ethoxide yielded only saturated ether, $C_2H_5OCF_2CHCl_2$ (1). Hine et al.[20] then reported that methoxide will catalyze hydrogen exchange of $CH_3OCF_2CHCl_2$ with methanol 10^4 faster than the elimination of HF. Therefore the carbanion $\{C_2H_5OCF_2CCl_2\}^{\ominus}$ formed by the reaction of $C_2H_5O^{\ominus}$ with $CF_2{=}CCl_2$, will take a proton from ethanol rather than eject fluoride to give a vinyl ether. This situation is ideal since the carbanion is generated from either direction; however, the chemistry is rather dull since the only reaction to occur is for proton transfer between carbon and oxygen.

Reaction of ethanolic ethoxide with 2-phenyl perfluoropropene (2),[21] is more interesting since a saturated ether (3) and two vinyl ethers (4) and (5) are formed, Scheme 2.

SCHEME 2

$$\underset{\underline{2}}{(C_6H_5)(CF_3)C{=}CF_2} \xrightarrow[-78^{o}]{EtO^{\ominus}} \underset{\underline{\text{INT-1}}}{\ominus\,C(C_6H_5)(CF_3)\text{-}CF_2OEt} \xrightarrow[k^{H}_{add}]{EtOH} EtO^{\ominus} + \underset{\underline{3}\ (15\%)}{C_6H_5CH(CF_3)CF_2OEt}$$

$$\downarrow k^{F}_{elim}$$

$$\underset{\underline{4}\ (76\%)}{(C_6H_5)(CF_3)C{=}C(OEt)(F)} \qquad \underset{\underline{5}\ (9\%)}{(C_6H_5)(CF_3)C{=}C(F)(OEt)} \qquad \underset{\underline{6}\ \text{(not observed)}}{(C_6H_5)(CF_2{=})C\text{-}CF_2OEt}$$

The carbanion INT-1 could be generated by reaction of ethanolic ethoxide with 3; however, this reaction is not competitive below room temperature. Thus the reaction as written in Scheme 2 need not include a back reaction from 3 to regenerate INT-1 under the reaction conditions, -78°C.

Is the same species formed from reaction of alkoxide with 2 or 3? A model for 3, $PhCH(CF_3)CF_2OCH_3$ (7), reacted with ethoxide in ethanol-O-d at 40°C until 20% $PhC(CF_3)=CFOCH_3$ (8) was formed, and recovered 7 was found to contain 3-4% deuterium.[22] Reaction of 2 under similar conditions yielded 15% 3-d, and 85% 4 and 5. The 85:15 split for fluoride loss vs. deuterium addition suggests the product ratio is independent of whether the intermediate was generated from 2 or 7. **Does INT-1 eject alkoxide to regenerate 2?** Product analysis from the reaction of 7 with ethoxide in EtOD indicated no 3, 4 or 5 was present, therefore $\{PhC(CF_3)CF_2OCH_3\}^{\ominus}$ (INT-2) does not eject methoxide. Reaction of methoxide with 2 in methanol results in formation of 34% 7 and 66% 8 at -78°. It follows that $k^F{}_{elim} > k^H{}_{add}$ for both INT-1 and INT-2.

The near unity PKIE, $k^H/k^D = 1.3$ at 40°C, associated with the ethanolic ethoxide-promoted dehydrofluorination of both 3 and 7, suggests that reaction occurs with substantial internal return. Therefore proton return to carbon is faster than loss of fluoride from the initially formed carbanion when alkoxide reacts with saturated ether, $k^H{}_{ret} > k^F{}_{elim}$. Ergo this carbanion cannot be the same as INT-1 or INT-2, and a second carbanion must be formed as shown in Scheme 3.[22] One intermediate, (H), is stabilized by a hydrogen bond, and the other, (F), has no contact stabilization from either a cation or solvent molecule. Internal return, k_{-1}, will be the low free energy process for the initial intermediate when alkoxide abstracts a proton from 3 or 7. The hydrogen bond must be broken, k_2, in a rate-limiting step to form intermediate F, which can partition to yield mixtures of vinyl or saturated ethers. An intermediate similar to F is formed from the rate-limiting step for reaction of alkoxide with alkene, k_N. Ejection of fluoride, $k^F{}_{elim}$, is faster than formation of H, k_{-2}, which is rate-limiting on the reaction pathway from F to 3 or 7. When reaction proceeds by Scheme 3, near unity PKIE is expected for hydron transfer in both directions.

SCHEME 3

$$
\begin{array}{ccccc}
 & & \uparrow k^{X}_{elim}\ (X = Cl, Br) & & \uparrow k^{X}_{elim}\ (X = F, Cl, Br) \\
\underset{\underline{R\text{-}h}}{X\text{-}C(OR)\text{-}C\text{-}H} + {}^{\ominus}OR & \underset{k^{H}_{-1}}{\overset{k^{H}_{1}}{\rightleftharpoons}} & \underset{\underline{H\text{-}h}}{X\text{-}C(OR)\text{-}C^{\ominus}\cdots H\text{-}OR} & \underset{k^{H}_{-2}}{\overset{k^{H}_{2}}{\rightleftharpoons}} & \underset{\underline{F}}{X\text{-}C(OR)\text{-}C^{\ominus}} + H\text{-}OR \\
 & & & & ROD \Downarrow\Uparrow ROH,\ k_{exch} \qquad RO^{\ominus},\ k_N:\ X\text{-}C{=}C \rightarrow \underline{F} \\
\underset{\underline{R\text{-}d}}{X\text{-}C(OR)\text{-}C\text{-}D} + {}^{\ominus}OR & \underset{k^{D}_{-1}}{\overset{k^{D}_{1}}{\rightleftharpoons}} & \underset{\underline{H\text{-}d}}{X\text{-}C(OR)\text{-}C^{\ominus}\cdots D\text{-}OR} & \underset{k^{D}_{-2}}{\overset{k^{D}_{2}}{\rightleftharpoons}} & \underset{\underline{F}}{X\text{-}C(OR)\text{-}C^{\ominus}} + D\text{-}OR \\
 & & \downarrow k^{X}_{elim}\ (X = Cl, Br) & & \downarrow k^{X}_{elim}\ (X = F, Cl, Br)
\end{array}
$$

If proton transfer between carbon and oxygen occurs with a highly asymmetric transition structure, there would be no need to suggest extensive internal return. Scheme 2 would be adequate, and Scheme 3 would not be necessary to explain the experimental results. Theoretical calculations of isotope effects associated with elimination reactions suggest that a C-H bond breaking of >95% is required in a transition structure to obtain k^H/k^D values as low as 1.5.[23] To differentiate between the two possible ways to interpret the experimental data, rates of all three isotopes of hydrogen are required. Albery and Knowles[24] warn that the rate constants must be determined to a high degree of accuracy to use deviations from an insensitive Swain-Schaad relationship to calculate internal return. Since exchange of $PhC^{i}H(CF_3)CF_2OMe$ competes with ethoxide-promoted dehydrofluorination, the accurate measurement of k^{T}_{elim} would be difficult.[25] An alternate method was used to address this question. Reaction of ethoxide with 2 in 1:1 mixtures of EtOH:EtOD resulted in a calculated isotope effect of k^H/k^D = 1.5 (-78°C) and 1.9 (20°) associated with the hydron transfer from ethanol to INT-1. The slight increase of PKIE with increasing temperature has been observed for reactions of other alkenes.[26] This behavior is not consistent with the concept of an asymmetric transition structure, and use of Scheme 3 is justified.

The leaving ability of fluoride has been studied by using a series of compounds, $PhCR_f{=}CF_2$, where R_f = CF_3 (2), CF_2CF_3 (9), and CHF_2 (10), and chloride can be included when R_f = CF_2Cl (11). Since alkoxide preferentially attacks $=CF_2$, halide leaving from R_f can be compared to hydron transfer from RO^iH to neutralize the carbanion or ejection of fluoride from $-CF_2OR$. The reaction of ethoxide with 2 (R_f = CF_3) results in no loss of allylic fluoride to form $PhC(CF_2OEt){=}CF_2$ (6), Scheme 2. When chloride is in the allylic position, 11, the only product is the allyl ether, 6. Allylic fluoride is displaced when R_f = C_2F_5, 9, but vinyl ethers are still the major products. Reaction of alcoholic alkoxide with 10 results in >98% (E)-$PhC(CF_2OR){=}CHF$ as the kinetic product. Table 1 summarizes the product distributions for these reactions. Also included are the reaction products for $PhCCl{=}CF_2$ (12), which can be compared to those for $CCl_2{=}CF_2$.

TABLE 1

Product Distribution from the Reaction[a]

$C_6H_5CR_f{=}CF_2$ + EtONa → $C_6H_5CHR_fCF_2OEt$(satd) + $C_6H_5CR_f{=}CFOEt$(vinylic) + $C_6H_5C(CF_2OEt){=}R_f$(allylic)

R_f	temp, °C	%satd	%vinylic	%allylic
CF_2Cl (11)	-78	0	0	100
CF_3 (2)	-78	15(34)[b]	85(66)[b]	0(0)[b]
CF_3 (2)	20	21(44)[b]	80(56)[b]	0(0)[b]
C_2F_5 (9)	-78	4	74	22
CF_2H (10)	-50		(<2)[c]	(>98)[b]
Cl (12)	0	40(49)[b]	60(51)[b]	

[a] Data from ref 22. [b] Values in parentheses are for reactions with MeONa/MeOH. [c] Thought to be $C_6H_5C(CF_2H){=}CFOMe$.

Note that reaction of 12 gives vinyl and saturated ethers. As with INT-1 and INT-2, $\{PhCClCF_2OR\}^{\ominus}$ ejects fluoride ion to yield both vinyl ethers, PhCCl=CFOR, at a rate competitive with proton transfer to form the saturated ether, $PhCHClCF_2OR$. Reaction of alcoholic alkoxide with $CCl_2{=}CF_2$ can not supply much information about relative energetics of the proton-transfer process since the intermediate, $ROCF_2CCl_2^{\ominus}$, only adds a proton. Reactions of the alkenes in Table 1 offer definite advantages over reactions that require the abstraction of a proton for investigations of the chemistry of carbanions in protic solvents. Carbanions are generated at lower temperatures and internal return will not be a problem. Transfer of proton from oxygen to carbon is fast,[27] but other chemical reactions appear to compete readily.

B. DEHYDROHALOGENATIONS

Scheme 3 indicates that beta chloride or bromide can leave from a hydrogen-bonded intermediate H. Table 2 summarizes some of the results that lead to this deduction. Ethoxide-promoted dehydrochlorination of $PhC^{i}HClCF_2Cl$ (14)[13] occurs $5x10^5$ faster than elimination of HF from $PhC^{i}HClCF_3$ (13).[28] These results can be compared to those of the corresponding β-phenethyl halides,[29] where elimination of HCl from 17 is only 81 times faster than the loss of HF from 16. Elimination rates of HBr relative to HCl are similar for both systems: (15/14) = 52 and (18/17) = 66. Since α-chlorine and β-fluorine increase benzylic hydrogen reactivity, it was not surprising that elimination of HCl occurs $7x10^4$ faster for 14 than for 17. Dehydrofluorinations are the anomaly since reaction of 13 is only 10 times faster than 16. Near unity PKIE indicates that internal return substantially decreases observed reactivity for the dehydrofluorination of 13 and the exchange of 13-d.[30] **Since there is no exchange of proton prior to loss of HCl, does dehydrochlorination of 14 occur by a concerted process?** Anomalous Arrhenius behavior of the isotope effect [k^H/k^D = 2.71 (50°), 2.73 (25°) and 2.77(0°) in EtOH and 2.30(50°), 2.29(25°) and 2.27(0°) in MeOH], coupled with different k^{35}/k^{37} for loss of HCl vs. DCl, suggests that reaction of 14 is not concerted, and a carbanion must be formed along the reaction pathway.[31]

TABLE 2

Rates, Activation Parameters, and PKIE for Alkoxide-Promoted Dehydrohalogenation and Exchange Reactions in Alcohols at 25°C.

Compound		ROH	k, M^{-1} s^{-1}	$\Delta H^{\ddagger}$, kcal	$\Delta S^{\ddagger}$, eu	k^H/k^D (°C)
$PhCH_2CH_2F$[a]	(16)	EtOH	4.34 $\times 10^{-8}$	25.5	-6	4.50 (50)[b]
$PhCHClCF_3$[c]	(13)	EtOH	4.40 $\times 10^{-7}$	29.7	12	1.04 (75)[d]
$PhCDClCF_3$[e]	(13-d)	EtOH	8.73 $\times 10^{-6}$	26.8	8	1.06 (75)[d]
$PhCH_2CH_2Cl$[a]	(17)	EtOH	3.52 $\times 10^{-6}$	23.2	-6	7.42 (25)[f]
$PhCH_2CH_2Br$[g]	(18)	EtOH	2.32 $\times 10^{-4}$	20.4	-7	7.53 (25)
$PhCHClCF_2Cl$[h]	(14)	EtOH	2.33 $\times 10^{-1}$	19.5	4	2.73 (25)
		MeOH	1.73 $\times 10^{-2}$	20.2	1	2.29 (25)
$PhCHBrCF_2Br$[i]	(15)	MeOH	9.09 $\times 10^{-1}$	17.1	-2	4.00 (25)

[a] Data from ref. 29a. [b] Measured in t-BuOH, ref. 29c. [c] Data from ref. 28. [d] Data from ref. 30. [e] Results for an exchange reaction. [f] Unpublished, K.S. Sweinberg. [g] Data from ref. 29b. [h] Data from ref. 31. [i] Data from ref. 13.

Scheme 3 is therefore a good model for these reactions. The hydrogen bond formed by the unique alcohol molecule generated by proton transfer can stabilize the carbanion, the proper alignment is maintained, and internal return can occur. **What is the energy necessary to break the hydrogen bond?** Values for the ΔH and the activation parameter $\Delta H_f^{\ddagger}$ have been reported for the following equilibrium:[32]

$$R-C\begin{matrix}\nearrow O\cdot\cdot H-O \searrow \\ \searrow O-H\cdot\cdot O \nearrow\end{matrix}C-R \underset{k_r}{\overset{k_f}{\rightleftharpoons}} R-C\begin{matrix}\nearrow O \quad \overset{H}{O} \searrow \\ \searrow O-H\cdot\cdot O \nearrow\end{matrix}C-R$$

For a series of aliphatic carboxylic acids, ΔH values range from 3 to 6 kcal mol^{-1}; however, corresponding $\Delta H_f^{\ddagger}$ values range from 9-12 kcal mol^{-1}. Therefore the barrier to break a hydrogen bond is substantially larger than the enthalpy of stabilization. The beta chloride in 14 could act as an intramolecular trap for a hydrogen-bonded intermediate if the energy of carbanion formation for 13 were similar to 14. The difference in $\Delta H^{\ddagger}$ for loss of HCl from 14 and exchange of 13-d would be a minimum for the barrier from H to F. The $\Delta\Delta H^{\ddagger}$, >7 kcal mol^{-1}, is within the values cited for the acid dimers.

The consequences of a hydrogen-bonded carbanion as the initial intermediate generated by proton transfer from carbon to oxygen have important mechanistic implications. When leaving groups are able to depart from H, experimental results would be similar to those of an E2 mechanism. The hydrogen bond inhibits diffusion controlled exchange and can preserve any stereochemistry at the carbanion site. Proton exchange with deuterated solvent is often used to check for carbanion formation prior to loss of a leaving group. When applied to the dehydrobromination of 18, no deuterium was found in recovered starting compound and an E2 mechanism was postulated.[33] Incorporation of deuterium can be taken as positive evidence for carbanion formation; however, the lack of exchange cannot rule out carbanion formation. It simply means that the loss of leaving group can occur much faster than the proton-transfer reaction to neutralize the carbanion.[34]

The PKIE for eliminations of $PhCHBrCF_2Br$ (15), $k^H/k^D = 4.00$, and $PhCHClCF_2Cl$ (14), $k^H/k^D = 2.29$, coupled with a sizable element effect, rate ratio 15/14 = 52, could be taken as evidence for a concerted reaction. Bunnett et al.[35] cleverly suggested the use of an element effect in place of leaving group isotope effects to study the mechanism of aromatic nucleophilic substitution reactions. Although the element effect was a good probe for our vinyl substitution reactions, we have reservations about using it in mechanism assignment for our dehydrohalogenations.[36] Hydrogen isotope effects associated with eliminations of 14 and 15 and chlorine isotope effects measured for 14 are inconsistent with an E2 mechanism.

What effect does β-halogen have on the rate of an elimination reaction? The series $PhCHBrCF_2Br$ (15), PhCHBrCFClBr (23) and $PhCHBrCFBr_2$ (22) varies a single beta halogen, F, Cl, and Br. Dehydrobromination is the only reaction product for 15 and 22, while reaction of 23 results in some dehydrochlorination as well as the elimination of HBr. Rates for these methoxide-promoted reactions are within a factor of four, and the largest difference

TABLE 3

Rates and Activation Parameters for Methanolic Sodium Methoxide Dehydrohalogenation of $C_6H_5CHClCFClX$ and $C_6H_5CHBrCFBrX$

Compound		k, M^{-1} s^{-1} 0°C	-25°C	ΔH‡ kcal	ΔS‡ eu
$PhCHClCF_2Cl$[a]	(14)	6.94 $\times 10^{-4}$	1.46 $\times 10^{-5}$	20.2	1
$PhCHClCFCl_2$[b]	(19)	6.22 $\times 10^{-4}$	1.25 $\times 10^{-5}$	20.5	2
PhCHClCFClBr[c]	(20)	2.84 $\times 10^{-2}$	1.08 $\times 10^{-3}$	17.0	-3
$PhCHBrCF_2Cl$[d]	(21)	6.11 $\times 10^{-4}$	1.10 $\times 10^{-5}$	21.0	4
$PhCHBrCF_2Br$[e]	(15)	5.87 $\times 10^{-2}$	2.35 $\times 10^{-3}$	16.8	-2
$PhCHBrCFBr_2$[f]	(22)	8.44 $\times 10^{-2}$	3.67 $\times 10^{-3}$	16.3	-4
PhCHBrCFClBr[f]	(23-A)	6.20 $\times 10^{-2}$	2.76 $\times 10^{-3}$	16.2	-5
	(23-B)	1.98 $\times 10^{-2}$	7.20 $\times 10^{-4}$	17.3	-3

[a]Data from ref. 32. [b]Unpublished, J.L. Fogarty. [c]Unpublished, H.F. Koch. [d]Unpublished, M.F. McEntee. [e]Updated data from ref. 13. [f]Unpublished, N.A. Touchette.

occurs for the two diastereomers 23-A and 23-B. Therefore the change of one beta halogen has little effect on the overall rate of reaction.[37]

Methoxide-promoted dehydrohalogenation of PhCHBrCFClBr (23) was investigated to probe the stereochemistry of elimination.[38] All attempted syntheses of 23 resulted in mixtures of the two diastereomers, which are not separated by gc, but can be analyzed by nmr. Isomer 23-A reacts at rates comparable to $PhCHBrCF_2Br$ (15), and 3 times faster than 23-B. Populations for each of the conformers of 23-A and 23-B were calculated from low temperature nmr measurements, **Figure 1**. Reaction of 23-A resulted in >99% (E)-PhCBr=CFCl (E-24) and <1% (Z)-PhCBr=CFBr (Z-25). The trans elimination product distribution agrees well with the rate ratios for 15/18 = 0.005 at -25°C. Loss of HCl from 23-B is greater then predicted: 3-6% (E)-PhCBr=CFBr (E-25) and 94-97% Z-24. The amount of E-25 was found to vary with temperature and hydrogen isotope (HCl vs. DCl).

Figure 1. Rotational Isomer Populations and Elimination Products for PhCHBrCFClBr

(R,R or S,S)-PhCHBrCFClBr (23-A)[a]

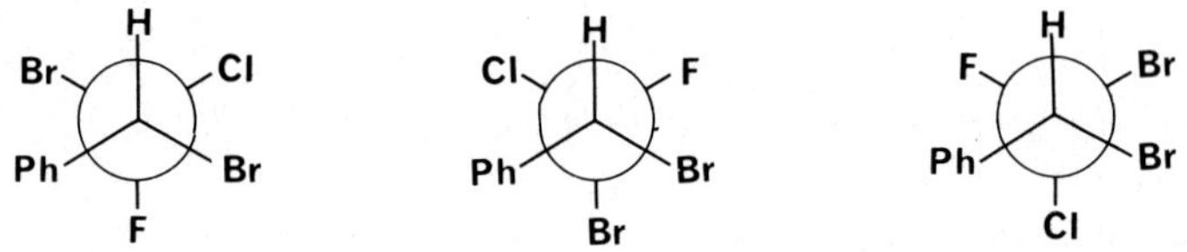

%	74	20	6
δppm:[b]	-65.1	-45.3	-48.7
J_{H-F}cps:	25	0	0
	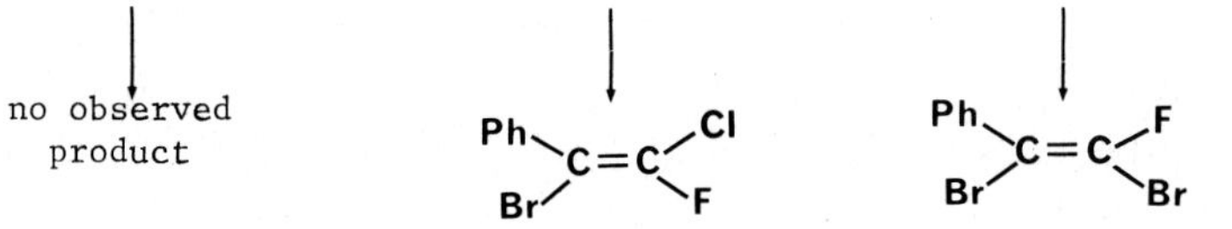 no observed product	>99%	<1%

(R,S or S,R)-PhCHBrCFClBr (23-B)[a]

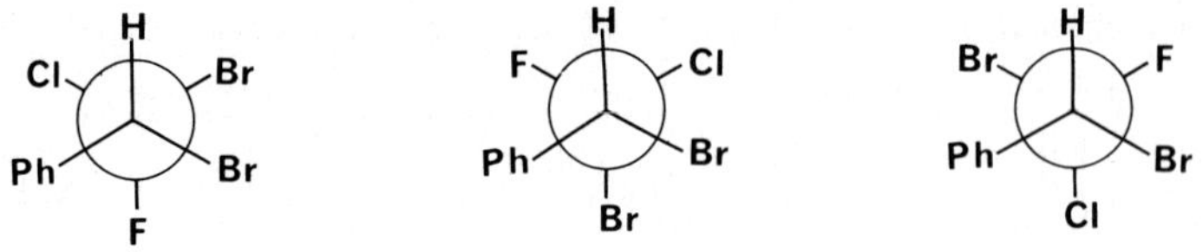

%	34	4	62
δppm:[b]	-64.7	-49.8	-44.4
J_{H-F}cps:	28	0	0
	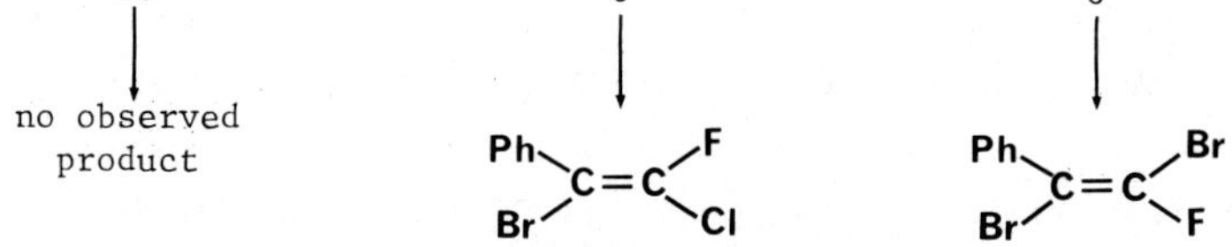 no observed product	94-97%	6-3%

[a]Data taken from spectra recorded at -125°C.

[b]Chemical shifts relative to $CFCl_3$.

The change of benzylic halogen from chlorine to bromine has only a small effect on the reaction rate: $PhCHClCF_2Cl$ (1.0) vs. $PhCHBrCF_2Cl$ (0.88); PhCHClCFClBr (1.0) vs. PhCHBrCFClBr (2.2 or 0.70).[39] However, product compositions can vary: $PhCHClCFCl_2$ results in about equal amounts of (E)- or (Z)-PhCCl=CFCl, but the elimination from $PhCHBrCFBr_2$ results in 85% (E)-PhCBr=CFBr and 15% (Z)-PhCBr=CFBr.

A frustration has been our inability to model the temperature dependence of PKIE for alkoxide-promoted eliminations of either 14 or 15 using the Koch-Dahlberg treatment.[14] Parameters that fit the k^H/k^D data do not model the Arrhenius behavior of k^D/k^T. If the anomalous behavior is not due to internal return, it could be attributed to reaction proceeding by two competing pathways: E2 vs. E1cB or syn- vs. anti-elimination. That loss of HCl and HBr from 23-A and 23-B occurs by stereospecific elimination rules out the syn vs. anti possibility. The E2 vs. E1cB does not look promising, but can not be ruled out at this time.

An alternative to Scheme 3 would require another step like a rotation that aligns the hydrogen-bonded electrons and leaving halide anti-periplanar. Populations of conformations with C-H and C-Br bonds anti-periplanar are 20% for 23-A, 4% for 23-B, and less than 1% for 15.[40] The rates of dehydrobromination resulting from these conformers would contribute only part of the observed rate constant. The most favorable conformer, 15, has C-H and C-F anti-periplanar, and does not result in product formation. A 120° rotation is needed to place the C-Br anti to the hydrogen-bonded electrons. That rotation competes with internal return, and results in different contributions to k_{obs} for $PhCDBrCF_2Br$ vs. $PhCHBrCF_2Br$. Similar arguments hold for the conformers of 23-A and 23-B. One could invoke the Curtin-Hammett principle[41] to argue against such an explanation. If conformers of a reactant are in highly mobile equilibrium, ground state populations should not be used to predict products. However, the activation energies for these reactions, 16-17 kcal mol^{-1}, are close to the rotational barriers, 9-15 kcal mol^{-1}, reported for some highly halogenated alkanes.[42]

TABLE 4

Rates of Protodetritiation, Activation Parameters, and PKIE Values for Exchange Reactions in Methanolic Sodium Methoxide

Compound	k, M^{-1} s^{-1} 25°(45°) C	ΔH‡ kcal M^{-1}	ΔS‡ eu	PKIE (°C) k^H/k^D	k^D/k^T
9-TFM-Fl[a]	9.3	15	-4	k^H/k^T = 28 ± 3 (-50)[j]	
$C_7F_{10}H_2$ (<u>26</u>)[b]	1.4 $x10^{-1}$	19	2	1.2 (-15)	
CF_3CCl_2D[c]	1.9 $x10^{-2}$	21	5		
C_6F_5H[d]	2.5 $x10^{-2}$	20	2		1.0(25)
$PhCH(CF_3)_2$[e] (<u>27</u>)	2.1 $x10^{-3}$ (2.8$x10^{-2}$)	24.1	10	1.1	1.0(25)
$(CF_3)_2CFD$(<u>28</u>)[f]	2.3 $x10^{-3}$	21	-1	1.4(20)[k]	
9-Ph-Fl[g]	1.7 $x10^{-3}$ (3.3$x10^{-2}$)	21.0	-1	6.4	2.5(25)
Fl[g]	3.1 $x10^{-5}$ (4.0$x10^{-4}$)	23.4	-1		2.3(25) 2.1(45) 1.8(100)
9-Me-Fl[g]	(2.2$x10^{-4}$)			5.2	2.3(45)
n-$C_7F_{15}D$(<u>29</u>)[f]	3.8 $x10^{-8}$	30	7	1.4(50)[k]	
Ph_3CH[h]	2 $x10^{-11}$	32	1	1.3	1.3(100)
$PhCH_3$[i]	2 $x10^{-16}$	38	-5		1.0(178)

[a]9-Trifluoromethylfluorene. M.F. McEntee, unpublished results. [b]1,4-dihydroperfluorobicyclo[2.2.1]-heptane. Ref. 44. [c]Ref. 20a [d]Ref. 17. [e]Ref. 30 [f]Ref. 43. [g]Ref. 10. Fl = fluorene, 9-Ph-Fl = 9-phenylfluorene, and 9-Me-Fl = 9-methylfluorene [h]Ref. 8a [i]Ref. 45. [j]Both H and T exchange were carried out in MeOD at temperatures between -41° and -51°C. [k]Corrected for KSIE = 2.6, see text for explanation.

C. HYDROGEN EXCHANGE REACTIONS

The ethoxide-catalyzed exchange of PhC^iHClCF_3 (13) with ethanol occurs with near unity PKIE at 75°C, Table 2. **Are near unity hydrogen isotope effects unusual for exchange reactions of C-H with alcohols?** Table 4 summarizes results for the exchange of some C-H compounds with methanolic methoxide. Only a quarter of the compounds listed result in the measurement of an appreciable PKIE: fluorene (Fl), 9-methylfluorene (9-Me-Fl), 9-phenylfluorene (9-Ph-Fl), and 9-trifluoromethylfluorene (9-TFM-Fl). The range of reactivity of the entire group is ca. 10^{16}, with 9-TFM-Fl the fastest and toluene the slowest. 9-TFM-Fl is $5.5x10^3$ faster than 9-Ph-Fl, which reacts as fast as 2-hydro-2-phenylperfluoropropane (27), and 2-hydroperfluoropropane (28). Although 9-Ph-Fl, 27, and 28 have the same reactivity at 25°C, the PKIE associated with the reactions differ significantly.

Experimental PKIE for all three isotopes of hydrogen have been reported for both 9-Ph-Fl and 9-Me-Fl.[10] Measurements of 9-Ph-Fl at 25°C give an observed $k^H/k^T = 16.0$ and $k^D/k^T = 2.50$, which allows the calculation of internal return factors[a] $a^H = 0.49$ and $a^T = 0.016$. At 45°C 9-Me-Fl has $k^H/k^T = 11.9$ and $k^D/k^T = 2.30$, which results in $a^H = 0.59$ and $a^T = 0.024$. These are examples of k^T as an accurate measurement for tritium transfer from carbon to alkoxide, but where moderate amounts of internal return reduce an observed rate for proton transfer. The kinetic solvent isotope effect (KSIE) was determined from detritiation rates in MeOD and MeOH to give $k^{OD}/k^{OH} = 1.8$ (9-Ph-Fl at 25°C) and = 2.2 (9-Me-Fl at 45°C).

Exchange rates with methanolic sodium methoxide are reported for all three isotopes of 2-hydro-2-phenylperfluorpropane (27): $k^H/k^T = 1.10$, $k^D/k^T = 1.05$, and $k^{OD}/k^{OH} = 2.63$ (using both MeOH and MeOD) at 25°C.[30] Similar results were obtained for ethanolic ethoxide: $k^H/k^T = 1.21$, $k^D/k^T = 1.08$, and $k^{OD}/k^{OH} = 2.74$. Since reaction has extensive internal return, even k^T will not measure

---

[a]The internal return factor, $a^H = k^H_{-1}/k_2$ and $a^T = k^T_{-1}/k_2$, for each labeled compound assumes a negligible isotope effect for the second step, $k^H_2 = k^D_2 = k^T_2$. To evaluate internal return factors the rates of reaction for all three isotopes are needed, and methods for calculation are given in reference 8.

the true rate of tritium transfer from carbon to alkoxide. The PKIE for the exchange rates of 27 are taken as the equilibrium hydrogen isotope effects for this reaction.

The data published[43] for methanolic sodium methoxide exchange of $(CF_3)_2CF^iH$ (28) must be carefully analyzed before using the reported rates to calculate PKIE associated with the reactions. Rates of $C_7F_{15}D$ and $C_7F_{15}T$ in MeOH resulted in $k^D/k^T = 1.4$ (70°). Eq. (9) was used to calculate a rate for $C_7F_{15}H$ (29) in MeOH, which was compared to k^H (in MeOD) to obtain KSIE, $k^{OD}/k^{OH} = 1.5$. This KSIE was used to calculate the reported k^H/k^D for 28.[b] If the KSIE measured for 27 (2.63) is used, the corrected k^H/k^D for 28 is 1.4 (20°) instead 2.4. The low value of the PKIE suggests extensive internal return.

Table 4 contains three compounds that have sp^3 carbon-hydrogen bonds {1,4-dihydroperfluorobicyclo[2.2.1]heptane (26)[44], 28, and (29)} and one, pentafluorobenzene (PFB),[17] with an sp^2 C-H bond that have near unity PKIE. **Why are the PKIE associated with the fluorenes different than those for these compounds?** One major difference between the two sets is that one can have extensive pi-electron delocalization while the other can not. Three other compounds {$PhCH(CF_3)_2$ (27), Ph_3CH, and $PhCH_3$} that result in near unity PKIE have benzylic hydrogens that exchange with methanol. Although this latter set is capable of pi-delocalization, these compounds appear to behave more like localized carbanions.

Experimental results for Ph_3CH ($k^H/k^T = 1.77$, $k^D/k^T = 1.34$, and $k^{OD}/k^{OH} = 2.29$ at 97.7°C) allows calculation of $k_1^H/k_1^D = 4.2$.[10] The PKIE for the exchange of toluene with methanolic methoxide at 178°C ($k^D/k^T = 1.0$)[45] suggests there is more internal return for that reaction than for the exchange of Ph_3CH. This does not hold when reaction occurs in cyclohexylamine:[46] PKIE for $PhCH_3$ are $k^H/k^T = 27.2$ and $k^D/k^T = 2.82$, for Ph_3CH are $k^H/k^T = 19.2$ and $k^D/k^T = 2.79$, and KSIE equals 1.57. These experimental isotope effects result in internal return factors: $a^H = 0.22$, $a^D = 0.02$, and $a^T = 0.005$ for toluene; $a^H \doteq 0.85$, $a^D = 0.06$, and $a^T = 0.02$

---

[b]This approach to calculation of KSIE was used by others in the early 1960's. The concept of internal return was proposed at that time and not too widely accepted. As a result, experimental sections must be read carefully before relying on values of KSIE reported in the text of papers.

for Ph_3C^iH at 25°C. A PKIE, $k_1^H/k_1^D = 11$, was calculated for the proton-transfer step for the reaction of both hydrocarbons with lithium cyclohexylamide in cyclohexylamine.

Carbonyl and nitro compounds are not included in Table 4 since charge will be localized on oxygen. The halogenated hydrocarbons rival the fluorenes in reactivity because their carbanions can be stabilized by field effects as well as by the interaction of the lone pair electrons with the σ^* orbital of a β-C-X.[47] The two trifluoromethyls of 27 increase the reactivity of the benzylic hydrogen by a factor of more than 10^{12}, while the two additional phenyl groups of Ph_3CH increase reactivity by 10^4. Our working hypothesis has been that benzylic carbanions behave as if they have little pi-delocalization in alcohol solvents. The next two sections will develope this concept further.

D. TRIMETHYL, ALKYLSILANES WITH METHOXIDE

Much work has been reported on proton-transfer reactions from a neutral carbon to oxide ions; however, few quantitative studies of the reverse process, transfer from hydroxyl to carbanion have been documented. Bockrath and Dorfman[48] used pulse radiolysis techniques to obtain rates of reaction for benzyl anion generated *in situ* with several alcohols and water in tetrahydrofuran (THF). The second order rates, $k \times 10^{-8}$ M^{-1} s^{-1}, at 297°K are: MeOH, 2.3 ± 0.3; EtOH, 1.4 ± 0.2 & EtOD, 1.2 ± 0.2; *t*-BuOH, 0.16 ± 0.02; H_2O, 0.53 ± 0.17. In all cases they were able to also obtain rates for the ion pair $PhCH_2^{\ominus}Na^{\oplus}$, which was more reactive than the free carbanion: (ion pair/free ion) is MeOH (25), EtOH (26), EtOD (18), *t*-BuOH (81) and H_2O (100).

They concluded from this study that:

(a) The magnitude of the different rate constants lies in the relative order to be expected for an acid-base reaction in which the acidity of the proton donor, as has been measured in either 2-propanol[49], or in DMSO solution,[50] is varied.

(b) Proton transfer from these donors, for free benzyl carbanion, is certainly far slower than diffusion controlled; even the rate constants for the ion paired species, while closer to the diffusion controlled limit, still depend upon the acidity of the proton donor.

(c) The free ion is less reactive than the ion pair by factors ranging from 20 to 100.

(d) The deuterium isotope effects, although small, are not negligible, indicating that O-H bond breaking is involved at the transition state of the reaction.

(e) The data indicate that the state of aggregation of the alcohol is not a factor in the kinetics, suggesting that there is a single reactive species.

The near unity k^H/k^D values agreed with previous studies on the reaction of hydroxylic species with organometallic compounds.[51] Known mixtures of ROH and ROD were added to the organometallic in an inert solvent. The H to D ratio in the resulting hydrocarbon was used to calculate k^H/k^D. The assumption had to be made that the organometallic compounds would be good models for carbanions. The kinetic studies seemed to support this assumption since PKIE measured for the free ions was similar to that for ion pairs. To study the behavior of free carbanions in protic solvents is more difficult.

The excellent experimental work reported by Eaborn et al.[52] was the first systematic investigation of this problem. Anions were generated *in situ* by the reaction of methoxide and an appropriate alkyltrimethylsilane in mixtures of MeOH and MeOD:

$$MeO^{\ominus} + Me_3SiR \longrightarrow MeOSiMe_3 + R^{\ominus} \quad (10)$$

$$MeO^{i}H + R^{\ominus} \longrightarrow MeO^{\ominus} + R^{i}H \quad (11)$$

Isotope effects are calculated from known ratios of MeOH:MeOD and the measured H:D in R^iH.[c] Kinetics were studied using a number of ring substituted benzyltrimethylsilanes, and this resulted in a $\rho = 4.7$.[53] The hydrogen isotope effects (k^H/k^D) associated with protonation of most benzylic anions are low (1.1-1.3) but increase steadily from p-CN (2.0), p-$PhSO_2$ (2.9), p-PhCO (7.0) to p-NO_2 (10).[52ac] Values of 10 were obtained for o-$NO_2PhCH_2^{\ominus}$, the 9-fluorenyl anion, and 9-methyl-9-fluorenyl anion; however, low values of 1.3 were calculated for the reaction of both $Ph_3C^{\ominus}$ and $Ph_2CH^{\ominus}$.[52c]

Eaborn et al.[52b] pointed out a discrepancy between results for PKIE associated with the methanolic methoxide-catalyzed exchange

---

[c]Problems were encountered in the measurement of accurate H:D ratios in the earlier work. The PKIE reported in 52b are corrected in 52c. The minor corrections did not alter the main concepts discussed in the earlier work.

of Ph_3C^iH (k_1^H/k_1^D = 4.2 at 100°C)[8a] compared to their near unity results for the protonation of the triphenylmethyl anion. After appropriate corrections, the predicted isotope effect, k_{-1}^H/k_{-1}^D, should be about 8 (25°) instead of the experimental value of 1.3. The experimental PKIE obtained by Streitwieser et al. for Ph_3C^iH was $(k^H/k^D)_{obs}$ = 1.32 at 97.7°C (k^H/k^T = 1.77 ± .01 and k^D/k^T = 1.34 ± .03) is low due to internal return: a^H = 25, a^D = 3.4 and a^T = 1.4. Scheme 3 can account for the apparent discrepancy as neither reaction has the proton-transfer step rate limiting. The results for protonation of the 9-methylfluorenyl anion and the methoxide-catalyzed exchange of 9-methylfluorene-9-*t* are in good agreement. For the fluorene reactions, the transfer of hydron between carbon and oxygen is rate limiting.

It was mentioned that initial studies of PKIE associated with proton transfer from hydroxyl groups to organometallic compounds were carried out by mixing solutions of the two reactants. The near unity hydrogen isotope effects were thought to arise from diffusion controlled reactions. The kinetic studies of Bockrath and Dorfman proved that the rates were below the diffusion limit; however, they resulted in the calculation of near unity isotope effects. Eaborn et al.[52b] probed this question further by reacting a solution of fluoren-9-yl-lithium in diethyl ether with an excess of 1:1 MeOH:MeOD. This resulted in an observed k^H/k^D of ca. 1.5, which is much lower than the value of 10 for reaction of 9-fluorenyl anion generated in methanol from the silane. They concluded that methanol in the vicinity of the organometallic is used up prior to diffusion, and results are due to mixing rather than diffusion controlled reaction rates.

Eaborn et al.[52c] initially analyzed their data using the widely accepted interpretation that PKIE values increase gradually until a maximum value is reached when the pK_a of the acid is equal to that of the conjugate acid of the base. After further studies, they noted a breakdown in this interpretation, and suggested the differences could be due to delocalization by conjugation vs. delocalization via σ bonds.[54] Reaction of fluoro-alkenes with alcoholic alkoxides also generates carbanions *in situ*, and the results reported for the benzyl anions obtained by reaction of methoxide with $XPhCH_2SiMe_3$ suggested a similar study for reaction of $XPhCH{=}CF_2$ with methanolic methoxide.[55]

E. PI-DELOCALIZED VS. LOCALIZED CARBANIONS

There are definite advantages to forming carbanions *in situ* by reaction of methanolic methoxide and β,β-difluorostyrenes. The reaction generating the carbanion is rate limiting, it occurs at a lower temperature than required for methoxide-catalyzed proton abstraction, and it eliminates problems associated with internal return. Methanol is an excellent trapping agent for carbanions, and only an intramolecular trap, such as β-halide ejection, is more efficient. The use of $XPhCH{=}CF_2$ (30-X) has the advantage of competing reactions: Intermediates (INT-3) can partition between proton-transfer from methanol to yield $XPhCH_2CF_2OMe$ (31-X), and the ejection of a β-fluoride to form $XPhCH{=}CFOMe$ (32-X).

$$\underset{\text{30-X}}{XPhCH{=}CF_2} \xrightarrow{MeO^{\ominus}} \underset{\text{INT-3}}{[\ XPhCHCF_2OMe\ ]^{\ominus}} \begin{cases} \xrightarrow{MeO^{i}H} \underset{\text{31-X}}{XPhCH^{i}HCF_2OMe} \\ \xrightarrow{-\ F^{\ominus}} \underset{\text{32-X}}{XPhCH{=}CFOMe} \end{cases}$$

Since reaction of methoxide with 30 is 10^3 faster than proton removal of 31-X or nucleophilic attack on 32-X, and fluoride ion will not react with 32-X in methanol, the products are stable to experimental conditions. Amounts of 31-X and 32-X form according to the values of k_{elim}, Scheme 4, and

$$k_{add} = k_f k_{prot} / (\ k_r + k_{prot}\ ). \qquad (12)$$

Calculation of the PKIE assumes that k_{elim} is the same in MeOH and MeOD:[56,57]

$$\frac{k^H}{k^D} = \frac{[\ \%\ \text{31-X} \ /\ \%\ \text{32-X}\ ]^{MeOH}}{[\ \%\ \text{31-X} \ /\ \%\ \text{32-X}\ ]^{MeOD}} \qquad (13)$$

SCHEME 4

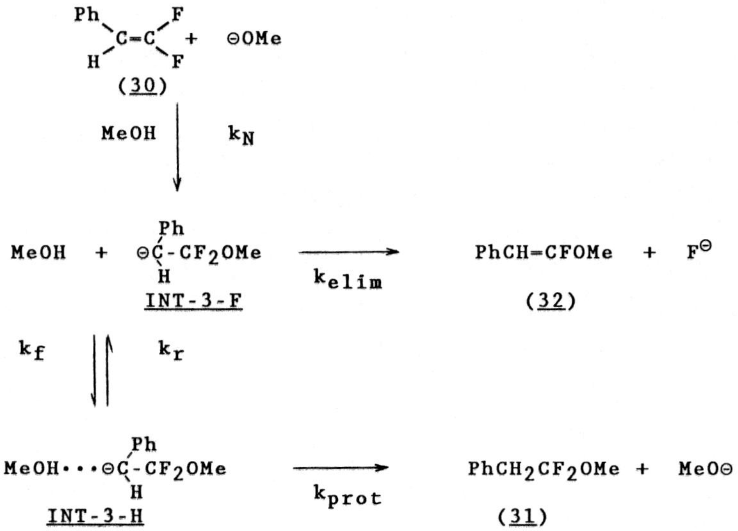

$$\underset{(\underline{30})}{PhHC{=}CF_2} + {}^{\ominus}OMe \xrightarrow[MeOH]{k_N}$$

$$MeOH + {}^{\ominus}C(Ph)(H)\text{-}CF_2OMe\ (\underline{INT\text{-}3\text{-}F}) \xrightarrow{k_{elim}} \underset{(\underline{32})}{PhCH{=}CFOMe} + F^{\ominus}$$

$$k_f \downarrow\uparrow k_r$$

$$MeOH\cdots{}^{\ominus}C(Ph)(H)\text{-}CF_2OMe\ (\underline{INT\text{-}3\text{-}H}) \xrightarrow{k_{prot}} \underset{(\underline{31})}{PhCH_2CF_2OMe} + MeO^{\ominus}$$

The PKIE (k^H/k^D) associated with the formation of both 31-m-NO_2 and 31-m-CF_3 increase slightly with increasing temperature: m-NO_2 is 1.20 (-50°), 1.24 (-25°), 1.29 (0°), 1.35 (25°) and 1.39 (50°) while m-CF_3 is 1.28 (0°), 1.34 (25°) and 1.40 (50°).

If $k_{prot} >> k_r$ (similar to reactions of 13 and 27), equation 12 reduces to $k_{add} = k_f$, and experimental PKIE are associated with the formation of INT-3-H from INT-3-F. A good model for these low isotope effects could be the proposal by Gold and Grist to explain the possible origins of KSIE.[58] They suggest the effects are due to multiple solvated methoxide ions that have a deuterium fractionation factor of 0.74. If the fractionation factor for INT-3-H is similar to methoxide, the measured PKIE values of 1.2 to 1.5 associated with protonation (k_{add}) would seem reasonable. It is not known if the fractionation factor will increase with temperature.

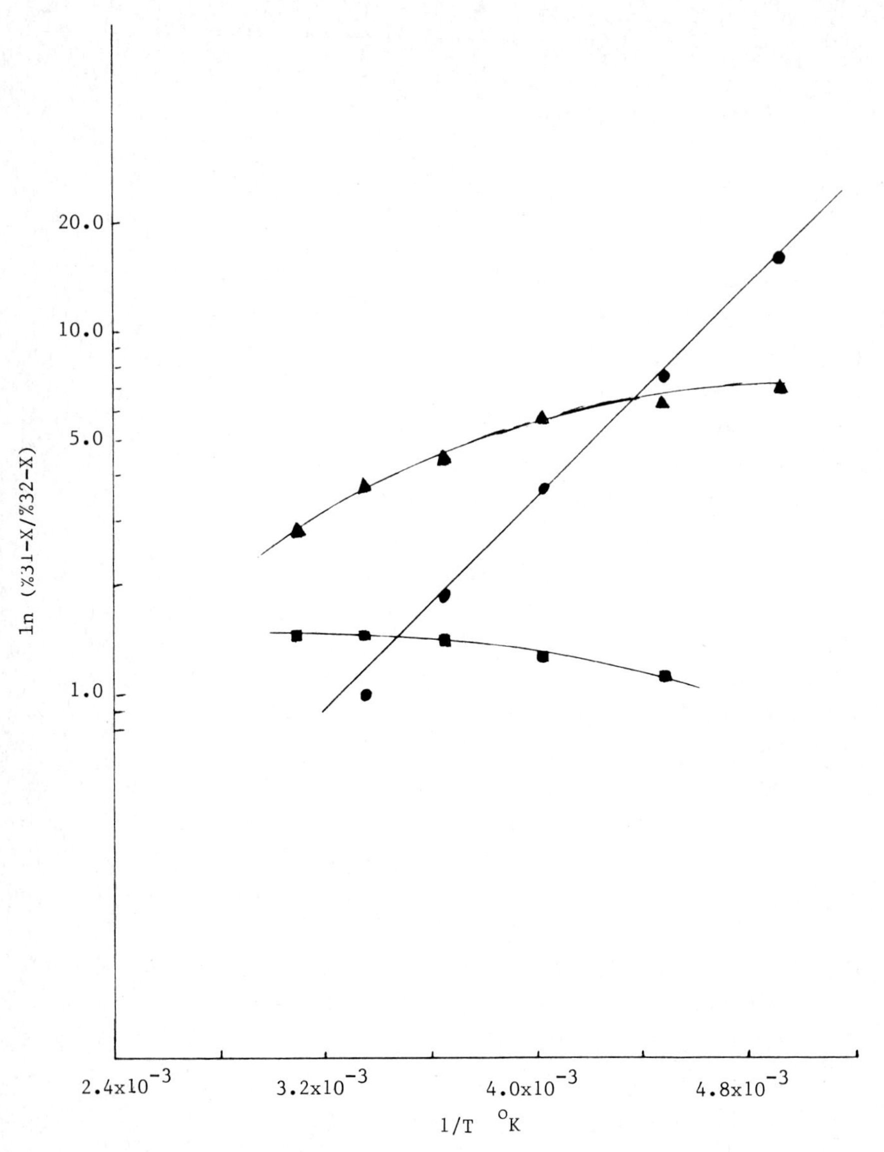

Figure 2. Plot of ln (%31-X/ %32-X) vs. 1/T for the partitioning of carbanions $XC_6H_4CHCF_2OMe^{\ominus}$ to form $XC_6H_4CH_2CF_2OMe$ (31-X) and $XC_6H_4CH{=}CFOMe$ (32-X). The circles (●) represent reactions of methanolic sodium methoxide with 30-p-NO_2, the triangles (▲) with 30-p-CN, and the squares (■) with 30-m-NO_2.

The PKIE associated with the protonation of INT-3-p-NO_2 shows more normal temperature behavior: 11.3 (-70°), 9.62 (-50°), 8.14 (-25°), 7.12 (0°) and 6.44 (25°).[26] However, the A^H/A^D of 1.9 is not normal, and suggests anomalous Arrhenius behavior of the type associated with internal-return mechanisms.[14] This behavior does suggest that neither k_f or k_{prot} is clearly rate limiting. The temperature behavior of the PKIE associated with protonation of INT-3-p-CN is similar to that for INT-3-m-NO_2 and INT-3-m-CF_3, but slightly larger: 1.33 (-70°), 1.36 (-50°), 1.53 (-25°), 1.69 (0°), 2.04 (25°) and 2.11 (50°). Since the isotope effects that increase with temperature have negative values for ($\Delta E_a{}^D$ - $\Delta E_a{}^H$), the magnitude of k^H/k^D must be due to A^H/A^D > 1.0. This is found as A^H/A^D is 2.2 (m-CF_3), 2.0 (m-NO_2) and 5.1 (p-CN).[d]

Comment must also be made regarding the temperature dependence of k_{add}/k_{elim} (Scheme 4) for reactions of methanolic methoxide with 30-m-NO_2, 30-p-CN, and 30-p-NO_2.[26] Figure 2 contains plots of ln (% 31-X/ % 32-X) vs. 1/T using the following data: m-NO_2 has 1.11(-50°), 1.26(-25°), 1.40(0°), 1.46(25°), 1.46(50°); p-CN has 6.81(-70°), 6.25(-50°), 5.77(-25°), 4.59(0°), 3.78(25°), 2.87 (50°); p-NO_2 has 16.0(-70°), 7.40(-50°), 3.61(-25°), 1.88(0°), 1.0(25°). The product distribution ratio for 30-p-NO_2 is linear between -70° and 0°C. The 25° point is ca. 15% below the line, and this can be attributed to competing loss of HF from 31-p-NO_2 to yield 32-p-NO_2. Since activation energies for elimination are much greater than for alkene reactions, the loss of some 31-p-NO_2 is only significant for the high temperature point. Both 30-p-CN and 30-m-NO_2[e] result in curved plots; however, with the same curvature occurring for reaction in MeOD, linear lines result from ln $(k^H/k^D)_{obs}$ vs. 1/T plots, when isotope effects are calculated using equation 13.

---

[d]Simple analysis of internal return will not predict the temperature dependence of the observed isotope effects. However, values at 25° can be rationalized if one assumes a $(k^H/k^D)_{prot}$ = 12 and $(k^H/k^D)_f$ = $(k^H/k^D)_r$ = 1.35. The $(k^H/k^D)_{obs}$ = 6.44 for formation of 31-p-NO_2 can be obtained if $[k_r = 10.9 \times k_{prot}]^H$ and $[k_r = 8.1 \times k_{prot}]^D$, and the $(k^H/k^D)_{obs}$ = 2.04 associated with formation of 31-p-CN would need $[k_r = 0.84 \times k_{prot}]^H$ and $[k_f = 0.62 \times k_{prot}]^D$ (see ref 26, p 4538).

[e]Results obtained for 30-m-NO_2 are quite similar for a variety of alkenes studied: $PhC(CF_3)=CF_2$, $PhCCl=CF_2$, and all 30-X other than the p-nitro and p-cyano.

Clearly intermediates for the three reactions behave quite differently toward protonation by methanol compared to the ejection of β-fluoride. Our isotope effects compare to those reported by Eaborn et al., which suggests that $[XPhCH_2]^{\ominus}$ and $[XPhCHCF_2OMe]^{\ominus}$ are similar in their reactivity toward methanol. The p-NO_2 compounds result in large isotope effects, and these anions will delocalize through the pi network to place the negative charge mainly on oxygen.[f] For proton transfer, the orbitals need to be properly aligned in an encounter complex, INT-3-H. If the barrier to reform INT-3-F is lower than that for proton transfer, significant isotope effects can result. The calculated ΔE_a between elimination and addition, 3.3 ± .1 kcal mol^{-1}, should give even larger amounts of addition product; however, a $\Delta S^{\ddagger}$ of ca 10 eu (A^{add}/A^{elim} = 0.004) favors elimination and offsets the big enthalpy difference. The ratio of addition to elimination for 30-m-NO_2 increases slightly with temperature, while the behavior of 30-p-CN is more like that of 30-p-NO_2. There are differences in E/Z vinyl ethers ratios for reaction of 30-p-NO_2 compared to 30-m-NO_2, and this will be discussed in the following section.

F. VINYL HALIDES WITH ALKOXIDES

Whether or not nucleophilic vinyl substitution reactions form carbanion intermediates or occur by a concerted mechanism is of practical and theoretical interest. Rappoport has reviewed the subject in depth.[59] There is good evidence for the formation of carbanions when fluoride is displaced: (1) carbanions are trapped by protonation from solvent;[22,60] (2) (E)- and (Z)-alkenes give the same product;[61] (3) fluoride is displaced by several orders of magnitude faster than chloride.[61] Favoring the single step mechanism is the complete retention of configuration observed in the the substitution of many vinyl chlorides or bromides.[59]

Modena et al. studied the reactions of methanolic methoxide and benzenethiolate with p-NO_2PhCH=CHX (33-X), X = F or Cl.[61,62] The reaction of Z-33-F or E-33-F gave only (E)-p-NO_2PhCH=CHOMe (E-34) or (E)-p-NO_2PhCH=CHSPh (E-35); however, substitution of chloride

---

[f] Of all the alkenes we have studied, only 30-p-NO_2 had colored solutions when reacting with methanolic methoxide.

by benzenethiolate occurs with complete retention.

p-NO_2Ph(H)C=C(F)H (Z-33-F) $\xrightarrow[\text{MeOH}]{\text{MeO}^{\ominus}}$ p-NO_2Ph(H)C=C(H)OMe (E-34) $\xleftarrow[\text{MeOH}]{^{\ominus}\text{OMe}}$ p-NO_2Ph(H)C=C(H)F (E-33-F)

(Z-33-F) (E-34) (E-33-F)

p-NO_2Ph(H)C=C(Cl)H (Z-33-Cl) $\xrightarrow[\text{MeOH}]{\text{PhS}^{\ominus}}$ p-NO_2Ph(H)C=C(SPh)H (Z-35)

(Z-33-Cl) (Z-35)

p-NO_2Ph(H)C=C(H)Cl (E-33-Cl) $\xrightarrow[\text{MeOH}]{\text{PhS}^{\ominus}}$ p-NO_2Ph(H)C=C(H)SPh (E-35) $\xleftarrow[\text{MeOH}]{^{\ominus}\text{SPh}}$ Z-33-F or E-33-F

(E-33-Cl) (E-35)

To explain the discrepancy, they suggested that chloride, a good leaving group, is displaced with kinetic control of the product forming step. Displacement of fluoride, a poorer leaving group, results in a thermodynamic control since bond rotation can occur prior to loss of $F^{\ominus}$.

Reactions of difluoroalkenes and methanolic methoxide occur with kinetic control of products. The exception is p-$NO_2PhCH{=}CF_2$ (30-p-NO_2), where addition to elimination ratios follow kinetic control, but ratios of E:Z vinyl ethers are equal from -70° to 0°. If negative charge in $(p\text{-}NO_2PhCHCF_2OMe)^{\ominus}$ is on oxygen, it would allow bond rotation prior to the elimination of fluoride. Reactions of PhC(CF_3)=CHF (36) with methanolic methoxide could give interesting results. If fluoride is displaced from 36 with retention, stereoconvergence found in reactions of 33-F can not be due to its poor leaving ability.

Reaction of both isomers of 36 resulted in the displacement of fluoride with complete retention; however, addition of methanol (3-11%) also was observed.[63]

Figure 3. Reaction of Z-PhC(CF_3) = CHF with Methoxide

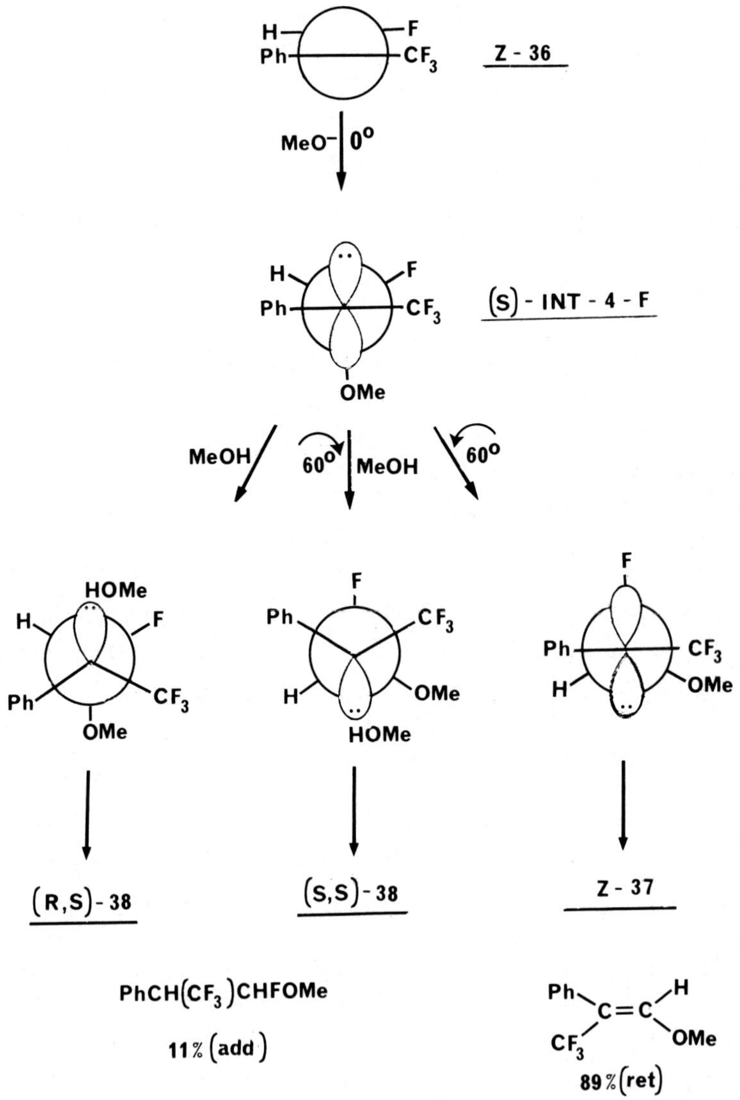
H
Ph
F
CF3
Z - 36
MeO−
0°
H
Ph
F
CF3
OMe
(S) - INT - 4 - F
MeOH
60°
MeOH
60°
HOMe
H
F
Ph
CF3
OMe
F
Ph
CF3
H
OMe
HOMe
F
Ph
CF3
H
OMe
(R,S) - 38
(S,S) - 38
Z - 37
PhCH(CF3)CHFOMe
11% (add)
Ph
H
C=C
CF3
OMe
89% (ret)

$$\begin{array}{c}Ph\\CF_3\end{array}\!\!>C=C<\!\!\begin{array}{c}F\\H\end{array}\ \xrightarrow[MeOH,\ 0^\circ]{^{\ominus}OMe}\ \begin{array}{c}Ph\\CF_3\end{array}\!\!>C=C<\!\!\begin{array}{c}OMe\\H\end{array}\ +\ PhCH(CF_3)CHFOMe$$

(E-36) 97% (E-37) 3% (38)

$$\begin{array}{c}Ph\\CF_3\end{array}\!\!>C=C<\!\!\begin{array}{c}H\\F\end{array}\ \xrightarrow[MeOH,\ 0^\circ]{^{\ominus}OMe}\ \begin{array}{c}Ph\\CF_3\end{array}\!\!>C=C<\!\!\begin{array}{c}H\\OMe\end{array}\ +\ PhCH(CF_3)CHFOMe$$

(Z-36) 89% (Z-37) 11% (38)

Formation of 38 argues against a concerted mechanism, and favors a two-step mechanism similar to that for reaction of $PhC(CF_3)=CF_2$ (2) with methanolic methoxide. More saturated ether 7 (44%) results from 2, and this is reasonable since loss of fluoride from -CHFOMe should be more favorable than from $-CF_2OMe$.[64] Therefore protonation of $\{PhC(CF_3)CF_2OMe\}^{\ominus}$ (INT-1) is more favorable than $\{PhC(CF_3)CHFOMe\}^{\ominus}$ (INT-4). The same ratio of diastereomers of 38 (45:55) is formed from either E-36 or Z-36. An explanation for the stereochemistry of this reaction is found in Figure 3. The first intermediate, INT-4-F has three possible options. It can (A) rehybridize and form the methanol hydrogen-bond that is anti-periplanar to C-OMe, (B) rotate clockwise 60°, rehybridize, and form a similar hydrogen-bond anti-periplanar to the C-H bond, or (C) rotate 60° anti-clockwise, placing C-F periplanar to the lone pair. The latter conformer will have fluoride in a position for leaving and result in the vinyl ether with retention. Since the isotope effects associated with reaction of 36 and 38 are similar to those for reactions of 2, 3, or 7, Scheme 3 should also define a mechanism for the reaction of 36 with methanolic methoxide.

Reaction of (E)-$PhC(CF_3)=CHCl$ with methanolic methoxide is 370 times slower than that of E-36, and occurs with 100% retention to form E-37 as the only product. This is consistent with formation of a carbanion and an inverse element effect, $k^{Cl}/k^{F} \ll 1.0$, is similar to other aromatic or vinyl nucleophilic substitutions. In light of these results, an earlier report that the reaction of $PhC(CF_3)=CFCl$ (39) with methanolic methoxide resulted in only 96% retention to give $PhC(CF_3)=CFOMe$ (8) seems odd.[65] It was also suggested that some substitution of fluoride occurred, but this

Figure 4. Reaction of (E)-PhC(CF_3) = CFCl with Methoxide

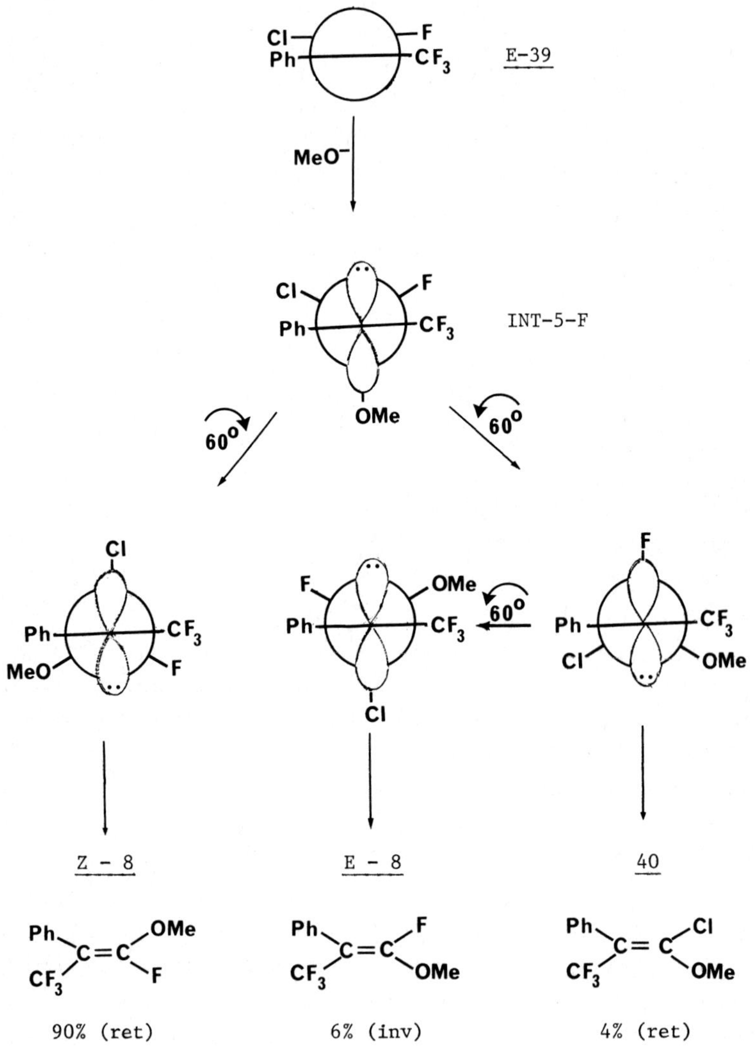

E-39

INT-5-F

Z - 8	E - 8	40
90% (ret)	6% (inv)	4% (ret)

was not experimentally documented.[66]

Ethoxide-promoted dehydrofluorination of $PhCHClCF_3$ is 5.3 x 10^5 slower than elimination of HCl from $PhCHClCF_2Cl$, Table 2. Thus $F^{\ominus}$ loss from $\{PhC(CF_3)CFClOMe\}^{\ominus}$ (INT-5) should not compete with $Cl^{\ominus}$ ejection; however, INT-5 has not yet formed a hydrogen-bond to MeOH. **Could fluoride ejection compete with chloride from an intermediate, INT-5-F, not stabilized by a hydrogen-bond?** This seemed important enough to reinvestigate the reactions of 39. The reaction of methanolic methoxide with 39 does eliminate fluoride (4%) in competition with chloride,[67] Figure 4. A 60° clockwise rotation of the initial intermediate, INT-5-F, results in a more stable rotomer with C-Cl periplanar to the lone pair electrons,[68] and loss of chloride gives 4. A counterclockwise rotation places C-F periplanar to the lone pair, and several options are open at this stage. Loss of fluoride yields $PhC(CF_3)$=CClOMe (40), and a second counterclockwise 60° rotation generates the other rotomer with C-Cl periplanar to the lone pair, which accounts for the 6% inversion product (5).

Scheme 3 could also account for these reactions. Chloride loss is the low energy process and will occur whenever the orbitals are properly aligned. Fluoride is capable of leaving when it is periplanar to the lone pair electrons, which results in substitution with retention. In competition with the ejection of $F^{\ominus}$ is the formation of a hydrogen-bonded intermediate. This is not stereospecific and occurs from either side. Fluoride will not leave from a hydrogen-bonded carbanion. If INT-5-F reacts in a similar way to INT-4-F, some INT-5-H should form. The absence of a saturated product is not surprising, since methoxide-promoted dehydrochlorination of $PhCH(CF_3)CFClOMe$ should be faster than the reaction of 39 with methoxide.

The stereochemistry of nucleophilic displacement reactions of alkenes has some renewed interest with the recent appearance of several theoretical calculations.[68,69] The qualitative results are in agreement with the conclusions reached in this chapter, but it is still to early to attempt to model actual product distributions.

IV. CARBON ACIDS

The interest in relative pK_a values of carbon acids is due to a desire to predict relative reactivities for reactions that are promoted or catalyzed by a base. In many instances, the rate of proton transfer from carbon to the base would be of greater value than an equilibrium measurement. Ritchie[70] once stated that, "those carbon acids whose conjugate bases have localized charge are predicted to have proton transfer rates considerably greater than acids of the same thermodynamic strength whose conjugate bases have delocalized charges. That is, saturated hydrocarbons, alkenes, alkynes and cycloalkanes whose conjugate bases are localized are expected to show 'kinetic acidities' greater than their thermodynamic acidities". An excellent example of this was mentioned at the end of section II: methoxide-catalyzed protodetritiation of pentafluorobenzene (PFB) with methanol is 15 times faster than that for 9-phenylfluorene (9-PhFl) at 25°C, Table 4. Relative pK_a values measured in cyclohexylamine (CHA) are 18.5 for 9-PhFl and 25.8 for PFB.[18]

What about rates for the actual proton-transfer steps for these compounds? The PKIE data for 9-PhFl suggests there is negligible internal return for the protodetritiation reaction, however, the isotope effect, $k^D/k^T = 1.0$, associated with the exchange of PFB implies there is extensive internal return. Therefore the actual proton transfer from carbon to oxygen occurs 50- 100 times faster than the observed protodetritiation rate. So much for relative pK_a data to predict relative rates of a base-catalyzed reaction. The pK_a values obtained in CHA are ion-pair values. **Will values measured in Me_2SO give a better correlation?** The pK_a for PFB can not be measured in Me_2SO since the anion is not stable.[71] The expected pK_a in Me_2SO would be greater than that determined in CHA, since the ΔpK_a of ca. 5 units between phenylacetylene (23) and 9-PhFl in CHA increases to about 11 pK_a units in Me_2SO.[1] For this reason, a pK_a in Me_2SO would be a worse indicator.

Values of pK_a cited for measurements in CHA and most of those reported for Me_2SO are obtained using an indicator method that utilizes carbon acids within several pK_a units of the compound to be determined. Therefore the indicator method compares energy levels of "free" carbanions. The reason phenyl acetylene has a lower pK_a in CHA than in Me_2SO is due to the stabilizing effect of the ion-pair gegen-ion.

The highly delocalized 9-PhFl$^{\ominus}$ will be more stable than the hydrogen-bonded carbanion, 9-PhFl$\ominus$•••HOMe, and as the proton is transfered from carbon to methoxide, the incipient intermediate is at a higher energy state. Prior to delocalization via the pi network, the carbon must rehybridize and a break the hydrogen-bond. This process has an energy barrier and is in competition with the internal-return reaction.

The proton transfer from PFB to methoxide can occur at a lower energy than that necessary for 9-PhFl; however, PFB$^{\ominus}$ doesn't gain stability as the free ion and needs significant additional energy input to break the hydrogen-bond. Internal return is the lowest free energy path, and this will wash out the PKIE associated with the proton-transfer step. The PFB anion has the electron pair in an sp^2 orbital, but a similar argument can be made for one that uses an sp^3 orbital. 1,4-Dihydroperfluorobicyclo[2.2.1]heptane (26) has a pK_a (CHA) of 22.3,[44] and undergoes protodetritiation at a rate 82 times faster than 9-PhFl at 25°C, Table 4. Both of the ions generated from PFB and 26 will have delocalized charge, but not through a pi network.[g] Regardless of the mechanism for this stabilization, it is still effective in the hydrogen-bonded intermediate.

Bordwell and Boyle[72] suggested that a localized carbanion is the initial intermediate resulting from a base-catalyzed proton removal from nitroalkanes. Therefore, electron delocalization to form a more stable intermediate would lag behind proton transfer. The PKIE results for the methanolic methoxide-catalyzed exchange reactions of 9-PhFl and 9-MeFl are consistent with an internal return mechanism. These results also suggest that delocalization of the electrons lags behind proton transfer. Bernasconi[73] deals with this problem and refers to it as the principle of imperfect synchronization.

---

[g]Schleyer and Kos[47c] discuss the importance of anionic hyperconjugation from a theoretical point, and disagree with the interpretation of calculations by Streitwieser et al.[47a] Both groups agree that there is a large stabilization due to perfluoro substituents, but disagree on the mechanism of stabilization.

V. CONCLUSIONS

A major goal for studies of chemical reaction mechanisms is to understand not only the detailed pathway of a reaction, but also the timing of various steps along that pathway. Proton-transfer reactions are of fundamental importance and many details of that process are still not known. The role of solvent reorganization associated with proton-transfer reactions is complex and under investigation.[74] Ahlberg and Thibblin reported enhanced PKIE due to ion-pair intermediates in base-promoted eliminations,[75] and temperature-independent PKIE.[76] Thibblin has some excellent work on hydrogen-bonded carbanion intermediates that include ArOH and quaternary ammonium ions as the hydrogen source.[77] One must take care in assigning "pK_a" values to organic molecules on the basis of isotopic exchange reactions, as well as ruling out the formation of a hydrogen-bonded carbanion as a reaction intermediate by saying there is "too great a pK_a difference between reactants".

A lack of exchange with solvent prior to elimination does not rule out the occurrence of a carbanion intermediate. Buncel and Bourns discuss this in depth in their paper on the mechanism of the ethoxide-promoted carbonyl elimination reaction of benzyl nitrate,[34] yet one still reads: the presence of a carbanion is ruled out due to a lack of exchange prior to elimination. They measured both hydrogen (k^H/k^D = 5.0 at 60°) and nitrogen (k^{14}/k^{15} = 1.0196 at 30°) isotope effects associated with that reaction, and concluded that the reaction occurred by a concerted mechanism rather than a two-step pathway. At that time (1960) any other interpretation of the data would certainly not have gotten past any reputable referee; however, similar results were obtained for the ethoxide-promoted elimination of PhC^iHClCH_2Cl (k^H/k^D = 4.2 and k^{35}/k^{37} = 1.00908 at 25°)[31] which appears to go by a two-step mechanism. The reaction of benzyl nitrate may be concerted, but the carbanion mechanism cannot be ruled out. This is a perfect example of words of caution given by Bunnett in the introduction to a chapter on "From Kinetic Data to Reaction Mechanism":[78]

> *"It is axiomatic that one cannot define a reaction mechanism absolutely. What one can do is to reject conceivable mechanisms that are not compatible with experimental evidence. When all conceivable mechanisms but one have been rejected, one is tempted to consider the mechanism to be established. However, such reasoning does not, and by definition cannot,*

take account of inconceivable mechanisms, those which have not occurred to the investigator or his contemporaries. At a later, especially in the light of advances in the meanwhile, one or more further alternative mechanisms may be recognized, and the mechanism previously thought to be the only tenable one may be found to be incompatible with new experimental evidence. Such has happened many times in the history of mechanistic studies. The conservative scientist therefore refers to the better-understood mechanism as 'generally accepted' or 'well recognized' rather than as 'established' or 'proven'."

Since polar C-F bonds can stabilize localized carbanions, there are advantages to using highly fluorinated compounds in studies of carbanions.[79] The increased reactivity of fluoroalkenes with nucleophiles allows for the generation of carbanions in protic solvents under rather mild conditions. This has allowed us to study the chemistry of carbanions, formed *in situ*, that are not stabilized by extensive π-delocalization, and compare it to that for highly π-delocalized anions. One should not study the chemistry of carbocations using only fluorocarbons, and for similar reasons, one should not predict the behavior of localized carbanions from compounds that result in π-delocalized carbanions.

ACKNOWLEDGEMENTS

For over a decade most of our work was supported by Research Corporation. Studies of alkene reactions was supported by the Petroleum Research Fund, and recent support has been by National Science Foundation grant no. CHE-8316219. It is a pleasure to acknowledge the hard work of the undergraduates, who carried out most of the experimental work, and the contributions of my senior co-authors Donald B. Dahlberg, Judith G. Koch, Gerrit Lodder and Duncan J. McLennan, who aided in the interpretation of the data and formulation of the ideas. We also greatfully acknowledge several NATO grants with Richard D. Chambers which have made possible discussions of many ideas regarding reaction mechanisms and fluoro-carbanions, as well as Sunderland football.

VI. REFERENCES

1. M.S. Matthews, J.E. Bares, J.E. Bartmess, F.G. Bordwell, F.J. Cornforth, G.E. Drucker, Z. Margolin, R.J. McCallum, G.J. McCollum, and N.R. Vanier, J.Am.Chem.Soc., **97** (1975) 7006.

2. (a) D.J. Cram, "Fundamentals of Carbanion Chemistry", Academic Press, New York 1965. (b) J.R. Jones, "The Ionization of Carbon Acids", Academic Press, New York 1973. (c) O.A. Reutov, I.P. Beletskaya, K.P. Butin, "C-H Acids", Pergamon Press, Oxford, England 1978. (d) E. Buncel, Carbanions. Mechanistic and Isotopic Aspects", Elsevier, Amsterdam 1975.

3. (a) A. Streitwieser, Jr., E. Juaristi, L.L. Nebenzahl, in "Comprehensive Carbanion Chemistry", edited by E. Buncel and T. Durst, Elsevier, Amsterdam 1980. (b) A. Streitwieser, Jr., Acc.Chem.Res., **17** (1984) 353.

4. (a) D.A.Bors, and M.J.Kaufman and A. Streitwieser, Jr., J.Am. Chem.Soc., **107** (1985) 6975. (b) S. Gronert and A. Streitwieser, Jr., J.Am.Chem.Soc., **108** (1986) 7016.

5. Albery (W.J. Albery, Annu.Rev.Phys.Chem. **31** (1980) 242) discusses this multistep approach to proton-transfer reactions and gives leading references.

6. (a) F.H. Westheimer, Chem. Rev., **61** (1961) 265. (b) L. Melander and W.H. Saunders, Jr., "Reaction Rates of Isotopic Molecules", Wiley-Interscience, New York, 1980.

7. (a) More O'Ferrall (R.A. More O'Ferrall, in "Proton-Transfer Reactions", edited by E.F. Caldin and V. Gold, Chapman & Hall, London 1975, pp 216-227) gives a detailed discussion and references. (b) R.L. Schowen, Prog.Phys.Org.Chem. **9** (1972) 286.

8. (a) A. Streitwieser, Jr., W.B. Hollyhead, G. Sonnichsen, A.H. Pudjaatmaka, C.J. Chang, and T.C. Kruger, T.C. J.Am.Chem. Soc., **93** (1971) 5096. (b) D.W. Boerth, and A. Streitwieser, Jr., J.Am.Chem.Soc., **103** (1981) 6443.

9. C.G. Swain, E.C. Stivers, J.F. Reuwer, and L.J. Schaad, J.Am. Chem.Soc. **80** (1958) 5885.

10. A. Streitwieser, Jr., W.B. Hollyhead, A.H. Pudjaatmaka, P.H. Owens, T.L. Kruger, P.A. Rubenstein, R.A. MacQuarrie, M.L. Brokaw W.K.C. Chu, and H.M. Niemeyer, J.Am.Chem.Soc., **93** (1971) 5088.

11. See reference 8, p. 5099.

12. F.G. Bordwell and W.J. Boyle, Jr., J.Am.Chem.Soc., **93** (1971) 512.

13. H.F. Koch, D.B. Dahlberg, M.F. McEntee, and C.J. Klecha, J.Am.Chem.Soc. **98** (1976) 1060.

14. H.F. Koch and D.B.Dahlberg, J.Am.Chem.Soc. **102** (1980) 6102.

15. R.P. Bell, Chem. Soc. Rev., **3** (1974) 513.

16. S.B. Kaldor and W.H. Saunders, Jr., J.Am.Chem.Soc., **101** (1979) 7594.

17. A. Streitwieser, Jr., J.A. Hudson, and F. Mares, J.Am.Chem. Soc., **90** (1968) 648.

18. A. Streitwieser, Jr., P.J. Scannon, and N.M. Neimeyer, J.Am. Chem.Soc., **94** (1972) 7936.

19. W.T. Miller, Jr., E.W. Fager, and P.H. Griswold, J.Am.Chem.Soc. **70** (1948) 431.

20. (a) J. Hine, R. Wiesboek, and R.G. Ghirardelli, J.Am.Chem.Soc. **83** (1961) 1219. (b) J. Hine, R. Wiesboek, and O.B. Ramsay, ibid. **83** (1961) 1222.

21. H.F. Koch and A.J. Kielbania,Jr., J.Am.Chem.Soc., **92** (1970) 729.

22. H.F. Koch, J.G. Koch, D.B. Donovan, A.G. Toczko and A.J. Kielbania, Jr., J.Am.Chem.Soc., **103** (1981) 5417.

23. W.H. Saunders, Jr., Chem.Scr., **8** (1975) 27.

24. W.J. Albery and J.R. Knowles, J.Am.Chem.Soc., **99** (197) 637.

25. Since k^H was measured in EtOH and k^D in EtOD, correction was made for the kinetic solvent isotope effect. This effect can be between $k^{OD}/k^{OH} = 2.0$ to 2.5 for systems that we have studied. An accurate value can only be assigned if k^T is known for both solvents.

26. H.F. Koch and A.S. Koch, J.Am.Chem.Soc., **106** (1984) 4536.

27. A.J. Kresge, Acc.Chem.Res., **9** (1975) 354.

28. H.F. Koch, D.B. Dahlberg, A.G. Toczko, and R.L. Solsky, J.Am.Chem.Soc., **95** (1973) 2029.

29. (a) C.H. DePuy and C.A. Bishop, J.Am.Chem.Soc., **82** (1960) 2535. (b) W.H. Saunders, Jr. and D.H. Edison, J.Am.Chem.Soc., **82** (1960) 138.

30. H.F. Koch, D.B. Dahlberg, G. Lodder, K.S. Root, N.A. Touchette, R.L. Solsky, R.M. Zuck, L.J. Wagner, N.H. Koch, M.A. Kuzemko, J.Am.Chem.Soc., **105** (1983) 2394.

31. H.F. Koch, D.J. McLennan, J.G. Koch, W. Tumas, B. Dobson and N.H. Koch, J.Am.Chem.Soc., **105** (1983) 1930.

32. (a) T. Sano, N. Tatsumoto, T. Niwa, and T. Yasunaga, Bull. Chem.Soc.Jpn., **45** (1972) 2669. (b) T. Sano, N. Tatsumoto, Y. Mende, and T. Yasunaga, *ibid*, **45** (1972) 2673.

33. P.S. Skell and C.R. Hauser, J.Am.Chem.Soc., **67** (1945) 1661.

34. E. Buncel and A.N. Bourns, Can.J.Chem., **38** (1960) 2457.

35. J.F. Bunnett, E.W. Garbisch, and K.M. Pruitt, J.Am.Chem.Soc. **79** (1957) 385.

36. Reference 31, p 1935.

37. It is interesting to note that in the series $YC_6H_4C^iHClCH_2X$, where X = Cl or Br, the effect of the β-halide is much different. After correcting for internal return, the magnitude of an element effect, k^{Br}/k^{Cl}, of about 35 comes mainly from a proton-transfer reaction and has little to do with the leaving group (R. Hage and D.J. Bogdan, unpublished results). Work is continuing on this problem.

38. K.S. Root, N.A. Touchette, J.G. Koch and H.F. Koch, Abst. of Papers, Euchem Conference on Mechanisms of Elimination Reactions, Assisi, Italy, Sept., 1977.

39. For other examples see reference 13, note 14.

40. J.G. Koch, unpublished results.

41. L.P. Hammett: "Physical Organic Chemistry", McGraw-Hill, New York, NY, 1970, p.119.

42. (a) D.S. Thompson, R.A. Newmark and C.H. Sederholm, J.Chem. Phy., **37** (1962) 411. (b) F.J. Weigert, M.B. Winstead, J.I. Garrels and J.D. Roberts, J.Am.Chem.Soc., **92** (1970) 7359.

43. S. Andreades, J.Am.Chem.Soc. **86** (1964) 2003.

44. A. Streitwieser, Jr., D. Holtz, G.R. Ziegler, J.O. Stoffer, M.L. Brokaw, and F. Guibe, J.Am.Chem.Soc. **98** (1976) 5229.

45. T.A. Keevil; Doctoral Dissertation, University of California, Berkeley, 1972.

46. A. Streitwieser, Jr., P.H. Owens, G. Sonnichsen, W.K. Smith, G.R. Ziegler, H.M. Niemeyer, and T.L. Kruger, J.Am.Chem.Soc., **95** (1973) 4254.

47. This topic was treated using theoretical calculations: (a) A. Streitwieser, C.M. Berke, G.W. Schriver, D. Grier, and J.B. Collins, Tetrahedron Suppl., **37** (1981) 345. (b) A. Pross, D.J. DeFrees, B.A. Levi, S.K. Pollack, L. Radom, and W.J. Hehre, J.Org.Chem., **46** (1981) 1693. (c) P.v.R. Schleyer and A.J. Kos, Tetrahedron, **39** (1983) 1141.

48. B.Bockrath and L.M.Dorfman, J.Am.Chem.Soc., **96** (1974) 5708.

49. J. Hine and M. Hine, J.Am.Chem.Soc., **74** (1952) 5266.

50. C.D. Ritchie, "Solute-Solvent Interactions," J.F. Coetzee and C.D. Ritchie, Ed., Marcel Dekker, New York, NY, (1969) p.230.

51. R. Wiberg, J.Am.Chem.Soc., **77** (1955) 5987; L.O. Assarson, Acta Chem.Scand., **12** (1958) 1545; Y. Pocker and J.H. Exner, J.Am. Chem.Soc., **90** (1968) 6764.

52. (a) R. Alexander, W.A. Asomaning, C. Eaborn, I.D. Jenkins, and D.R.M. Walton, J.Chem.Soc.,Perkin Trans. 2, (1974) 490. (b) C. Eaborn, D.R.M. Walton, and G.J. Seconi, *ibid* (1976) 1857. (c) D. Macciantelli, G. Seconi, and C. Eaborn, *ibid* (1978) 834.

53. R.W.Bott, C.Eaborn, and T.W.Swaddle, J.Chem.Soc. (1963) 2342.

54. (a) G.Seconi, C.Eaborn and A.J.Fischer, J.Organomet.Chem., **177** (1979) 129. (b) G. Seconi, C. Eaborn, and J.G. Stamper, *ibid* **204** (1981) 153.

55. H.F. Koch, J.G. Koch, N.H. Koch and A.J. Koch, J.Am.Chem. Soc., **105** (1983) 2388.

56. This assumption has validity since a fractionation factor of 1.0 has been reported for fluoride and H_2O and D_2O.[57]

57. W.J. Albery in "Proton-Transfer Reactions"; E.F. Caldin and V. Gold, Eds.; Chapman and Hall; London, 1975; p 283.

58. V. Gold and S. Grist, J.Chem.Soc. B (1971) 2282.

59. Z. Rappoport, Acc.Chem.Res., **14** (1981) 7.

60. G.Marchese and F.Naso, La.Chim.E.Ind.(Milano) **53** (1971) 760.

61. G.Marchese, F.Naso and G.Modena, J.Chem.Soc.(B) (1969) 290.

62. G.Marchese, F.Naso and G.Modena, J.Chem.Soc. (B) (1968) 958.

63. H.F. Koch, J.G. Koch and S.W. Kim, Abst. of Papers for VI IUPAC Conference on Physical Organic Chemistry, Bull.Soc.Chim. Belg., **91** (1982) 431.

64. H.F. Koch, W. Tumas and R. Knoll, J.Am.Chem.Soc., **103** (1981) 5423.

65. D.J. Burton and H.C. Krutzch, J.Org.Chem., **36** (1971) 2351.

66. H.C. Krutzch, Doctoral Dissertation, University of Iowa, Iowa City, 19

67. H.F. Koch, J.G. Koch and M.J. Barnes, Abst. of Papers for VII IUPAC Conference on Physical Organic Chemistry, Auckland, NZ, August 20-24, 1984.

68. (a) Y. Apeloig and Z. Rappoport, J.Am.Chem.Soc., **101**(1979) 5095. (b) R.D. Bach and G.J. Wolber, J.Am.Chem.Soc., **106** (1984) 1401.

69. D. Cohen, R. Bar and S.S. Shaik, J.Am.Chem.Soc., **108** (1986) 231.

70. C.D. Ritchie, J.Am.Chem.Soc., **91** (1969) 6479.

71. F.G. Bordwell, private communication.

72. F.G. Bordwell and W.J. Boyle, Jr., J.AM.Chem.Soc., **97** (1975) 3447.

73. C.F. Bernasconi, Tetrahedron, **41** (1985) 3219.

74. Bernasconi is active in this field and published an excellent summary (C.F. Bernasconi, Pure Appl.Chem. **54** (1982) 2335) with pertinent references.

75. A. Thibblin, S. Bengtsson and P. Ahlberg, J.Chem.Soc. Perkin Trans. II (1977) 1569.

76. A. Thibblin, I. Onyido and P. Ahlberg, Chem.Scr., **19** (1982) 145.

77. (a) A. Thibblin, J.Am.Chem.Soc., **105** (1983) 853. (b) A. Thibblin, J.Chem.Soc. Chem.Comm. (1984) 92.

78. J.F. Bunnett, in "Techniques of Chemistry, Vol. VI, Investigation of Rates and Mechanisms of Reactions, Part I" edited by E.S. Lewis, Wiley-Interscience, New York 1974.

79. R.D. Chambers, Chapter 5 in this volume.

SUBJECT INDEX

Ab initio methods, 6-9

Absorption spectra of anions,
 uv-visible, 226

Acetyl anion, 54

Acetylide dianion, 72

Acid-base behavior upon photoexcitation,
 226, 228

Acidity, photoexcited carbanions, 229

Activation parameters for
- -, alkoxide-promoted dehydrohalogenated reactions, 331, 333, 336
- -, alkoxide-catalyzed hydrogen exchange reactions, 331, 336

Acyl organometallics, electrochemically
 induced reductive elimination, 160

Alkoxyvinyl anions, 52

Alkyl anions, 47

Alkyl halides,
- -, reduction of, 108, 130
- -, stereochemistry of electroreduct, 133

Alkyllithiums, electrooxidation of, 109

Alkylmagnesiums, electrooxidation of,
 109

Alkyñyl anions, 54-57

Alkyltrimethylsilanes reactions with
 methanolic methoxide, 340

Allyl anion, 57-58
- -, geometry, 57
- -, molecular orbitals, 224, 235
- -, photochemistry
  - -, 2-aryl-1,3-diphenyl-, 240
  - -, 1,3-diphenyl-2-(4-chlorophenyl)-, 240
  - -, 1,3-diphenyl-2-(4-cyanophenyl)-, 240
  - -, 1,3-diphenyl-, 228, 236-238
  - -, 1,3-diphenyl-2-phenylsulfonyl, 246
  - -, 1,3-diphenyl-2-phenylthio-, 246
  - -, 1,3-dicyano-, 240
  - -, 1,2,3-triphenyl-, 225
  - -, 1,1,2,3-tetraphenyl-, 245
- -, photoprotonation, 236, 244
- -, proton affinity, 58
- -, rotational barrier, 58
- -, substituted, conformation preferences, 59
- -, 2-substituted, 246
- -, tō cyclopropyl anion cyclisation, 224-5, 240-1, 243

Amphielectronic radicals, 115

Amylidenedisodium, 175

Anion inversion barriers, 16-19

Anion proton affinities, 14-16

Anion radicals, charge dispersal in,
 105

Anion radicals, protonation of, 104

Anion vibrational frequencies, 23-24

Anionic excited states, 228

Anions, photochemistry, 224

Anthracene
- -, 9,10-diphenyl, 102,104
- -, photoaddition, 248
- -, radical anion, 248

Antiaromatic anion, 66-68, 40

Antiaromaticity, electrochemical
 evidence for, 114

Aromatic anions, 67-68

Aromatic hydrocarbons, reduction
 potentials of, 101, 142

Aromatic multiply-charged anions,
 73-74

Arrhenius behavior of
- -, PKIE, 325
- -, protonation vs. β-fluoride

loss for carbanion intermediates, 344, 346
2-Aryl-1,3-propenyl anions, 240
Arylation, photoinitiated, 256
Arylhalides, electroreduction of, 134
Arylmethyl anions, 227
Asymmetric induction, electrochemical, 147

Band bending, 128
Basis functions, 5
Basis sets, 6-8
-, diffuse augmented, 7
-, extended, 7
-, minimal, 6
-, polarization, 7
Basicity of photoexcited anions, 229
Bathochromic shift upon carbanion formation, 226, 249
1,2-Benzanthracene, 102
Benzene derivatives
-, fluorinated, kinetic acidities, 283
Benzil-benzilic acid rearrangement
-, fluorinated derivatives, 309
Benzo[3,4]fluorenene, 252
Benzophenone as polymerization inhibitor, 233, 241
Benzyl anion, 64, 227
Benzyl radical, electron affinity of, 232
Bianthrone, 116, 162
Bicycloalkanes, fluorinated, 275
Bicycloalkylidine derivatives, fluorinated, 296
Bicyclo[3.3.0]octadienediyl dianion, 73
Bicyclo[5.1.0]octa-3-4-dienyl anion, 247
4-Biphenylyl, 239
4-Biphenylyldiphenylmethyl anion, 235
4-Biphenylylmethyl anion, absorption, 227
Birch reduction, 62-63
Bisphenols, 261
t-Butylacetylene, 109

Carbanion intermediates generated by reaction of
-, alcoholic alkoxide and alkenes, 326, 342, 346
-, alcoholic alkoxide and saturated compounds, 323, 327, 337, 352
-, methanolic methoxide and Me_3SiR, 340
Carbanions, oxidation peak potentials of, 110
Carbanion stabilities, effect of F and perfluoroalkyl, 272
Carbazole, 126
Carbenes
-, by carbanion irradiation, 261-2
-, fluorinated, 310
Carbonyl anions, 53-54
Carbonyls
-, electroreduction of, 145
-, reductive dimerization of, 145
Carboxylic acid derivatives, electroreduction of, 153
Cations
-, migratory aptitude in electrogenerated, 123
-, reduction of, 115
-, ring expansion of electrochemically generated, 123
Cesium benzenide, 176
9-(2-Chlorophenyl)fluorenyl anion, carbene formation, 251
Chrysene, 102
Cinnamylmagnesium bromide, 243, 246
cis,cis,cis,cis-1,3,5,7-cyclononatetraene, 254
cis-stilbene, isomerization, 239

Configuration interaction, 6
Copper derivatives, fluorinated, 311
Correlated wavefunctions, 8-9
Correlation energy, 8
Coronene, 102
Counter cation, effect of, on redox potentials, 103
Crystal structures, 177-192
-, of benzophenonedilithium complex with THF and tmeda, 188
-, of bis(dimethoxylethanelithium) pentalenide, 179-180
-, of bis(pentamethyldiethylenetriaminelithium) stilbenide, 182
-, of bis(potassium-diglyme) 1,3,5,7-tetramethylcyclooctateraenide, 190
-, of bis[tri(tetrahydrofuran)sodium] p-terphenylide, 191-192
-, of bis(tetramethylethylenediaminelithium)acenaphthylenide, 181-182
-, of bis(tetramethylethylenediaminelithium)anthracenide, 180-181
-, of bis(tetramethylethylenediaminelithium) $\Delta^{9,9'}$bifluorenide, 182
- of bis(tetramethylethylenediaminelithium) 1,2-diphenylbenzocyclobutadienide, 183
-, of bis(tetramethylethylenediaminelithium) trans, trans-1,4-diphenylbutadienide, 183
-, of bis(tetramethylethylenediaminelithium) hexatrienide, 183-184
-, of bis(tetramethylethylenediaminelithium) naphthalenide, 180
-, of bis(tetramethylenediaminelithium) o-quinodi(phenylmethanide), 184-185
-, of bis(tetramethylethylenediaminelithium) o-quinodi(trimethylsilylmethanide), 184-185
-, of bis(tetramethylethylenediaminelithium) stilbenide, 182
-, of bis(tetramethylethylenediamine lithium) tribenzylidenemethanide 186-187
-, of 5,5-di(lithio-diethyl ether)-2,2,8,8-tetramethyl-3,6-nonadiyne, 186-187
-, of dilithioferrocene-pentamethyldiethylenetriamine, 188-189
-, of 3,6-di(lithio-THF)-2,2,7,7-tetramethyl-3,4,5-octatriene, 185
-, of 2,2'di(lithio-tmeda)biphenyl, 179
-, of 1,3-di(lithio-tmeda)dibenzyl ketone, 187
-, of 1,4-di(lithio-tmeda)-1,4-diphenyl-2-butyne, 185-186
-, of dipotassium cyclooctatetraenide·diglyme, 190-191
-, of dirubidium cyclooctatetraenide·diglyme, 191
-, of disodiumacetylenide, 190
-, of methylcesium, 189
-, of methyllithium, 177
-, of methyllithium-tetramethylethylenediamine, 177
-, of methylpotassium, 189
-, of methylrubidium, 189
-, of methylsodium, 189
-, of phenyllithium-diethyl etherate, 178
-, of phenyllithium-pentamethyldiethylenetriamine, 178-179
-, of phenyllithium-tretramethylethylenediamine, 178
Current-potential curve, 96
Curtin-Hammett principle, 335
Cyanide anion, 55-56

-, geometry, 13
-, proton affinity, 15
Cyanocyclopropyl anion, 48
p-Cyanofluorobenzene, 137
Cyanomethyl anion, 30
-, substituted, 30-31
p-Cyanophenolate, photoreduction,263
Cyano radical, electron affinity, 21, 56
Cyclic voltammetry, 97
Cyclisation reactions, induced by F^-, 306
Cycloalkenyl anions, 61-62
Cycloalkyl anions, 47-49
Cyclobutadienes, fused naphthoquinone, 113
Cyclobutanecarboxylic acid, 122
Cyclobutyl anion, 49
Cycloheptatrienyl anion, 68
Cycloheptatrienyl trianion, 74
Cyclohexadienyl anion, 63, 68-69
-, stabilization energies, 63-64
-, substituted, 62-64
Cyclononatetraenyl anion, 249, 254
Cyclononatetraenyl dianion, 69
Cyclooctatetraene
-, fluorinated, 286
-, reduction, 287
Cyclooctatetraene dianion, 73
Cyclooctatetraenyl dianion, 253
Cyclooctatrienyl anion, 248
Cyclopentadiene, 109, 126
Cyclopentadienyl anions, 68,
-, photolysis, 227, 249
Cyclopropa-acenaphthene, 242
Cyclopropanation, 44
Cyclopropenyl anion, 66-67
-, open-chain isomers, 66
Cyclopropenyl radical, electron affinity, 67
Cyclopropyl anion, 48, 224, 240-242, 246,
-, electrocyclic transformation, 49
Cyclopropylmethyl anion, 31-32

Dehydrohalogenation reactions
-, alkoxide-promoted dehydrobromination, 330, 332
-, alkoxide-promoted dehydrochlorination, 330, 332
-, alkoxide-promoted dehydrofluorination, 330
-, effect of α-halide on rate of, 335
-, effect of β-halide on rate of, 332
-, stereochemistry, 333
Delocalized anions, 286
-, lag of delocalization behind proton transfer, 353
-, PKIE associated with hydrogen exchange reactions, 338
-, proton transfer from MeOH to, 341, 345, 347
Dianion, bis-malonitrilobiphenyl, 112
Dianion, bis-malonitrilopyrene, 112
Dianion, m-bis-i-propylbenzene, 112
Dianions, 175-221
-, electrooxidation of, 112
-, protonation of, 104
9,10-Dibromoanthracene, 135
1,2-Dibromocyclohexane, 116, 133
1,2-Dibromo-1,2-difluoroethane, 132
Dicesium biphenylide, 176
9,10-Dichloroanthracene, 135
9,9-Dichlorofluorene, 134
1,3-Dicyanoallyl anion, 240
Dicyclopentenyl, 249
Dienolates, 257
Diffuse-argmented basic sets, 7

Difluoromethyl anion, 33
Dihydrofluorene, 252
Dihydroindenes, 254
Dimer formation, from carbanion
photolysis, 251, 260
1,1-Dimethoxyethylene, 252
Dimethyl sulfoxide radical anion, 234
8,8-Dimethyl-2,4,6-Cyclooctatrienyl
anion, photocyclization, 247
2,3-Dimethyl-1,3-butadiene, 252
8,8-Dimethylbicyclo[5.1.0]octa-3,4-
dienyl anion, 247
1,3-Diphenyl-2-phenylsulfonylpropenyl
anion, 246
1,3-Diphenyl-2-phenylthiopropenyl
anion, 246
1,3-Diphenylallyl anion, 238
1,3-Diphenylallylsodium, 237
1,7-Diphenylheptadienyl, 227, 247
1,3-Diphenylindenyl anion, 245
1,3-Diphenylisoindenylidene, 252
Diphenylmethane, 109
Diphenylmethyl anion, 35
-, absorption spectrum, 227
1,9-Diphenylnonatetraenyl, 227, 247
1,5-Diphenylpentadienyl, 227, 247
1,3-Diphenylpropenyl, 227-228
-, anions, 236
Disilylmethyl anion, 35
Disodium carboxylates, 176
Dithioesters, electroreduction of, 155

Elcb elimination, 44
Electrical double layer, 96
Electrochemical cell, 95
Electrogenerated bases, 138
Electrogenerated nucleophiles, 140
Electron affinities, 20-23, 100
-, of radicals, 232
Electron correlation, 8
Electron ejection, 226, 233, 236-7, 260
Electron transfer, 233, 236, 248
Electron-hole pair, 128
Electronic effects
-, of F and perfluoroalkyl, 276
Electrostatic models for organodial-
kali metal compounds, 193-198
Element effect applied to dehydro-
halogenation reactions, 332
Emission, from irradiated carbanions,
226, 228
Enolate anions, 58-59
Enolate photolysis, 255, 257
Enthalpy of activation ($\Delta H^{\neq}$)
-, alkoxide-catalysed hydrogen
exchange reactions, 331, 336
-, alkoxide-promoted dehydrohalogenation reactions, 331, 333, 336
-, hydrogen-bond breaking in
carboxylic acid dimers, 331
Entropy of activation ($\Delta S^{\neq}$)
-, alkoxide-catalyzed hydrogen
exchange reactions, 331, 336
-, alkoxide-promoted dehydrohalogenation reactions, 331, 333, 336
Epoxidation, 44, 244
Ethyl anions, 35-38
-, comparison of α- and β-substituted, 46,
-, conformation, 37
-, 1,2-hydrogen shift, 38
-, inversion-rotation, 38
-, proton affinity, 38
-, stability, 36
-, α-substituted, 45-46
-, β-substituted, 39-45
-, β-substituted, conformational
preferences, 39-41
-, β-substituted, hyperconjugation,
39-45
-, β-substituted, stabilization
energies, 40-41

Ethylene dianion, 72
Ethyl phenylglyoxylate, 148
Ethyl radical, electron affinity, 36
Ethyl vinyl ether, addition compound, 252
Ethynyl anion, 54-55
-, 1,2-hydrogen shift, 55
-, proton affinity, 54
-, sila-analogues, 56
-, substituted, 55
-, vibrational frequencies, 24
Ethynyl radical, electron affinity, 21, 54-55
1-Ethynylvinyl anion, 52
2-Ethynylvinyl anion, 52
Excited states of carbanions, 224
Extended basic sets, 7

Fermi level, 99
Fluoradene, 251
Fluorenyl anion, 227, 249, 251
-, substituted, 111
Fluorescence of photoexcited carbanions, 228, 230, 236, 238, 251
Fluoride ion
-, exchange processes induced, 299
-, generation of observable carbanions, 297-298
-, induced processes, 292, 301
-, leaving ability from beta carbanions, 329
-, reactions with cyclic systems, 295-296
-, reactions with fluorinated alkenes, 293
-, reactions with hexafluoro-3-butyne, 294
-, sources, 300, 306
Fluorinated alkenes
-, cyclic, 284, 286
-, electronic effects in, 288
-, formation of carbanions, 297
-, nucleophilic attack, 287
-, oligomerisation, 293
-, reaction with CN^-, 307
-, reaction with F^-, 293
Fluorinated arenes
-, nucleophilic attack, 289
Fluorine, effect on carbanion stabilities, 272
Fluorocyclopropyl anion, 48
β-Fluoroethyl anion, 42
Fluoromethyl anion, 32-33
α-Fluorovinyl anion, 52
β-Fluorovinyl anion, 52
Formyl anion, 54
Formylmethyl anion, 58-59
1-Formylvinyl anion, 51
Förster equation, 225
Franck-Condon factors, 231
Friedel Crafts reactions, negative, 301
Frontier orbitals, 249, 292
-, nucleophilic attack, 289

Geminal dihalides, reduction of, 134

2-Halo-1,3-diphenylindenyl, 252
2-Haloindenyl anion, 252
Hartree-Fock method, 8
Heavy atom isotope effects
-, chlorine, 354
-, hydrogen isotope effect with chlorine in elimination reactions, 330
-, nitrogen, 354
Hexachlorocyclopentadiene, 132
Hexafluoro-2-butyne, reaction with F^-, 294
Hexafluorocyclobutene
-, formation of ylides, 294, 295
-, reaction with F^-, 294
-, reaction with pyridine, 295

Hexafluoropropene, nucleophilic attack, 288, 307
Hexamethylphosphoramide, 103
Homoaromatic anions, 68-71
Homoconjugation, 70
HOMO-LUMO energies, effect of F and perfluoroalkyl, 289
Huckel molecular orbital, 225, 235, 245, 250, 255
Hydride addition
-, to acetylene, 50-51
-, to ethylene, 38
Hydrogen bonds
-, associated with carboxylic acid dimers, 331
-, consequence of,
-, comparing equilibrium to kinetic acidities, 352
-, for proton transfer from ROH to carbanions, 341, 343, 347
-, to carbanion intermediates in dehydrohalogenation reactions, 330, 332
-, to carbanion intermediates to hydrogen exchange reactions, 328, 332
-, estimate of energy to carbanions, 332
Hydrogen, deuterium exchange
-, in fluorinated bicycloalkanes, 275
-, in fluorinated systems, 272
-, in nitroalkanes, 272, 273
Hydrogen exchange reactions
-, activation parameters for, 336
-, as a probe for E1cB mechanism, 330, 332, 354
-, effects of internal return on, 325, 337, 341
-, relative to dehydrofluorination, 326, 327, 331
1,2-Hydrogen shift
-, ethyl anion, 38
-, ethynyl anion, 55
-, vinyl anion, 51-52
Hydrogen transfer from RO^iH to carbanions
-, competition with elimination of β-fluoride, 326, 329, 339, 344-5,
-, PKIE associated with, 329, 339, 341, 345, 347
1-Hydropyrenyl monoanion, 176
Hyperconjugation, 39-45
Hypsochromic shift of ion pairs, 230, 249

Imine anions, conformational preferences, 60-61
Indenone, 150
Indenyl anion, 227, 249
Indirect electroreductions, 160
Internal return
-, calculation of $a^{iH} = k_{-1}/k_2$, 337, 338, 341
-, calculations using deviations from Swain-Schaad, 324, 328
-, effect of kinetic acidities, 325
-, kinetic expressions for, 323
Intersystem crossing in photoexcited carbanions, 232
Intramolecular charge transfer, 225, 255
Inversion barriers, 16-19
o-Iodoaniline, 136
Ion-pair, 103, 111, 226, 228, 230, 251
Ionicity and carbanion photo-rearrangement, 246
Irreversibility, 98
Isocyanocyclopropyl anion, 48
Isocyanomethyl anions, 31

Isogyric reactions, 22
Isomerization of photoexcited carbanions, 233, 236
Isopropyl anion, 47

Ketyl radical anions, 145, 153
Kinetic acidities
-, bicyclic systems, 275
-, effects of internal return on, 325
-, fluorinated benzenes, 283
-, fluorinated cyclopentadienes, 286
-, fluoroalkanes, 272
-, nitroalkane derivatives, 272, 273
-, use to assign pK_a values, 352, 354
Kinetic solvent isotope effect (KSIE), 337, 338
Kolbe electrolysis, 121

Lactic acid, 125

Mercaptomethyl anion, 33
Methanimidamide anion, 59
Methoxyazocine, 117
9-Methoxyfluorene, 139
Methyl anion, 25
-, inversion barrier, 16, 19, 25
-, proton affinity, 15-16, 25
-, vibrational frequencies, 24, 25
Methyl anions, substituted,
-, geometries, 25-29
-, inversion barriers, 25-29
-, stabilization energies, 25-29
Methylene dianion, 72
N-Methylpyrrole, 126
Methyl radical,
-, electron affinity, 21, 25, 232
-, from irradiation in DMSO, 234
1-Methylvinyl anion, 62
2-Methylvinyl anion, 62
Minimal basis set, 6
Mixed Kolbe couplings, 122
MNDO calculations, 242, 246
Molecular orbital calculations,
-, assessment of, 10-24
-, electron affinities, 20-23
-, geometries, 10-14
-, inversion barriers, 16-19
-, proton affinities, 14-16
-, vibrational frequencies, 23-24
Molecular orbital energy, from half-wave potent, 99
Molecular orbital theory,
-, ab initio, 6-9
-, semiempirical, 6, 9-10
Møller-Plesset perturbation theory, 9
Multiply-charged anions, aromatic, 73-74
Multiply-charged carbanions, 72-74

Naphtholate excited states, 229, 263
Naphthols, photoinduced proton transfer, 260
Naphthoquinone, 113
Naphthyl, as substituent, 239, 240
Negative hyperconjugation, 39-45, 277, 339, 353
-, calculations, 280
Nitranions, photochemistry, 243
Nitrobenzene, 151
p-Nitrobenzenediazonium, 136
p-Nitrobromobenzene, 135
2-Nitrobutane, 119
Nitro compounds, electroreduction of, 150
Nitrocyclopropyl anion, 48
Nitroketones, electrocyclization of, 152
Nitromethane, derivatives, acidity, 272, 273
Nitromethyl anion, 29

o-Nitrophenylsulphones, electro-cyclization of, 152
p-Nitrotoluene, 153
Non-bonding molecular orbital (NBMO), 225, 227, 235
7-Norbornadienyl anion, 49
7-Norbornenyl anion, 49
7-Norbornyl anion, 49
Nucleophiles, electrogenerated, 140
Nucleophilic addition to
 -, alkenes, 38, 44
 -, alkynes, 50
Nucleophilic aromatic substitution, 63
 -, effect of F as a substituent, 289
 -, in fluorinated arenes, 289
Nucleophilic vinylic substitution, 44

Observable anions, fluorinated, 297, 300
Observable carbanions,
 -, chemical shifts in, 297-298
 -, fluoride ion exchange, 299
 -, generation, 298
 -, quenching, 299
 -, sigma complexes, 300
Olefins, electroreduction of, 144
Oligomerisation,
 -, fluorinated alkenes, 293
 -, hexafluoro-2-butyne, 294
 -, tetrafluoroethylene, 293
Optical activity, loss of, in electro-oxidation, 124
Organodialkali metal compounds, 175-221
Organomercurials, as radical sources, 256
Organometallic compounds,
 -, di-lithio derivatives, 308, 309
 -, fluorinated, 307
 -, PKIE associated with neutralization by MeOH, 341
 -, relative stabilities, 307
Organometallics, reductive cleavage of, 158
Oxazirine oxide, from nitranion irradiation, 243
Oxetenyl anion, 67
Oxidation peak potentials, effect of substituent, 111

Pentadienyl anions, 61
 -, methyl-substituted, 61
Pentafluoropyridine, 303
Perfluoroalkyl, electronic effects, 276
Perfluoropropene, nucleophilic attack, 288
Perylene, 102, 129
 -, as triplet quencher, 233
Phenalenyl anion, 241, 242
 -, photodetachment spectrum, 231
Phenolate excited states, 229, 255, 260
Phenols, irradiation, 260
Phenylacetylene, 119
9-Phenylfluorene pK_a, 228
9-Phenylfluorenyl anion, 227, 239
1-Phenylindene, 249
1-Phenylindenyl anion, 250
Phenylphenolate, irradiation, 260
Photodetachment, 231, 234
Photodimers, 240
Photoejection from carbanions, 230, 238, 240, 255
Photoelectrochemistry, anionic, 127
Photoexcitation of carbanions, 225, 230, 235
Photohydrolysis, 263
Photomethylation, 234, 251
Photooxidation on semiconductor powders, 125
Photorearrangement of carbanions, 226
Photosensitization, carbanions, 254

pK_a
-, of hydrocarbons, 109
-, of photoexcited carbon acid, 228, 229
-, of radical cations, 115
-, thermodynamic cycle for determination of, 106
Polarization basis sets, 7
Polyenyl anions, irradiation, 227
Polyfluoroalkylation reactions, 301
Potential energy surfaces, 4-5
Predictive capabilities of quantum chemical methods, 10-24
Primary kinetic isotope effect (PKIE),
-, associated with,
-, alkoxide-catalyzed hydrogen exchange, 331, 336, 337
-, alkoxide-promoted dehydrobromination, 332, 333
-, alkoxide-promoted dehydrochlorination, 331, 333
-, alkoxide-promoted dehydrofluorination, 323, 331
-, proton transfer from ROH to carbanion, 325, 340, 342, 345
-, of hydrogen coupled with heavy atom isotope effects, 330, 354
-, of internal return on, 324, 325
-, increasing with increasing temperature, 329, 343, 345
Principle of imperfect synchronisation, 353
Propargyl anion, 64-65
-, diphenyl-substituted, 65
-, nitrogen, oxygen and silicon analogues, 65
-, substituted, 64
Prop-1-ynyl anion, 56, 65
n-Propyl anion, 42, 47
Proton affinities, 14-16
β-Proton exchange, 44
Protonation of photoexcited allyl anions, 236, 244
Pyrene dianion, 176
Pyrene tetraanion, 176
Pyrones, electroreduction of, 149
Quantum chemical methods, 5-10
Quantum mechanical calculations upon
-, allyllithium, 203
-, dilithioacetylene, 206-207
-, 2,2'-dilithiobiphenyl, 199
-, dilithiobutadienes, 203-204
-, 1,4-dilithiobutane, 208-209
-, 1,3-dilithio-2,2-dimethylpropane, 207-208
-, 1,1-dilithioethane, 210
-, 1,2-dilithioethane, 207
-, dilithioethylenes, 205-206
-, 1,3-dilithiopropane, 207-208
-, 1,3-dilithiopropylene, 205
-, α,α'-dilithio-o-xylene, 200
-, dilithium cyclobutadienide, 203-204
-, dilithium 1,4-di-t-butyldiacetylenide, 201-202
-, dilithium pentalenide, 200
-, 1,4-diphenyl-1,3-butadiene dianion, 200-201
-, hexalithiocarbon, 213
-, hypervalent organolithium compounds, 213
-, organodialkali metal compounds, 198-214
-, pentalithiocarbon, 213
-, pentalithiomethonium ion, 213
-, tetralithioallene, 211-212
-, tetralithiomethane, 212-213
-, trilithiomethane, 212-213
Quasi-reversibility, 98

Radiationless decay, carbanions, 236
Radical anions
-, aryl halides, bond cleavage in, 135
-, conformation equilibration of, 117
-, dimerization of, 135, 140
-, disproportionation equilibria of, 102
Radical coupling, 118, 126
Radicals
-, cross-coupling of, 120
-, reduction of, 107, 108
Rearrangements, induced by F^-, 303
Redox umpolung, 94
Reduction potential of carbanion substituent, 238
Reductive eliminations, 156
Resonance and photodetachment spectrum, 231
Resonance-stabilized carbanions, 264
Second harmonic ac voltammetry, 108
Semi-conductors, band positions in, 128
Semi-conductor powders, photooxidation on, 125
Semi-empirical methods, 9-10
Sigma complexes, 291-300
Silylated methyl anions, 34-35
Silylethyl anion, 42
Silymethyl anion, 34-35
Skeletal rearrangements, induced by F^-, 303
Solar energy conversion, by carbanion photolysis, 254
Solvated electrons, 145
Spiroisoindene, 252
SRNI reaction, 234, 235, 248, 249, 256, 257-259
Stabilization energies
-, substituted cyclohexadienyl anions, 63-64
-, α-substituted ethyl anions, 46
-, β-substituted ethyl anions, 40-41
-, substituted methyl anions, 46
Stereochemistry of
-, cyclopropanation, 44
-, dehydrohalogenation reactions, 333
-, Elcb elimination, 44
-, epoxidation, 44
-, fluorocarbanions, 285
-, fluorovinyl anions, 285
-, nucleophilic addition to alkenes, 44
-, nucleophilic vinyl substitution reactions, 347, 349, 351
-, β-proton exchange, 44
Stilbene, electrocatalyzed geometric isomerization, 118
Substituted allyl anions, 59-60
Substituted cyclohexadienyl anions, 62-64
α-Substituted ethyl anions, 39-45
β-Substituted ethyl anions, 45-46
Substituted ethynyl anions, 55
Substituted methyl anions, 25-29
Substituted vinyl anions, 52
Sulfonium ions, reductive coupling of, 162
Swain-Schaad relationship, 324, 328

tert-butylmercury chloride, 234, 256
Tetrafluoroethylene, reaction with F^-, 293, 303, 307
Tetraphenylcyclopentadienide, 114
1,1,3,3-Tetraphenylpropenyl anion, photomethylation, 245
Theoretical studies
-, β-C-X bond stabilization of carbanions, 339, 353

-, fluorinated anions, 280
-, nucleophilic vinyl substitution reactions, 351
Thienyl anions, 53
Thiopyrones, electroreduction of, 149
Toluenes, substituted, acidities, 64
Tosylates, electroreduction of, 157
Trialkylgermanium halides, 131, 159
Triarylmethyl anions, 234
Triazines
-, formation of sigma complexes, 300
-, reaction with F^-, 300
-, trifluoro, 298
2,2,2-Trifluoroethyl anion, 42
Trifluoromethyl anion, 33
Trifluoromethyl substituent, photo-hydrolysis, 263
Trimethylenemethyl anion, 32
1,2,3-Triphenylallyl anion, charge density, 225
Triphenylcyclopropenium, 115
Triphenylcyclopropenyl anion, 67
Triphenylmethane, 109
Triphenylmethyl anion, 227, 232, 234, 235
Triphenylmethyl radical spin density, 232
2,4,6-Triphenylpyridinium, 156
Triplet state, carbanions, 232
Trisilylmethyl anion, 35
Two electron oxidations, 124

Unsaturated fluorocarbons, nucleophilic attack, 287, 289

Vibrational frequencies, 23-24
Vicinal dihalides, reduction of, 132
Vinylacetylene
-, deprotonated, 57
-, dideprotonated, 73
Vinyl anions, 49-52
-, fluorinated, 282
-, 1,2-hydrogen shift, 51-52
-, inversion barrier, 19-50
-, nitrogen and phosphorus analogues, 53
-, proton affinity, 50
-, stereochemistry, 285
-, substituted, 52-53
Vinyl radical, electron affinity, 50

Wolff rearrangement, 262

Y delocalisation, 73

Ylids
-, fluorinated, 294, 312
-, generation from haloalkanes, 314

Zinc derivatives
-, fluorinated, 312
-, reactions of perfluoroalkylzinc derivatives, 312